Catalysts Deactivation, Poisoning and Regeneration

Catalysts Deactivation, Poisoning and Regeneration

Special Issue Editors

Luciana Lisi
Stefano Cimino

MDPI • Basel • Beijing • Wuhan • Barcelona • Belgrade

Special Issue Editors
Luciana Lisi
Istituto di Ricerche sulla Combustione
IRC-CNR
Italy

Stefano Cimino
Istituto di Ricerche sulla Combustione
IRC-CNR
Italy

Editorial Office
MDPI
St. Alban-Anlage 66
4052 Basel, Switzerland

This is a reprint of articles from the Special Issue published online in the open access journal *Catalysts* (ISSN 2073-4344) from 2018 to 2019 (available at: https://www.mdpi.com/journal/catalysts/special_issues/CDPR)

For citation purposes, cite each article independently as indicated on the article page online and as indicated below:

LastName, A.A.; LastName, B.B.; LastName, C.C. Article Title. *Journal Name* **Year**, *Article Number*, Page Range.

ISBN 978-3-03921-546-1 (Pbk)
ISBN 978-3-03921-547-8 (PDF)

Contents

About the Special Issue Editors

Luciana Lisi is a Senior Researcher at the Institute of Research on Combustion of the National Research Council (CNR) in Italy. She is co-responsible for a research group active in the field of materials and catalytic processes for energy and environment. She holds a Ph.D. in Chemical Engineering and has authored more than 100 papers in ISI journals focused on partial and total oxidation of light hydrocarbons; steam and dry reforming of hydrocarbons, catalytic conversion of biomass; post-combustion treatment of exhaust gases from power plants and automotive engines; biogas purification; synthesis and characterization of transition metal oxides and noble metals based catalysts. In 2018 she joined the Editorial Board of *Catalysts*.

Stefano Cimino is a Senior Researcher at the Institute of Research on Combustion of the National Research Council (CNR) in Italy where he is co-responsible for the Heterogeneous Catalysis group. He graduated cum laude in 1997 and in 2001 he obtained a Ph.D. in Chemical Engineering from the University of Napoli Federico II. Since then he has worked in the field of heterogeneous catalysis with a specific emphasis on catalytic processes for sustainable development and environmental protection. He has authored more than 70 articles in international journals and he has filed 6 international patents licensed to industrial partners. His areas of interest include catalyst deactivation, poisoning, and regeneration, catalytic and advanced combustion systems, structured and multifunctional catalytic reactors, advanced catalysts for partial oxidation, steam/dry/tri-reforming and methanation; gas purification (SCR DeNOx, VOC, CO, Hg, and H2S removal). Since 2018, Stefano Cimino has been a member of the Editorial Board of *Catalysts*.

Editorial

Catalyst Deactivation, Poisoning and Regeneration

Stefano Cimino * and Luciana Lisi *

Istituto Ricerche Sulla Combustione, CNR, 80125 Napoli, Italy
* Correspondence: stefano.cimino@cnr.it (S.C.); l.lisi@irc.cnr.it (L.L.)

Received: 25 July 2019; Accepted: 5 August 2019; Published: 5 August 2019

Catalyst life-time represents one of the most crucial economic aspects in most industrial catalytic processes, due to costly shut-downs, catalyst replacements and proper disposal of spent materials. Not surprisingly, there is considerable motivation to understand and treat catalyst deactivation, poisoning and regeneration, which causes this research topic to continue to grow [1]. The complexity of catalyst poisoning obviously increases along with the increasing use of biomass/waste-derived/residual feedstocks [2,3] and with requirements for cleaner and novel sustainable processes, such as those implementing a catalytic assisted chemical looping approach [4,5].

This Special Issue provides insight for several specific scientific and technical aspects of catalyst poisoning and deactivation, proposing more tolerant catalyst formulations and exploring possible regeneration strategies. In particular, 14 research articles focus on heterogeneous catalysts by investigating thermal [6–8], physical [9,10] and chemical [11–19] deactivation phenomena, and also exploring less conventional poisons related to the increasing use of bio-fuels [17]. Some regeneration strategies [11,16], together with solutions to prevent or limit deactivation phenomena [7,9,11,16], are also discussed. Eventually, one review paper [20] analyzes the rich chemistry of rhodium/phosphine complexes, which are applied as homogeneous catalysts to promote a wide range of chemical transformations, showing how the in situ generation of the active species, as well as the reaction of the catalyst itself with other components in the reaction medium, can lead to a number of deactivation phenomena.

More in detail, the effect of the gas flow rate on the formation of hotspots during the Catalytic Partial Oxidation of logistic fuels on Rh-based monoliths for the on-board production of syngas is investigated in [6]. Solutions to prevent the irreversible thermal deactivation of Ni/ZrO_2 catalysts during the Dry Reforming of Methane are proposed in [7] by inhibiting the transition of ZrO_2 into its monoclinic phase via modification with La. For the same reaction, it is well known that coke deposition is often the main cause of the deactivation of Ni-based catalysts—in [10], the authors report on the beneficial addition of barium to NiLa-based catalysts obtained from perovskite precursors to depress coke formation.

Novel La-oxysulfate/oxysulfide oxygen carrier materials promoted with small amounts of Co, Mn or Cu are presented in [8], assessing their reactivity and stability during the cyclic operation of a Chemical Looping Combustion process fueled with either hydrogen or methane.

Experimental data for the ageing and deactivation effects in real-life catalysts are generally scarce in the open literature, so the insight on industrial catalysts operated in full-scale systems in [9,11,15] is an important contribution. In particular, transient operations of the industrial reactor, such as shutdowns with insufficient air purging, are identified to cause an unusual deactivation behavior of a commercial V_2O_5/TiO_2 catalyst used for the production phthalic anhydride, as a consequence of excessive vanadium reduction and coke deposition [9]. A case study [11] on the deactivation of a commercial V_2O_5-WO_3/TiO_2 monolith catalyst operated for 18,000 h in a SCR (DeNOx) unit with tail-end configuration treating the exhaust gases of a MW incinerator highlights the formation and surface deposition of ammonium (bi)sulphates, which occurred due to the low reaction temperature, even though the inlet feed was rather clean after a desulphurization unit with $CaO/Ca(OH)_2$ and fabric filters. Furthermore, one research article [15] reports a detailed characterization of vehicle-aged oxidation catalysts from heavy-duty diesel and natural gas engines, showing the different effects of thermal ageing and chemical poisons (SO_2, phosphorous, zinc, silicon) on those commercial catalysts containing Pt or Pd dispersed on alumina.

Sulphur, mainly as SO_2, remains one of the main poisons for catalytic systems treating exhaust gases from combustion processes in stationary and mobile applications, since it forms highly stable sulphates. The individual and combined decomposition of aluminum and palladium sulphates, components of DOC, TWC and MOC, is presented in [12]. The same authors also address the role of sulphates formed on the support in restoring the Pd active species under different atmospheres and under simulated exhaust gas [13]. The deactivating effect of SO_2 and the formation of undesired by-products during the NH_3 SCR over a $MnFe/TiO_2$ catalyst for low-medium temperature is analyzed in [14] by investigating the mechanism of the main and side reactions occurring during the SCR process. On the other hand, in [16] the authors study regeneration strategies (oxidation/reduction) for a Co-Zn zeolite catalyst that was severely poisoned during the SCR of NOx with propane in the presence of SO_2 as a result of coke deposition and the formation of sulphates.

In addition, alkali and alkali earth metals are often responsible for the severe deactivation of SCR catalysts, especially when treating exhaust from the combustion of renewable fuels. Indeed, those elements can potentially come from bio-fuels or urea solutions in diesel engines [17] or even from fossil fuels in thermal power plants burning carbon [18]. Accordingly, in [18] the authors investigate the impact of the deposition of sulphur containing sodium salts onto a V_2O_5-WO_3/TiO_2 SCR catalyst with special regards to the NO removal rate as well as to the oxidation of SO_2 to SO_3. Moreover, as reported in [17], relatively high contents of sodium affect the hydrothermal stability of a Cu/SSZ-13 zeolite catalyst for the NH_3-SCR of a diesel exhaust.

Notably, water is often identified as causing catalyst poisoning, associated with the oxidation of the active metal, acceleration of sintering and even leaching of the active metals. In [19] the authors investigate the role of water addition during the hydrogenation of octanal over two copper-based catalysts and demonstrate that water can indeed promote process selectivity without affecting conversion.

Finally, the guest editors would like to express their deepest gratitude to all authors for their valuable contributions, as well as to Ms. Milly Chen and to all the staff of the editorial office for their significant support that made this Special Issue possible.

Conflicts of Interest: The authors declare no conflict of interest.

References

1. Argyle, M.; Bartholomew, C. Heterogeneous Catalyst Deactivation and Regeneration: A Review. *Catalysts* **2015**, *5*, 145–269. [CrossRef]
2. Lange, J.P. Renewable Feedstocks: The Problem of Catalyst Deactivation and its Mitigation. *Angew. Chem. Int. Ed.* **2015**, *54*, 13187–13197. [CrossRef] [PubMed]
3. Kim, S.; Tsang, Y.F.; Kwon, E.E.; Lin, K.-Y.A.; Lee, J. Recently developed methods to enhance stability of heterogeneous catalysts for conversion of biomass-derived feedstocks. *Korean J. Chem. Eng.* **2019**, *36*, 1–11. [CrossRef]
4. Hu, J.; Galvita, V.V.; Poelman, H.; Marin, G.B. Advanced chemical looping materials for CO utilization: A review. *Materials* **2018**, *11*, 1187. [CrossRef] [PubMed]
5. Dou, B.; Zhang, H.; Song, Y.; Zhao, L.; Jiang, B.; He, M.; Ruan, C.; Chen, H.; Xu, Y. Hydrogen production from the thermochemical conversion of biomass: Issues and challenges. *Sustain. Energy Fuels* **2019**, *3*, 314–342. [CrossRef]
6. Batista, R.; Carrera, A.; Beretta, A.; Groppi, G. Thermal Deactivation of Rh/α-Al_2O_3 in the Catalytic Partial Oxidation of Iso-Octane: Effect of Flow Rate. *Catalysts* **2019**, *9*, 532. [CrossRef]

7. Al-Fatesh, A.S.; Arafat, Y.; Ibrahim, A.; Kasim, S.O.; Alharthi, A.; Fakeeha, A.H.; Abasaeed, E.A.; Bonura, G.; Frusteri, F. Catalytic Behaviour of Ce-Doped Ni Systems Supported on Stabilized Zirconia under Dry Reforming Conditions. *Catalysts* **2019**, *9*, 473. [CrossRef]
8. Cimino, S.; Mancino, G.; Lisi, L. Performance and Stability of Metal (Co, Mn, Cu)-Promoted $La_2O_2SO_4$ Oxygen Carrier for Chemical Looping Combustion of Methane. *Catalysts* **2019**, *9*, 147. [CrossRef]
9. Richter, O.; Mestl, G. Deactivation of Commercial, High-Load o-Xylene Feed VO_x/TiO_2 Phthalic Anhydride Catalyst by Unusual Over-Reduction. *Catalysts* **2019**, *9*, 435. [CrossRef]
10. Gomes, R.; Costa, D.; Junior, R.; Santos, M.; Rodella, C.; Fréty, R.; Beretta, A.; Brandão, S. Dry Reforming of Methane over NiLa-Based Catalysts: Influence of Synthesis Method and Ba Addition on Catalytic Properties and Stability. *Catalysts* **2019**, *9*, 313. [CrossRef]
11. Cimino, S.; Ferone, C.; Cioffi, R.; Perillo, G.; Lisi, L. A Case Study for the Deactivation and Regeneration of a V_2O_5-WO_3/TiO_2 Catalyst in a Tail-End SCR Unit of a Municipal Waste Incineration Plant. *Catalysts* **2019**, *9*, 464. [CrossRef]
12. Kinnunen, N.M.; Nissinen, V.H.; Hirvi, J.T.; Kallinen, K.; Maunula, T.; Keenan, M.; Suvanto, M. Decomposition of Al_2O_3-Supported $PdSO_4$ and $Al_2(SO_4)_3$ in the Regeneration of Methane Combustion Catalyst: A Model Catalyst Study. *Catalysts* **2019**, *9*, 427. [CrossRef]
13. Kinnunen, N.M.; Kallinen, K.; Maunula, T.; Nissinen, V.H.; Keenan, M.; Suvanto, M. Fundamentals of Sulfate Species in Methane Combustion Catalyst Operation and Regeneration—A Simulated Exhaust Gas Study. *Catalysts* **2019**, *9*, 417. [CrossRef]
14. Lee, T.; Bai, H. Byproduct Analysis of SO_2 Poisoning on NH_3-SCR over $MnFe/TiO_2$ Catalysts at Medium to Low Temperatures. *Catalysts* **2019**, *9*, 265. [CrossRef]
15. Kanerva, T.; Honkanen, M.; Kolli, T.; Heikkinen, O.; Kallinen, K.; Saarinen, T.; Lahtinen, J.; Olsson, E.R.L.; Vippola, M. Microstructural Characteristics of Vehicle-Aged Heavy-Duty Diesel Oxidation Catalyst and Natural Gas Three-Way Catalyst. *Catalysts* **2019**, *9*, 137. [CrossRef]
16. Pan, H.; Xu, D.; He, C.; Shen, C. In Situ Regeneration and Deactivation of Co-Zn/H-Beta Catalysts in Catalytic Reduction of NO_x with Propane. *Catalysts* **2019**, *9*, 23. [CrossRef]
17. Wang, C.; Wang, J.; Wang, J.; Wang, Z.; Chen, Z.; Li, X.; Shen, M.; Yan, W.; Kang, X. The Role of Impregnated Sodium Ions in Cu/SSZ-13 NH_3-SCR Catalysts. *Catalysts* **2018**, *8*, 593. [CrossRef]
18. Xiao, H.; Dou, C.; Shi, H.; Ge, J.; Cai, L. Influence of Sulfur-Containing Sodium Salt Poisoned V_2O_5–WO_3/TiO_2 Catalysts on SO_2–SO_3 Conversion and NO Removal. *Catalysts* **2018**, *8*, 541. [CrossRef]
19. Govender, A.; Mahomed, A.S.; Friedrich, H.B. Water: Friend or Foe in Catalytic Hydrogenation? A Case Study Using Copper Catalysts. *Catalysts* **2018**, *8*, 474. [CrossRef]
20. Alberico, E.; Möller, S.; Horstmann, M.; Drexler, H.-J.; Heller, D. Activation, Deactivation and Reversibility Phenomena in Homogeneous Catalysis: A Showcase Based on the Chemistry of Rhodium/Phosphine Catalysts. *Catalysts* **2019**, *9*, 582. [CrossRef]

Article

Thermal Deactivation of Rh/α-Al_2O_3 in the Catalytic Partial Oxidation of Iso-Octane: Effect of Flow Rate

Roberto Batista, Andrea Carrera, Alessandra Beretta * and Gianpiero Groppi *

Laboratory of Catalysis and Catalytic Processes, Dipartimento di Energia, Politecnico di Milano, via La Masa 34, 20156 Milano, Italy; roberto.batistadasilva@polimi.it (R.B.); andrea.carrera@polimi.it (A.C.)
* Correspondence: alessandra.beretta@polimi.it (A.B.); gianpiero.groppi@polimi.it (G.G.); Tel.: +39-02-23993284 (A.B.); +39-02-23993264 (G.G.)

Received: 20 May 2019; Accepted: 12 June 2019; Published: 14 June 2019

Abstract: Catalytic partial oxidation (CPO) of logistic fuels is a promising technology for the small-scale and on-board production of syngas (H_2 and CO). Rh coated monoliths can be used as catalysts that, due to Rh high activity, allow the use of reduced reactor volumes (with contact time in the order of milliseconds) and the achievement of high syngas yield. As the CPO process is globally exothermic, it can be operated in adiabatic reactors. The reaction mechanism of the CPO process involves the superposition of exothermic and endothermic reactions at the catalyst inlet. Thus, a hot spot temperature is formed, which may lead to catalyst deactivation via sintering. In this work, the effect of the flow rate on the overall performance of a CPO-reformer has been studied, using iso-octane as model fuel. The focus has been on thermal behavior. The experimental investigation consisted of iC8-CPO tests at varying total flow rates from 5 to 15 NL/min, wherein axially resolved temperature and composition measurements were performed. The increase of flow rate resulted in a progressive increase of the hot spot temperature, with partial loss of activity in the entry zone of the monolith (as evidenced by repeated reference tests of CH_4-CPO); conversely, the adiabatic character of the reformer improved. A detailed modelling analysis provided the means for the interpretation of the observed results. The temperature hot spot can be limited by acting on the operating conditions of the process. However, a tradeoff is required between the stability of the catalyst and the achievement of high performances (syngas yield, reactants conversion, and reactor adiabaticity).

Keywords: CPO reactor; effect of flow rate; deactivation; iso-octane; Rh catalysts

---

1. Introduction

Nowadays, the industrial sector (mining, manufacturing, agriculture, construction, and others) accounts for the largest share in energy consumption all around the world. According to IEA, the transportation sector ranks at the second position in terms of energy consumption and projections show that, in the 2015–2040 period, its demand for energy will grow more quickly than the industrial field, reaching 1%/year, 0.3% higher than the industrial rate [1]. To supply this ever-increasing demand, while coping with the commitment to mitigating CO_2 emissions, fuel cell and hydrogen technology can be a key player [2,3]. The final goal of a green energy market is the full exploitation of renewable energy sources (with H_2 production via water electrolysis); however, the development of a decentralized H_2-production and supply chain based on small scale processors represents a realistic transition strategy [4–7]. Small-scale reformers have also been proposed for the on-board applications of H_2 (fueling of auxiliary power units based on fuel cells, the injections in the combustion chamber, and the regeneration of catalytic traps) in view of an improvement of the vehicle efficiency [8–11].

Natural gas, LPG, and liquid hydrocarbons can be converted catalytically into hydrogen-rich steams by steam reforming (SR) and catalytic partial oxidation (CPO). The use of noble metal-based catalysts is an important aspect of the process intensification, since higher activity allows for smaller catalyst inventory and faster dynamic response. Furthermore, such catalysts reduce the risk of coke formation with respect to non-precious metals as Ni and Fe [12,13]. Operating at very short contact times mitigates the cost issues associated with the adoption of precious metal catalysts. Among noble metals, Rh was reported to provide the highest activity and lower the tendency to coke formation at typical CPO conditions [4,14,15].

Concerning the reactor design, steam reforming of methane is an already consolidated industrial technology based on multi-tubular reactors but the necessity of a large energy input due to its high endothermicity makes the reactors hardly scalable down to small sizes (1–10 kW) of interest for distributed applications [2,16–19]. Instead, the catalytic partial oxidation (CPO) of hydrocarbons is a more flexible technology as it is globally exothermic and can be carried out in simple adiabatic structured reactors that are easily scalable.

The autothermal operation of the so-called short contact time CPO reformers has been successfully demonstrated by the pioneering and extensive work of Lanny Schmidt and coworkers, who have shown the obtainment of high syngas yields via partial oxidation of gaseous and vaporized liquid hydrocarbons over Rh washcoated foams [20–23]. The results from the Minnesota group have been largely confirmed in the years by several groups [24–27]. Besides, the development of advanced experimental and modelling tools has significantly contributed to the comprehension of the transport and chemical phenomena that govern the performance of CPO reformers [28]. Basini and co-workers have addressed a comprehensive analysis of the reduction of investment costs and energy consumption, the flexibility towards feedstock composition and product capacity, and the simplicity of technical and operational processes [29].

In previous works, the authors have reported the results of recent studies on the autothermal CPO of model hydrocarbons, representative of logistic fuels: iso-octane (iC8), a model for gasoline; and n-octane, a model for diesel [19,30]. The measurement of axially resolved temperature and concentration profiles and the engineering analysis of the reactor by the means of mathematical modelling have shown that the CPO of logistic fuels is a more severe process than the CPO of light hydrocarbons, being characterized by a higher peak surface temperature and the onset of gas-phase reactions leading to the formation of coke precursors; both factors can significantly contribute to accelerate catalyst deactivation by sintering and coking [17,19,30].

In this work, the effect of the input load on the performance of an iC8-CPO reformer was investigated by both experimental and modelling approaches. Flow rate is a key parameter of the reformer performance; it affects the reaction pathways, the output product yield, and the extent of heat dissipations. In turn, these factors can significantly impact the thermal behavior of the reactor and, consequently, the catalyst stability. At this scope, experiments and calculations were performed for a 400/7 CPSI cordierite honeycomb monolith, coated with a 2 wt% Rh/ α-Al_2O_3 active phase.

2. Results and Discussion

2.1. Conversion and Selectivity Performances

Experiments of iC8-CPO were performed at constant feed composition (iC8, Air, N_2 with 3% iC8 and C/O = 0.9) and varying total flow rate from 5 to 15 NL/min; Figure 1 reports the integral results of the experiments in terms of reactant conversions and product yields.

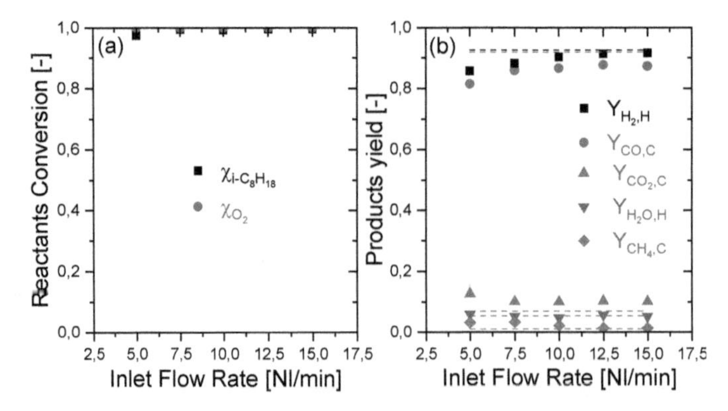

Figure 1. Effect of flow rate on the integral performance of the catalytic partial oxidation (CPO) reactor: (**a**) reactants conversion (χ); (**b**) products yield ($Y_{i,j}$, i = product and j = reference atom balance). Feed composition: iC8 = 3%, air with C/O = 0.9 and N_2 complement. Symbols = experiments; dashed lines = calculated adiabatic equilibrium.

It was verified that O_2, the limiting reactant, was fully converted under all the conditions. Except in the case of 5 NL/min, iso-octane was also completely converted since its conversion was not limited by thermodynamics. At increasing flow rate, the selectivity and yield of H_2 and CO increased, progressively approaching the expected equilibrium values under adiabatic conditions (dotted lines in Figure 1). The yields of CH_4, CO_2, and H_2O, instead, moderately decreased and tended to the calculated equilibrium values as inlet flow increased. This result might appear counter-intuitive, considering the indirect-consecutive nature of CO and H_2 formation and the expected negative effect of reducing the contact time on the formation of terminal products. However, the axial evolution of temperature profiles changed considerably at increasing flow rates; the measurements obtained during the iso-octane experimental campaign are presented in Figure 2, where thin lines represent the measurements obtained by the thermocouple (representative of the gas-phase temperature), while thicker lines represent the measurements obtained from the optical-fiber/pyrometer system (representative of the emitting surface temperature).

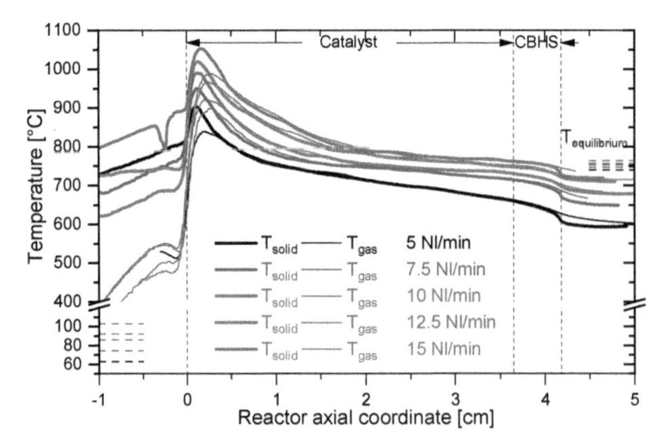

Figure 2. Experimental temperature profiles varying the inlet flow rate. Feed composition: iC8 = 3%, Air with C/O = 0.9 and N_2 complement (T_{solid} measured with an optcal fiber and T_{gas} measured with a thermocouple).

The temperature of the catalyst surface and of the gas phase measured along the entire axial coordinate increased significantly with the increase of flow rate. Several factors have a role in this trend, including operational, thermodynamic, and kinetic factors.

First, the inlet temperature increased with the flow rate from a value of 62 °C (at 5 NL/min) to 103 °C (at 15 NL/min) because of the enhanced heat exchange between the pre-heating cartridge and the gas flow with increasing flow rate. Thus, the adiabatic equilibrium temperature also increased; the single values calculated for the various experiments are reported as short dotted bars at the right-hand side of Figure 2. Secondly, as better shown in Figure 3a, the measured outlet temperature increased more markedly than the adiabatic temperature; thus, the difference between the outlet adiabatic equilibrium temperature and the outlet measured temperature decreased with the increase of the flow rate.

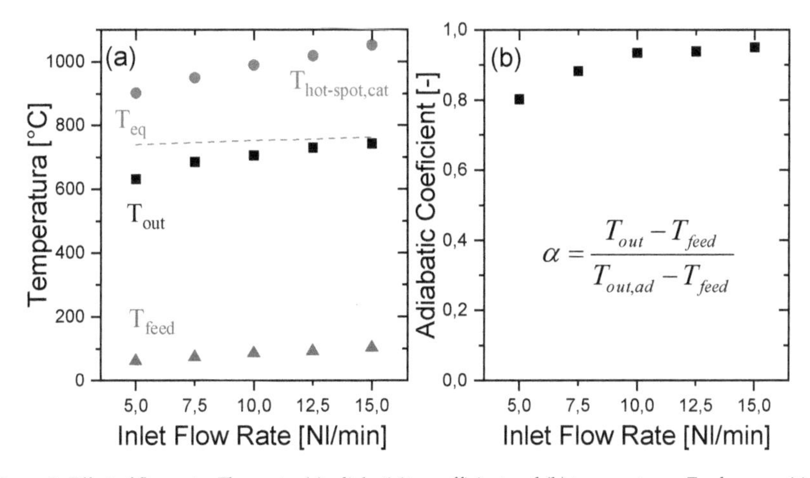

Figure 3. Effect of flow rate. Flow rate: (**a**) adiabaticity coefficient and (**b**) temperatures. Feed composition: iC8 = 3%, air with C/O = 0.9 and N_2 complement.

In other words, at increasing load, the reactor better approached the adiabatic behavior. This effect can be quantified through the definition of an adiabaticity coefficient, expressing the ratio between the measured temperature rise across the CPO reactor and ideal temperature rise for the fully adiabatic reactor, as follows:

$$\alpha = \frac{T_{out} - T_{feed}}{T_{out,ad} - T_{feed}} \quad (1)$$

As shown in panel (b) of Figure 3, the adiabaticity coefficient increased significantly with the flow rate, passing from a value of 80% in the case of 5 NL/min, to 93% in the case of 10 NL/min, and finally to 95% in the case of 15 NL/min. This trend reveals the impact of heat dispersions on the thermal balance of the reactor or, in other words, the relative impact of heat dispersion over heat load. The data clearly show that although heat dispersions expectedly grew on absolute basis due to the progressive increase of the reactor temperature, the ratio between heat dispersion and the inlet enthalpy flux entering the reactor decreased. The criticality of obtaining a full adiabatic behavior at the lab scale is well known, and this is especially true when dealing with miniaturized systems, given the high surface-to-volume ratio; thus, the experiments were extremely important to verify the sensitivity of the system to a key parameter as input flow. It was concluded that at total flows above 10 NL/min, the CPO reactor can be treated as fully adiabatic.

Lastly, it is observed that another important phenomenon was the increase and enlargement of the hot spot region at increasing flow rate (as shown in Figure 2 and highlighted in Figure 3, panel (a)), which cannot be explained by the above-mentioned factors. A modelling analysis was thus performed to understand more deeply the kinetic effects involved in the temperature profile, and its dependence on flow rate.

2.2. Modeling Analysis

To gain insight into the correlation between inlet flow rate and hot spot temperatures, the reactor was simulated, assuming a perfect adiabatic behavior. The predicted gas-phase and solid phase temperature profiles are reported in Figure 4.

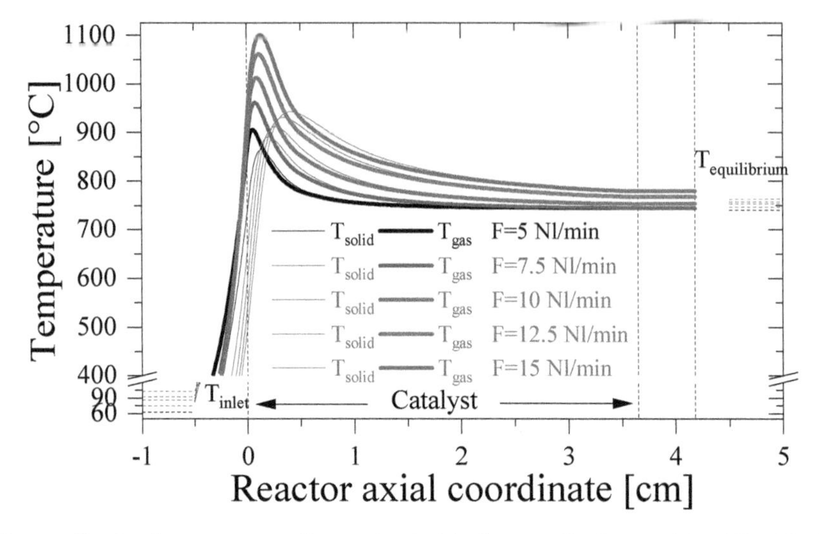

Figure 4. Simulated temperature profiles varying the inlet flow rate. Feed composition: iC8 = 3%, air with C/O = 0.9 and N_2 complement. Very good agreement with the experimental results was obtained, since the calculations showed a progressive increase of the whole temperature profiles and an especially important increase of temperatures in the hot spot at the monolith entrance.

Notably, a progressive enlargement of the hot spot is predicted; in fact, the decline of the temperature downstream of the maximum becomes more gradual at increasing flow, such that at any flow rate, the consumption of O_2 is more rapid than the consumption of i-C8 and the formation of CO and H_2 starting from the very entrance of the monolith. Thus, the heat release occurs across a shorter distance than heat consumption, which originates from the hot spot at the entrance.

The simulated concentration profiles are reported in Figure 5.

Panels (a) and (b) present a progressive extension of the iso-C_8H_{18} and O_2 consumption zones with inlet flow increase. In fact, a higher flow rate corresponds to a higher velocity of the gas phase inside the reactor; thus, there is an expected delay of consumption of the reactants. In particular, the O_2—consumption length (Figure 5b) grows from 0.25 cm (5 NL/min) to 0.75 cm (15 NL/min). This region is the so-called oxy-reforming zone, where the hot spots develop as the result of the balance between exothermic reactions responsible for O_2 consumption (mainly H_2 oxidation) and endothermic reactions responsible for the fuel consumption (iC8 steam reforming to CO and H_2) [31].

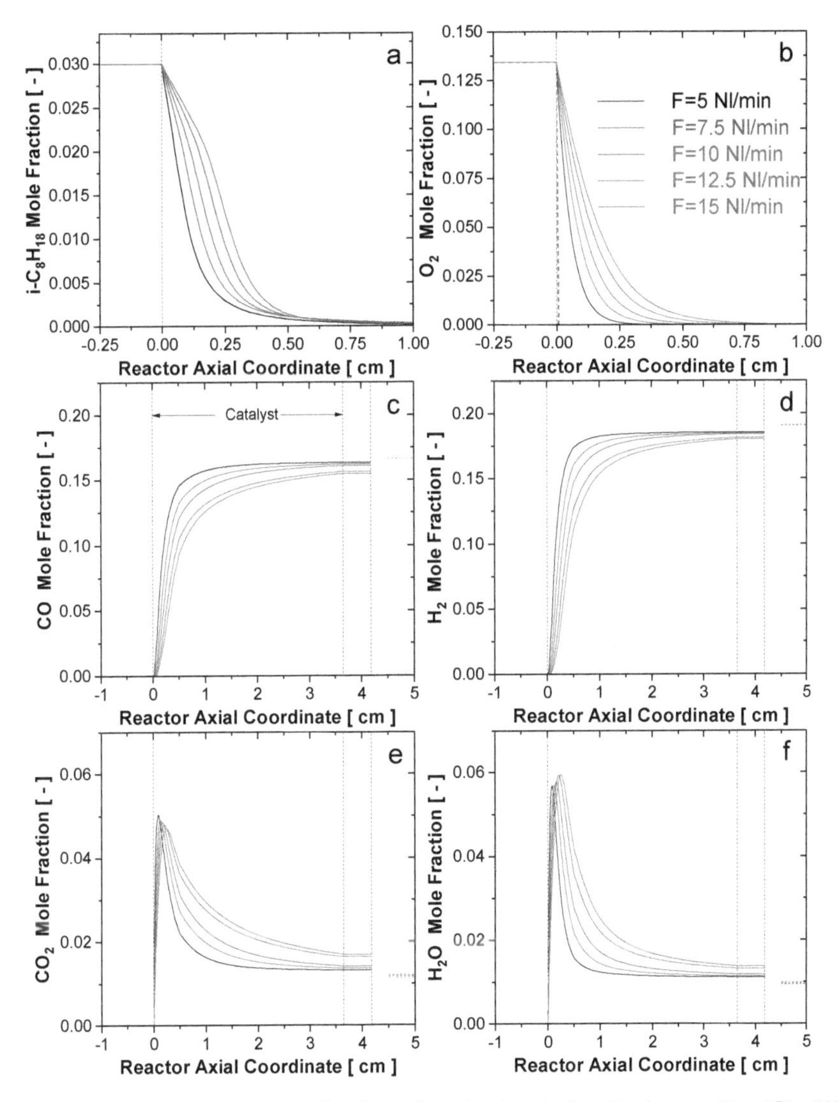

Figure 5. Simulated concentration profiles obtained varying the inlet flow. Feed composition: iC8 = 3%, air with C/O = 0.9 and N_2 complement.

However, the balance between exothermic and endothermic reactions can change. This is more clearly shown in Figure 6, where the conversion profiles of the reactants are plotted in the various flow conditions; taking the coordinate 0.25 cm as a reference, here the oxygen conversion moves from 98% at 5 NL/min to 78% at 15 NL/min. On the other hand, iso-octane conversion moves from 88% of 5 NL/min to 53% of 15 NL/min. Thus, the exothermic contribution increases over the endothermic one and temperatures grow consequently. A change of selectivity is also produced, leading to an increased concentration of H_2O and CO_2 a decreased concentration of CO and H_2.

Such an unbalancing of exothermic and endothermic contributions is fuel-specific, being related to the slow diffusivity of i-C8, which enhances the consecutive nature of the surface process [2], and to its high gas phase reactivity, which results in the onset of homogeneous reactions upon an ignition delay. The onset of gas phase reactions is progressively shifted downstream on increasing the flow rate, as evidenced by the change of slope of iC8 conversion curves in Figure 6. In addition to this fuel-specific effect, there is also a general trend associated with the increase of gas velocity in CPO processes: at increasing importance of convection, conduction is less effective in smoothing the surface hot spot [32].

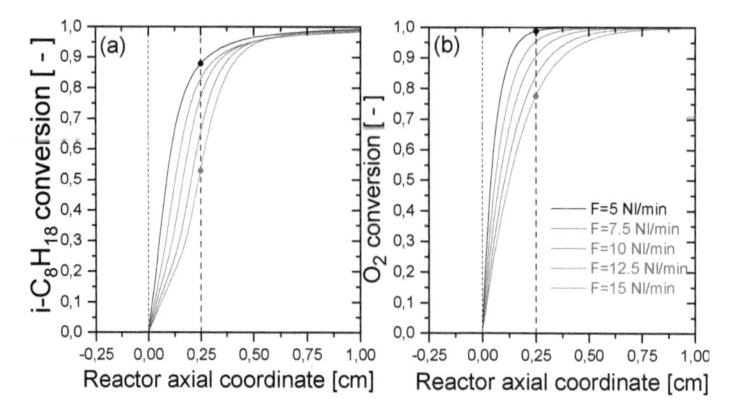

Figure 6. Effect of flow rate: (**a**) i-C_8H_{18} conversion and (**b**) O_2 conversion. Feed composition: iC8 = 3%, air with C/O = 0.9 and N_2 complement.

2.3. Catalyst Stability

The effect on the catalyst stability of performing iC8 experiments at increasing flow rate, and thus the effect of exposing the catalyst to progressively temperature increase, was verified by systematically repeating methane CPO tests; these were carried out on the fresh catalyst and after every iso-octane CPO test.

The reactor integral performance was measured, and the results are reported in Table 1, in terms of reactant conversion and product selectivity. Negligible differences were observed between the experiment on fresh catalyst and the following tests, and a close approach to thermodynamic equilibrium was found.

Table 1. Methane CPO: reactants conversion and products selectivity.

Table	χCH_4	χO_2	σH_2	σCO	σCO_2	σH_2O
	[-]	[-]	[-]	[-]	[-]	[-]
equilibrium	0.86	1.00	0.92	0.86	0.14	0.08
fresh catalyst	0.84	1.00	0.90	0.84	0.16	0.10
after 5 NL/min	0.84	1.00	0.90	0.85	0.15	0.10
after 7.5 NL/min	0.84	1.00	0.90	0.85	0.15	0.10
after 10 NL/min	0.84	1.00	0.90	0.85	0.15	0.10
after 12.5 NL/min	0.84	1.00	0.90	0.85	0.15	0.10
after 15 NL/min	0.84	1.00	0.90	0.84	0.16	0.10

More sensitive data were, however, obtained from the axially resolved temperature measurements shown in Figure 7.

The outlet temperature remained aligned with the adiabatic equilibrium, but changes of the temperature profiles were observed in the entering zone. In fact, the fresh catalyst showed a hot spot of temperature of about 780 °C and flattening of the solid and gas temperature profiles (which indicates where the system approaches the thermodynamic equilibrium) at about 1.5 cm from the entrance. Test after test, the maxima measured by the thermocouple and the pyrometer, as well as the axial extension of the hot spot, increased. After the iC8 experiment at 15 NL/min, the hot spot temperature measured by the optical fiber amounted to 856 °C, while the flattening of the solid and gas-phase temperatures was observed in correspondence with the coordinate of 2.5 cm.

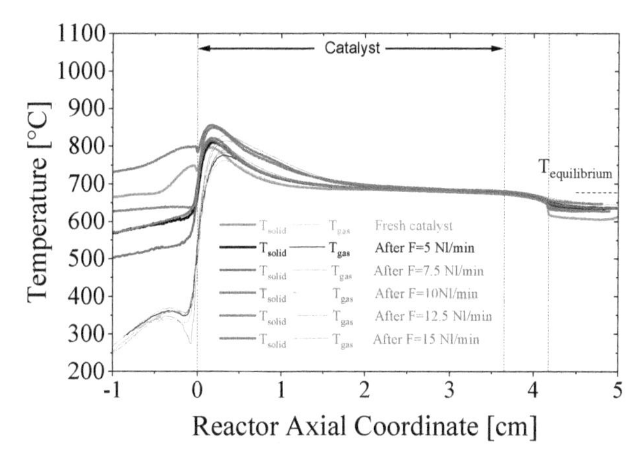

Figure 7. Effect of flow rate variation on the catalyst stability. CH_4 = 27.3%, air with C/O = 0.9, and T_{in} = 25 °C, F = 10 NL/min. (T_{solid} measured with an optcal fiber and T_{gas} measured with a thermocouple).

This is the clear evidence of a progressive deactivation of the catalyst, likely due to sintering of Rh clusters after exposure to temperature exceeding 900 °C in the iC8 experiments. The phenomenology of deactivation has been discussed in previous works from this and other research groups [14,32,33]. Since O_2 consumption in the oxidation reactions is fully mass transfer controlled, while CH_4 consumption via steam reforming is more chemically controlled, a loss of surface sites will preferentially affect the steam reforming reaction, with a consequent loss of the heat consumption rate and thus an increase of temperatures in the oxy-reforming zone.

Despite the deactivation probed by the temperature measurements, the evidence that equilibrium conversion and syngas yield were still reached in every CH_4 CPO test confirms that the monolith was sufficiently oversized for keeping a stable methane CPO application.

Taking into consideration all the results obtained, methane CPO tests confirm that, when feeding a logistic fuel such as iso-octane, a flow rate of 10 NL/min is a relatively safe condition that is able to preserve catalyst stability and avoid an important deactivation via sintering.

3. Materials and Methods

3.1. Catalyst Synthesis, Structural and Morphological Characterizations

The catalyst evaluated in this work is a sample of 400/7 cordierite honeycomb monolith, coated with a 2 wt% Rh/ α-Al_2O_3 active phase. The Rh/α-Al_2O_3 catalytic powder was synthesized by incipient wetness impregnation of α-Al_2O_3 with $Rh(NO_3)_3$ solution, followed by drying at 120 °C overnight. Surface area and pore size volume of the catalyst powders, respectively, 5 m^2/g and 0.21 mL/g, were determined by N_2 adsorption–desorption at 77 K with the BET method using a Micromeritics TriStar 3000 instrument. Rhodium content, 1.71% w/w, was determined by ICP-MS using a XSeries instrument.

Rh dispersion on the catalyst powders was estimated by hydrogen pulse chemisorption using a TPD/R/O 1100 Thermo Fischer Instrument. A pre-treatment was performed consisting of an initial reduction with a 5% H_2/Ar flow (50 Ncm^3/min) from room temperature up to 500 °C (heating rate 7 °C/min). After 1h at 500 °C, the sample was cooled down to 40 °C in pure Argon. The chemisorption was performed with 20–30 pulses (0.86 Ncm^3) of the same diluted H_2 mixture. Ageing experiments under reaction atmosphere were performed in order to evaluate the catalyst stability using mild

(4% CH_4, C/O = 0.9, N_2 to balance) and severe (27% CH_4 in air with C/O = 0.9) conditions, ramping the temperature from R.T to 850 °C (heating rate 10 °C/min, 850 °C hold for 4h) at GHSV = 80,000 NL/kg/h. The results of chemisorption on the fresh and aged catalysts are presented in Table 2, showing a progressive decrease of Rh dispersion from 69% for the as prepared catalyst to 7% under severe ageing conditions, which resemble those achieved in the adiabatic reactor.

Table 2. Catalyst morphological characteristics.

Catalyst	Rh Load	Rh Dispersion (%)		
	% w/w	Fresh	Aged (4% CH_4)	Aged (27% CH_4)
Rh/α-Al_2O_3	1.71	69	23	7

After drying, the catalyst powders were suspended with water and nitric acid before undergoing a 24h-long ball milling. The coating proceeded through the dip-coating technique. Further details on synthesis methodology can be found elsewhere [34].

Aiming to verify the presence of heat losses along the reactor, the monolith was partly uncoated, in the rear part, forming a continuous back heat shield (CBHS), being completely coated at the entrance and avoiding a continuous front heat shield, thus reducing the hot spot temperature.

Table 3 reports the characteristics of the catalyst used in this work.

Table 3. Catalyst specifications.

Catalyst	L_{cat}	LCBHS	m_{cat}	t_{cat}	Void Fraction
	[cm]	[cm]	[g]	[μm]	[-]
Rh/α-Al_2O_3	3.65	0.53	0.63	9.43	0.714

L_{Cat} = catalyst length; LCBHS = continuous back heat shield length; m_{cat} = catalyst deposited mass; t_{cat} = catalyst thickness.

The catalysts were inserted in a quartz pipe between a FeCrAlloy 15 ppi foam and an inert cordierite monolith inserted in such a way to guarantee downstream position to the back-heat shield.

3.2. Experimental Setup

A capillary (ID 320 μm), carefully allocated in a central channel of the catalyst, was capable to slide along the axial coordinate, as the reactor is equipped with spatially resolved sampling apparatus. Two different types of capillaries are employed in alternance: a (ID = 200 μm, OD = 350 μm), opened at its tip, allowing collecting gas samples and a second capillary (OD = 500 μm), closed at its tip, used to host the 250 μm K-Type thermocouple or a 45° ground optical fiber. Optical fiber and thermocouple are used to collect, respectively, the temperatures of the solid and gas phase. An Agilent MicroGC 3000A equipped with two columns (Plot U and a Molecular Sieve, Agilent, Santa Clara, CA, USA) was used to determine the gas phase concentration of different species. Plot U was operated at 60 °C, in order to get better resolution for light hydrocarbons (C1–C4) and then at 160 °C in order to get peaks of small tailing effect. N_2 was chosen as internal standard. As Plot U is not capable to separate N_2 from CH_4, O_2, H_2, and CO, the Molecular Sieve was used as well.

The effect of the inlet flow rate on CPO of iC8 was conducted with a progressive increase of the total flow from 5 NL/min to 15 NL/min. The inlet concentration of the hydrocarbon was 3%, and the C/O ratio was kept constant and equal to 0.9. The goal of this campaign is to evaluate the adiabaticity of the reactor and the activity of the catalyst. The stability of the catalyst has been evaluated, by performing methane CPO tests on the fresh catalyst and after each iso-octane experiment.

3.3. Reactor Modelling

A mathematical 1D, dynamic, fixed bed, heterogeneous, single channel reactor model was used for the reactor design and for the analysis of experimental results. Its development was presented in previous works [16,34].

The model accounts for axial convection and diffusion, gas–solid transport term, solid conduction, and mass and energy balances for both solid and gas phase (Table 4).

Table 4. Model equations.

Gas Phase	
Mass Balance	$\frac{\partial \omega_i}{\partial t} = -\frac{G}{\rho_g \varepsilon}\frac{\partial \omega_i}{\partial z} - \frac{a_v}{\varepsilon} k_{mat,i}\left(\omega_i - \omega_{wall,i}\right) + \frac{\mathfrak{D}_{i,mix}}{\varepsilon}\frac{\partial^2 \omega_i}{\partial z^2} + MW_i \sum_{j=1}^{NR_g} \upsilon_{i,j} r_j^{homo}$
Enthalpy Balance	$\frac{\partial T_g}{\partial t} = -\frac{G}{\rho_g \varepsilon}\frac{\partial T_g}{\partial z} - \frac{a_v h (T_g - T_s)}{\varepsilon \rho_g \hat{c}_{p,g}} - \sum_{j=1}^{NR_g} \Delta H_{r,j} r_j^{homo}$
Solid Phase	
Mass Balance	$0 = a_v \rho_g k_{mat,i}\left(\omega_i - \omega_{wall,i}\right) + \frac{MW_i \rho_w \alpha}{\rho_s \hat{S}_{att}} \eta \sum_{j=1}^{NR_s} \upsilon_{i,j} r_j^{het}$
Enthalpy Balance	$\frac{\partial T_s}{\partial t} = \frac{a_v h (T_g - T_s)}{(1-\varepsilon)\rho_s \hat{c}_{p,s}} + \frac{\frac{\partial}{\partial z}\left(k_{ax}^{eff} \frac{\partial T_s}{\partial z}\right)}{(1-\varepsilon)\rho_s \hat{c}_{p,s}} + \frac{\alpha}{(1-\varepsilon)\rho_s \hat{c}_{p,s} \hat{S}_{att}} \eta \sum_{j=1}^{NR_s} \Delta H_{r,j} r_j^{het}$
Boundary Conditions	
Reactor Inlet	$\omega_i\|_{z=0} = \omega_{i,feed}$ $\quad T_g\|_{z=0} = T_{feed}$ $\quad \frac{\partial T_s}{\partial z}\Big\|_{z=0} = 0$
Reactor Outlet	$\frac{\partial \omega_i}{\partial z}\Big\|_{z=L} = 0$ $\quad \frac{\partial T_s}{\partial z}\Big\|_{z=L} = 0$
Initial Conditions	$\omega_i(z,0) = 0$ $\quad T_g(z,0) = T_{feed}$ $\quad T_s(z,0) = 650\ ^\circ C$

This model is an extension of a previous, already validated methane CPO model, now taking into consideration a catalytic kinetic scheme for i-C_8H_{18} conversion [30]. The gas phase kinetics are also included, as reported in [19].

The kinetic scheme is presented in Table 5.

Table 5. Heterogeneous reaction mechanism for iso-octane CPO [30,35,36].

Reaction Name and Chemical Equation	Rate Equation [mol/atm/g_{cat}/s]	k_i@873K [mol/atm/g_{cat}/s]	$E_{activation}$ [kJ/mol]	Ref.
CH_4 oxidation $CH_4 + 2\,O_2 \rightarrow CO_2 + 2\,H_2O$	$r_{CH_4}^{OX}\Big\|_{873K} = \frac{k_{CH_4}^{OX} p_{CH_4}}{1+K_{H_2O}^{ads} p_{H_2O}} \sigma_{O_2}$	1.030×10^{-1}	91.96	[33]
CH_4 steam reforming $CH_4 + H_2O \rightarrow CO + 3\,H_2$	$r_{CH_4}^{SR}\Big\|_{873K} = \frac{k_{CH_4}^{SR} p_{CH_4}}{1+K_{O_2}^{ads} p_{O_2} + K_{CO}^{ads} p_{CO}} \sigma_{H_2O}\left(1 - \eta_{CH_4}^{SR}\right)$	1.027×10^{-1}	91.80	[33]
CO methanation $CO + 3\,H_2 \rightarrow CH_4 + H_2O$	$r_{CO}^{meth}\Big\|_{873K} = k_{CO}^{meth} p_{H_2} \sigma_{CO}\left(1 - \eta_{CO}^{meth}\right)$	1.500×10^{-3}	30.00	[26]
Water Gas Shift $CO + H_2O \rightarrow CO_2 + H_2$	$r_{WGS}\|_{873K} = k_{WGS} p_{H_2O} \sigma_{CO} (1 - \eta_{WGS})$	6.831×10^{-3}	74.83	[33]
Reverse Water Gas Shift $CO_2 + H_2 \rightarrow CO + H_2O$	$r_{RWGS}\|_{873K} = k_{RWGS} p_{CO_2} \sigma_{H_2} (1 - \eta_{RWGS})$	1.277×10^{-2}	62.37	[26]
H_2 oxidation $H_2 + 1/2\,O_2 \rightarrow H_2O$	$r_{H_2}^{OX}\Big\|_{873K} = k_{H_2}^{OX} p_{H_2} \sigma_{O_2}$	2.666×10^{3}	61.65	[33]
CO oxidation $CO + 1/2\,O_2 \rightarrow CO_2$	$r_{CO}^{OX}\Big\|_{873K} = k_{CO}^{OX} p_{CO} \sigma_{O_2}$	1.937×10^{1}	76.07	[33]
iso-C_8H_{18} total oxidation iso-$C_8H_{18} + 25/2\,O_2 \rightarrow 8\,CO_2 + 9\,H_2O$	$r_{i-C_8H_{18}}^{OX}\Big\|_{873K} = \frac{k_{i-C_8H_{18}}^{OX} p_{i-C_8H_{18}}}{1+K_{H_2O}^{ads} p_{H_2O}} \sigma_{O_2}$	4.600×10^{-1}	80.00	[34]

Table 5. *Cont.*

Reaction Name and Chemical Equation	Rate Equation $[mol/atm/g_{cat}/s]$	k_i@873K $[mol/atm/g_{cat}/s]$	$E_{activation}$ [kJ/mol]	Ref.
iso-C_8H_{18} steam reforming iso-$C_8H_{18} + 8\,H_2O \rightarrow 8\,CO + 17\,H_2$	$r^{SR}_{i-C_8H_{18}}\Big\|_{873K} = \frac{k^{SR}_{i-C_8H_{18}} p_{i-C_8H_{18}}}{1+K^{poison}_{i-C_8H_{18}} \frac{p_{i-C_8H_{18}}}{p_{H_2O}}} \sigma_{H_2O}\left(1-\eta^{SR}_{i-C_8H_{18}}\right)$	7.500×10^{-2}	69.00	[34]
$r_j^{chemical\ reaction} = r_j^{chemical\ reaction}\Big\|_{873\ K} \exp\left[-\frac{E_{att,\ j}}{R}\left(\frac{1}{T}-\frac{1}{873}\right)\right]$				
Adsorption		$K_i^{0,ads}$ **@873K [1/atm]**	$\Delta H_{adsorption}$ **[kJ/mol]**	**Ref.**
O_2		5.461×10^{0}	−72.83	[33]
CO		2.114×10^{2}	−37.15	[33]
H_2O		8.974×10^{0}	−57.48	[33]
Poisoning term		$K_i^{0,\ poisoning}$ **@873K [-]**	$\Delta H_{adsorption}$ **[kJ/mol]**	**Ref.**
i-C_8H_{18}		6.000×10^{0}	−26.00	[34]
n-C_8H_{18}		6.000×10^{0}	−26.00	[34]
$r_j = K_j^0 \exp\left[-\frac{E_{att,\ r_j}}{R}\left(\frac{1}{T}-\frac{1}{873}\right)\right]$				

4. Conclusions

Rhodium supported catalysts, operated in adiabatic reactors, are effective in the small-scale production of synthesis gas from liquid fuels. These features make this process attractive for the development of a compact reformer. Rhodium-based catalysts are able to efficiently reform iso-octane into synthesis gas.

Experiments and model simulations confirm the indirect reaction mechanism, which is responsible for the presence of an oxy-reforming zone in the first part of the catalyst and a reforming zone further downstream.

It has been observed that the increase of the inlet flow rate promotes the adiabaticity of the reactor, but it leads to a higher catalyst hot spot temperature, due to the higher inlet enthalpic flux. The catalyst stability has been evaluated by performing a methane CPO after each octane test. The optimal inlet flow rate has been set to 10 NL/min, the setting that guarantees the best compromise between the adiabaticity of the reactor and the stability of the catalyst.

Author Contributions: The research was completed through the cooperation of all authors. All authors were responsible for the study of concept, data acquisition and interpretation. G.G. and A.B. were responsible for the project's conceiving and design, being also responsible for drafting and revising the manuscript.

Funding: This research was funded by MIUR—Ministero dell'Istruzione, dell'Università e della Ricerca (Italy) within the PRIN—2015 program, HERCULES project; and by MISE—Ministero dello Sviluppo Economico (Italy) within the Industria 2015 program, MICROGEN 30 project.

Conflicts of Interest: The authors declare no conflict of interest.

References

1. U.S. Energy Information Administration. *International Energy Outlook 2017*; U.S. Energy Information Administration: Washington, DC, USA, 2017.
2. Carrera, A.; Beretta, A.; Groppi, G. Catalytic Partial Oxidation of Iso-Octane over Rh/α-Al_2O_3 in an Adiabatic Reactor: An Experimental and Modeling Study. *Ind. Eng. Chem. Res.* **2017**, *56*, 4911–4919. [CrossRef]
3. Yvonne, R.; Simon, L.; Johannes Pfister, C.D. *Fuel Cells and Hydrogen for Green Energy in European Cities and Regions. A Study for the Fuel Cells and Hydrogen Joint Undertaking*; Sederanger: Frankfurt, Germany, 2018.
4. Farrauto, R.J.; Liu, Y.; Ruettinger, W.; Ilinich, O.; Shore, L.; Giroux, T. Precious Metal Catalysts Supported on Ceramic and Metal Monolithic Structures for the Hydrogen Economy. *Catal. Rev.* **2007**, *49*, 141–196. [CrossRef]

5. Heck, R.M.; Gulati, S.; Farrauto, R.J. The Application of Monoliths for Gas Phase Catalytic Reactions. *Chem. Eng. J.* **2001**, *82*, 149–156. [CrossRef]
6. Farrauto, R.; Hwang, S.; Shore, L.; Ruettinger, W.; Lampert, J.; Giroux, T.; Liu, Y.; Ilinich, O. New Material Needs for Hydrocarbon Fuel Processing: Generating Hydrogen for the PEM Fuel Cell. *Annu. Rev. Mater. Res.* **2003**, *33*, 1–27. [CrossRef]
7. Groppi, G.; Tronconi, E. Honeycomb Supports with High Thermal Conductivity for Gas/Solid Chemical Processes. *Catal. Today* **2005**, *105*, 297–304. [CrossRef]
8. Specchia, S. Fuel Processing Activities at European Level: A Panoramic Overview. *Int. J. Hydrog. Energy* **2014**, *39*, 17953–17968. [CrossRef]
9. Kraaij, G.J.; Specchia, S.; Bollito, G.; Mutri, L.; Wails, D. Biodiesel Fuel Processor for APU Applications. *Int. J. Hydrog. Energy* **2009**, *34*, 4495–4499. [CrossRef]
10. Specchia, S.; Tillemans, F.W.A.; van den Oosterkamp, P.F.; Saracco, G. Conceptual Design and Selection of a Biodiesel Fuel Processor for a Vehicle Fuel Cell Auxiliary Power Unit. *J. Power Sources* **2005**, *145*, 683–690. [CrossRef]
11. Kolb, G.; Baier, T.; Schürer, J.; Tiemann, D.; Ziogas, A.; Specchia, S.; Galletti, C.; Germani, G.; Schuurman, Y. A Micro-Structured 5 kW Complete Fuel Processor for Iso-Octane as Hydrogen Supply System for Mobile Auxiliary Power Units Part II—Development of Water–Gas Shift and Preferential Oxidation Catalysts Reactors and Assembly of the Fuel Processor. *Chem. Eng. J.* **2008**, *138*, 474–489. [CrossRef]
12. Hou, Z.; Chen, P.; Fang, H.; Zheng, X.; Yashima, T. Production of Synthesis Gas via Methane Reforming with CO_2 on Noble Metals and Small Amount of Noble-(Rh-) Promoted Ni Catalysts. *Int. J. Hydrog. Energy* **2006**, *31*, 555–561. [CrossRef]
13. le Saché, E.; Santos, J.; Smith, T.J.; Centeno, M.A.; Arellano-Garcia, H.; Odriozola, J.A.; Reina, T.R. Multicomponent Ni-CeO_2 Nanocatalysts for Syngas Production from CO_2/CH_4 Mixtures. *J. CO2 Util.* **2018**, *25*, 68–78.
14. Tavazzi, I.; Beretta, A.; Groppi, G.; Maestri, M.; Tronconi, E.; Forzatti, P. Experimental and Modeling Analysis of the Effect of Catalyst Aging on the Performance of a Short Contact Time Adiabatic CH_4-CPO Reactor. *Catal. Today* **2007**, *129*, 372–379. [CrossRef]
15. Fichtner, M.; Mayer, J.; Wolf, D.; Schubert, K. Microstructured Rhodium Catalysts for the Partial Oxidation of Methane to Syngas under Pressure. *Ind. Eng. Chem. Res.* **2001**, *40*, 3475–3483. [CrossRef]
16. Maestri, M.; Beretta, A.; Groppi, G.; Tronconi, E.; Forzatti, P. Comparison among Structured and Packed-Bed Reactors for the Catalytic Partial Oxidation of CH_4 at Short Contact Times. *Catal. Today* **2005**, *105*, 709–717. [CrossRef]
17. Qi, A.; Wang, S.; Ni, C.; Wu, D. Autothermal Reforming of Gasoline on Rh-Based Monolithic Catalysts. *Int. J. Hydrog. Energy* **2007**, *32*, 981–991. [CrossRef]
18. Costa, D.S.; Gomes, R.S.; Rodella, C.B.; da Silva, R.B.; Fréty, R.; Teixeira Neto, É.; Brandão, S.T. Study of Nickel, Lanthanum and Niobium-Based Catalysts Applied in the Partial Oxidation of Methane. *Catal. Today* **2018**. [CrossRef]
19. Carrera, A.; Pelucchi, M.; Stagni, A.; Beretta, A.; Groppi, G. Catalytic Partial Oxidation of n-Octane and Iso-Octane: Experimental and Modeling Results. *Int. J. Hydrog. Energy* **2017**, *42*, 24675–24688. [CrossRef]
20. Nogare, D.D.; Degenstein, N.J.; Horn, R.; Canu, P.; Schmidt, L.D. Modeling Spatially Resolved Data of Methane Catalytic Partial Oxidation on Rh Foam Catalyst at Different Inlet Compositions and Flowrates. *J. Catal.* **2011**, *277*, 134–148. [CrossRef]
21. Panuccio, G.J.; Williams, K.A.; Schmidt, L.D. Contributions of Heterogeneous and Homogeneous Chemistry in the Catalytic Partial Oxidation of Octane Isomers and Mixtures on Rhodium Coated Foams. *Chem. Eng. Sci.* **2006**, *61*, 4207–4219. [CrossRef]
22. Wanat, E.C.; Venkataraman, K.; Schmidt, L.D. Steam Reforming and Water–Gas Shift of Ethanol on Rh and Rh–Ce Catalysts in a Catalytic Wall Reactor. *Appl. Catal. A Gen.* **2004**, *276*, 155–162. [CrossRef]
23. Degenstein, N.; Subramanian, R.; Schmidt, L. Partial Oxidation of n-Hexadecane at Short Contact Times: Catalyst and Washcoat Loading and Catalyst Morphology. *Appl. Catal. A Gen.* **2006**, *305*, 146–159. [CrossRef]
24. Donazzi, A.; Maestri, M.; Michael, B.C.; Beretta, A.; Forzatti, P.; Groppi, G.; Tronconi, E.; Schmidt, L.D.; Vlachos, D.G. Microkinetic Modeling of Spatially Resolved Autothermal CH_4 Catalytic Partial Oxidation Experiments over Rh-Coated Foams. *J. Catal.* **2010**, *275*, 270–279. [CrossRef]

25. Maestri, M.; Beretta, A.; Faravelli, T.; Groppi, G.; Tronconi, E.; Vlachos, D.G. Two-Dimensional Detailed Modeling of Fuel-Rich H_2 Combustion over Rh/Al_2O_3 Catalyst. *Chem. Eng. Sci.* **2008**, *63*, 2657–2669. [CrossRef]
26. Batista da Silva, R.; Brandão, S.T.; Lucotti, A.; Tommasini, M.S.; Castiglioni, C.; Groppi, G.; Beretta, A. Chemical Pathways in the Partial Oxidation and Steam Reforming of Acetic Acid over a $Rh-Al_2O_3$ catalyst. *Catal. Today* **2017**, *289*, 162–172. [CrossRef]
27. Donazzi, A.; Livio, D.; Beretta, A.; Groppi, G.; Forzatti, P. Surface Temperature Profiles in CH_4 CPO over Honeycomb Supported Rh Catalyst Probed with in Situ Optical Pyrometer. *Appl. Catal. A Gen.* **2011**, *402*, 41–49. [CrossRef]
28. Beretta, A.; Donazzi, A.; Groppi, G.; Maestri, M.; Tronconi, E.; Forzatti, P. Gaining Insight into the Kinetics of Partial Oxidation of Light Hydrocarbons on Rh, through a Multiscale Methodology Based on Advanced Experimental and Modeling Techniques. *Catalysis* **2013**, *25*, 1–49.
29. Iaquaniello, G.; Antonetti, E.; Cucchiella, B.; Palo, E.; Salladini, A.; Guarinoni, A.; Lainati, A.; Basini, L. Natural Gas Catalytic Partial Oxidation: A Way to Syngas and Bulk Chemicals Production. In *Natural Gas—Extraction to End Use*; InTech: London, UK, 2012.
30. Pagani, D.; Batista, R.; Silva, D.; Moioli, E.; Donazzi, A.; Lucotti, A.; Tommasini, M.; Castiglioni, C.; Brandao, S.T.; Beretta, A.; et al. Annular Reactor Testing and Raman Surface Characterization of the CPO of i-Octane and n-Octane on Rh Based Catalyst. *Chem. Eng. J.* **2016**, *294*, 9–21. [CrossRef]
31. Beretta, A.; Groppi, G.; Carrera, A.; Donazzi, A. *Analysis of the Impact of Gas-Phase Chemistry in Adiabatic CPO Reactors by Axially Resolved Measurements*; Academic Press: Cambridge, MA, USA, 2017; pp. 161–201.
32. Beretta, A.; Groppi, G.; Lualdi, M.; Tavazzi, I.; Forzatti, P. Experimental and Modeling Analysis of Methane Partial Oxidation: Transient and Steady-State Behavior of Rh-Coated Honeycomb Monoliths. *Ind. Eng. Chem. Res.* **2009**, *48*, 3825–3836. [CrossRef]
33. Cimino, S.; Lisi, L.; Russo, G. Effect of Sulphur during the Catalytic Partial Oxidation of Ethane over Rh and Pt Honeycomb Catalysts. *Int. J. Hydrog. Energy* **2012**, *37*, 10680–10689. [CrossRef]
34. Livio, D.; Donazzi, A.; Beretta, A.; Groppi, G.; Forzatti, P. Experimental and Modeling Analysis of the Thermal Behavior of an Autothermal C_3H_8 Catalytic Partial Oxidation Reformer. *Ind. Eng. Chem. Res.* **2012**, *51*, 7573–7583. [CrossRef]
35. Donazzi, A.; Beretta, A.; Groppi, G.; Forzatti, P. Catalytic Partial Oxidation of Methane over a 4% $Rh/\alpha-Al_2O_3$ catalyst. Part I: Kinetic Study in Annular Reactor. *J. Catal.* **2008**, *255*, 241–258. [CrossRef]
36. Pagani, D.; Livio, D.; Donazzi, A.; Beretta, A.; Groppi, G.; Maestri, M.; Tronconi, E. A Kinetic Analysis of the Partial Oxidation of C_3H_8 over a 2% Rh/Al_2O_3 Catalyst in Annular Microreactor. *Catal. Today* **2012**, *197*, 265–280. [CrossRef]

Article

Catalytic Behaviour of Ce-Doped Ni Systems Supported on Stabilized Zirconia under Dry Reforming Conditions

Ahmed Sadeq Al-Fatesh [1,*], Yasir Arafat [1,*], Ahmed Aidid Ibrahim [1], Samsudeen Olajide Kasim [1], Abdulrahman Alharthi [2], Anis Hamza Fakeeha [1], Ahmed Elhag Abasaeed [1], Giuseppe Bonura [3] and Francesco Frusteri [3]

1 Chemical Engineering Department, College of Engineering, King Saud University, P.O. Box 11421 Riyadh, Saudi Arabia; aidid@ksu.edu.sa (A.A.I.); sofkolajide2@gmail.com (S.O.K.); anishf@ksu.edu.sa (A.H.F.); abasaeed@ksu.edu.sa (A.E.A.)
2 College Sciences and Humanities, Prince Sattam Bin Abdulaziz University, Al-Kharj 11942, Saudi Arabia; a.alharthi@psau.edu.sa
3 CNR-ITAE, Istituto di Tecnologie Avanzate per l'Energia "Nicola Giordano", Via S. Lucia sopra Contesse 5, 98126 Messina, Italy; giuseppe.bonura@itae.cnr.it (G.B.); francesco.frusteri@itae.cnr.it (F.F.)
* Correspondence: aalfatesh@ksu.edu.sa (A.S.A.-F.); engryasir@ksu.edu.sa (Y.A.); Tel.: +966114676859 (A.S.A.-F. & Y.A.); Fax: +966114678770 (A.S.A.-F. & Y.A.)

Received: 19 April 2019; Accepted: 16 May 2019; Published: 22 May 2019

Abstract: Ni supported on bare and modified ZrO_2 samples were synthesized using the incipient wet impregnation method. The *t*-ZrO_2 phase was stabilized by incorporation of La_2O_3 into ZrO_2. Moreover, the influence of CeO_2-doping on the physico-chemical and catalytic properties under CO_2 reforming conditions was probed. The characterization data of the investigated catalysts were obtained by using XRD, CO_2/H_2-TPD, BET, TPR, TPO, TGA, XPS and TEM characterization techniques. In the pristine Ni/Zr catalyst, the *t*-ZrO_2 phase transformed into the monoclinic phase. However, upon support modification by La_2O_3, significant effects on the physicochemical properties were observed due to the monoclinic-to-tetragonal ZrO_2 phase transformation also affecting the catalytic activity. As a result, superior activity on the La_2O_3 modified Ni/Zr catalyst was achieved, while no relevant change in the surface properties and activity of the catalysts was detected after doping by CeO_2. The peculiar behavior of the Ni/La-ZrO_2 sample was related to higher dispersion of the active phase, with a more pronounced stabilization of the *t*-ZrO_2 phase.

Keywords: CO_2 reforming; Ni-catalyst; syngas; tetragonal zirconia; phase stabilization

---

1. Introduction

Modern civilization is confronted with two major challenges: dwindling energy resources [1,2] and global warming caused by greenhouse gases (mainly CH_4 and CO_2) [3,4]. The dry reforming of the methane (DRM) reaction for the conversion of greenhouse gases into a valuable synthesis gas (H_2 and CO) product is a potential contestant for confronting both these challenges simultaneously. Syngas is recognized as the building block for the production of H_2 and liquid fuels such as olefins, paraffins, methanol, oxygenates and aromatics in the petrochemical industry, employing the Fischer–Tropsch synthesis process [5,6].

It is an acknowledged fact that DRM is hampered by some side reactions, such as the Boudouard reaction (Equation (2)), water gas shift reaction (Equation (3)), and methane decomposition

(Equation (4)), which are considered the major reactions leading to the deactivation of catalysts as a result of coke deposition [7].

$$CH_4 + CO_2 \leftrightarrow 2CO + 2H_2 \qquad \Delta H^0 = +261 \text{ kJ mol}^{-1} \tag{1}$$

$$2CO \leftrightarrow C + CO_2 \qquad \Delta H^0 = -171 \text{ kJ mol}^{-1} \tag{2}$$

$$H_2 + CO_2 \leftrightarrow H_2O + CO \qquad \Delta H^0 = +41 \text{ kJ mol}^{-1} \tag{3}$$

$$CH_4 \leftrightarrow 2H_2 + C \qquad \Delta H^0 = +75 \text{ kJ mol}^{-1} \tag{4}$$

Coke formation represents a serious problem in DRM. Therefore, noble metal catalysts like Ir, Pt, Rh, Ru and Pd, based on anti-coking properties, were used to constrain such phenomena [8]. In spite of their marvelous characteristics, noble metals are rare and expensive, and end up with a higher cost-benefit ratio. Therefore, the application of noble metals is no longer commercially feasible. On the other hand, Ni-based catalysts are cheap and abundant, and also demonstrate an interesting catalytic performance for the DRM reaction. Unfortunately, nickel particles undergo thermal sintering and tend to promote coke formation over the catalyst surface, leading to catalyst deactivation. Therefore, besides the relevance of active metal components, the role of support material is crucial in inhibiting (or at least limiting) the formation of coke. This can be accomplished by: modifying the electronic properties of the catalyst through a control of the metal-support interaction [9]; improving the oxygen storage capacity [10]; or controlling the size of active metal particles [11]. Consequently, several oxide supports, like Al_2O_3, SiO_2, ZrO_2, CeO_2 and La_2O_3 have been employed to enhance the catalytic activity and stability of Ni catalysts. Among different oxides, ZrO_2 is extensively employed as a functional support, possessing several unique properties like redox properties, acid-base bi-functional properties, thermal and mechanical stability, high ionic conductivity and oxygen transport properties, making it a favourable support for reforming reactions [12,13]. Moreover, the partial CO_2 activation, as well as the strong anchoring effects endowed by Zr^{4+}, may effectively boost the DRM reaction. However, it was found that Ni/ZrO_2 catalysts undergo serious coke formation over their surface, leading to fast deactivation and, subsequently, reactor plugging [14,15].

The deactivation of the catalyst may occur for several reasons, for instance, commonly available ZrO_2 possesses a rather smaller surface area (viz. <50 m^2/g) in comparison to its counterparts [16,17]. Moreover, ZrO_2 is characterized by three polymorph structures, named as the monoclinic phase (*m*-ZrO_2, room temperature–1175 °C), tetragonal phase (*t*-ZrO_2, 1175–2370 °C), and cubic phase (*c*-ZrO_2, 2370–2680 °C) [18]. The *c*-ZrO_2 is poorly stable at room temperature, which restricts its large scale application in catalysis as compared to *m*-ZrO_2 and *t*-ZrO_2. Furthermore, *t*-ZrO_2 is found to have a better performance than *m*-ZrO_2 [19]. In addition, *t*-ZrO_2 is widely utilized as a support for Ni catalysts in CH_4 reforming processes, even if its thermodynamic instability represents a problem to overcome. Rezaei et al. proposed that the surface energy difference between monoclinic and tetragonal phases may induce the *t*-ZrO_2 phase to be thermodynamically stable for tiny crystals [20]. Consequently, in order to exploit the excellent properties associated with ZrO_2 support, it is essential to find a way to prepare high surface area ZrO_2 and stabilizing *t*-ZrO_2, which will obviously contribute to enhancing the catalytic performance.

Recently, heteroatom oxide compounds (e.g., Y_2O_3, CeO_2, TiO_2 and K_2O) have been reported to stabilize *t*-ZrO_2 in catalytic processes [19,21]. When La_2O_3 is assigned to modify the support for the dry reforming reaction, the ability of support to activate CO_2 was enhanced by means of $La_2O_2CO_3$ formation. In this way, it discards the accumulated carbon through oxidation of coke at the Ni-La_2O_3 interface [22,23].

Furthermore, in order to promote the oxygen storage capacity of ZrO_2, as well as to enhance the thermal stability of Ni-based catalysts, CeO_2 was proposed as a promoter (in fact, ceria demonstrates outstanding redox properties as a result of a quick transition between Ce^{4+}/Ce^{3+}).

The purpose of this study was to exploit the aforementioned peculiar properties of ZrO_2 support, trying to circumvent the limitations associated with its instability and low surface area. To achieve this objective, high surface area ZrO_2 supported Ni catalysts were prepared using the incipient wet-impregnation method. Subsequently, ZrO_2 support was modified by heteroatom oxides, like La_2O_3, with an intention that it might stabilize the *t*-ZrO_2 phase. In addition, 1% CeO_2 promoter was also incorporated to enhance the oxygen storage capacity of catalysts. Consequently, the unique properties achieved as an outcome of the synergistic effect of the ceria promoter and high surface area ZrO_2 support modifiers (La_2O_3) were studied to stabilize *t*-ZrO_2 and their eventual influence on the catalytic performance.

2. Results and Discussion

2.1. Physicochemical Features of the Ceria Promoted Ni/x-Zr (x = 0, La_2O_3) Catalysts

Figure 1a illustrates the X-ray diffraction patterns of fresh catalyst samples. The diffraction peaks of NiO can be featured on the diffractograms at 2θ = 37.3° and 43.3° corresponding to the [101] and [012] reflections, respectively.

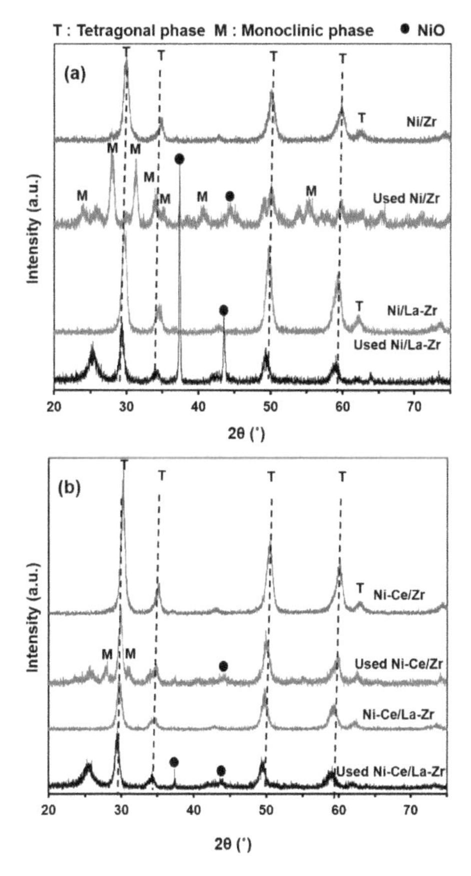

Figure 1. X-ray diffraction patterns of fresh and used (**a**) Ni/x-Zr (x = 0, La_2O_3) and (**b**) Ni-Ce/x-Zr catalysts.

According to the literature, monoclinic zirconia (*m*-ZrO_2) was found at 2θ≈24.0°, 28.2°, 31.5°, 34.2°, 34.4°, 35.3° and 40.7°, while tetragonal zirconia (*t*-ZrO_2) appeared at 2θ≈30.0°, 34.8°, 35.1°,

50.0° and 59.4° [24]. Based on this evidence, it can be observed that "pure" Ni/Zr catalyst predominantly consists of *t*-ZrO_2. After incorporation of La_2O_3, the peaks related to *t*-ZrO_2 became sharper and more pronounced. It implies that La_2O_3 contributed to enhance the stability of the *t*-ZrO_2 phase of the Ni/Zr catalyst. Moreover, it is clear that the crystalline La_2O_3 phase (2θ = 28° and 49°, JCPDS: 01-089-4016) was not individually identified on the diffractograms, indicating that La_2O_3 was highly dispersed and incorporated into the ZrO_2 lattice. Furthermore, when the Ni/Zr catalyst was promoted using 1% CeO_2 (see Figure 1b), more intense diffraction patterns attributing to *t*-ZrO_2 appeared for the Ni-Ce/Zr catalyst. However, when Ni/La-Zr was promoted by 1% CeO_2, the intensity of diffraction peaks corresponding to *t*-ZrO_2 was considerably reduced for the Ni-Ce/La-Zr catalyst.

The textural properties of Ni/x-Zr (x = 0, La_2O_3) and CeO_2 promoted fresh catalysts are shown in Figure 2 and Table 1. It is evident that the incorporation of lanthanum as a ZrO_2 modifier in the catalyst composition led to a significant enhancement of surface area and porosity with respect to the reference Ni/Zr sample.

Table 1. Properties and H_2 consumption during reduction of NiO species of fresh Ni/x-Zr and Ni-Ce/x-Zr (x = 0, La_2O_3) catalysts (calcined at 600 °C).

SAMPLE	BET (m^2/g)	P.V. (cm^3/g)	P.D. (nm)	H_2 Consumption (μmol/g)
La-Zr	67	0.247	4.0	-
Ni/Zr	39	0.111	11.3	1049.3
Ni/La-Zr	62	0.246	15.8	814.2
Ni-Ce/Zr	41	0.139	13.3	945.5
Ni-Ce/La-Zr	65	0.242	15.0	782.2

In particular, the Ni/La-Zr sample presented the highest cumulative pore volume (P.V., 0.246 cm^3/g), accounting for an average pore diameter (P.D.) just smaller than 16 nm. However, the undoped Ni/Zr sample exhibited the lowest extension of surface area (39 m^2/g) along with the lowest porosity (0.111 cm^3/g).

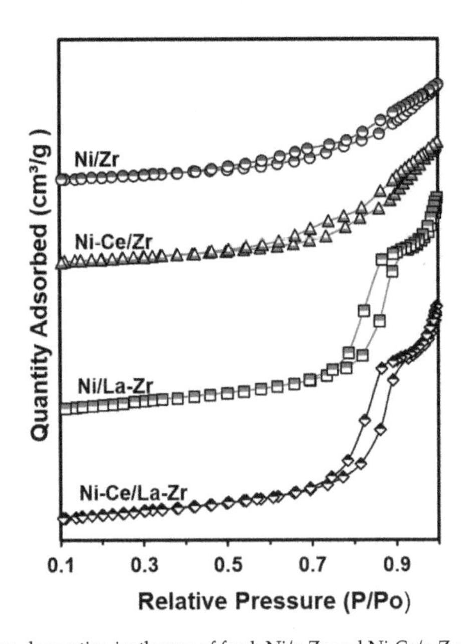

Figure 2. N_2 adsorption–desorption isotherms of fresh Ni/x-Zr and Ni-Ce/x-Zr (x = 0, La_2O_3) catalysts.

Moreover, the N_2 adsorption–desorption branch of the Ni/Zr sample exemplifies a typical type IV isotherm, characteristic for mesoporous solids with a H_4-type hysteresis loop. The doping of ZrO_2 with La_2O_3 dramatically affected the hysteresis loop, transforming it into a H_1-type hysteresis loop, associated with a cylindrical pore geometry as well as relatively high uniformity in pore size. Consequently, the La_2O_3 modified catalyst possessed a high surface area and a mesoporous structure. After promotion with 1% CeO_2, no significant change in the textural properties of the catalysts was observed either in quantitative (see Table 1) or qualitative (see Figure 2) terms.

The reducibility and metal-support interaction of the Ni/Zr system after the support modification was evaluated by TPR analysis. Figure 3 illustrates the reduction profile of Ni/x-Zr (x = 0, La_2O_3) and Ni-Ce/x-Zr (x = 0, La_2O_3) samples.

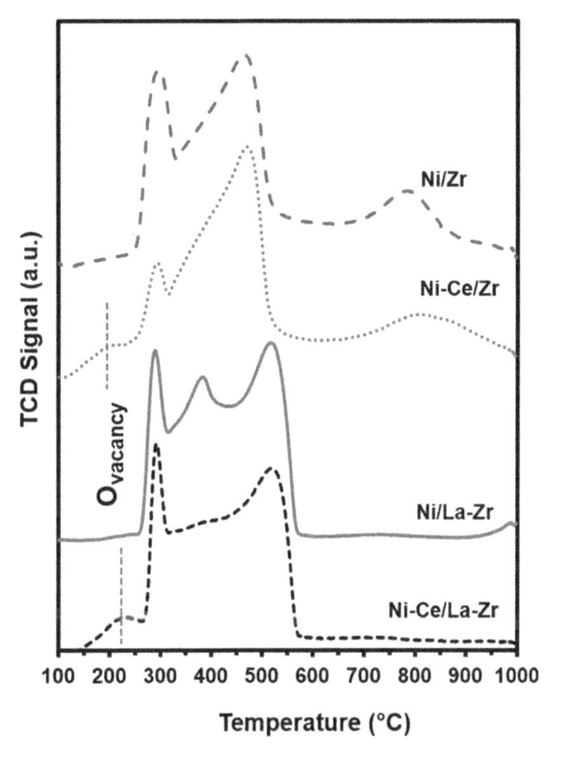

Figure 3. TPR profiles of fresh Ni/x-Zr and Ni-Ce/x-Zr (x = 0, La_2O_3) catalysts.

It is evident from the TPR profile that the modification of ZrO_2 support by the addition of La_2O_3 has a considerable influence on the reduction profiles, being in all cases characterized by the occurrence of multiple reduction peaks, spanning the range 100–1000 °C, and ascribed to different nickel species with different degrees of interaction with the support. A first reduction peak observed in the range 260–290 °C with a shoulder at some lower temperature in the case of CeO_2 promoted catalysts, may be assigned to the reduction of surface oxygen species [25,26]. The second reduction peak can be referred to the reduction of NiO grafted onto the support by the weak interaction. The reduction peak at a higher temperature (>750 °C) may be attributed to the development of NiO–ZrO_2 solid solutions [27]. It is interesting to observe that the modification of ZrO_2 with La_2O_3 considerably improved the reduction kinetics, considering that NiO was completely reduced (NiO→Ni) below 550 °C owing to the inhibition of formation of NiO–ZrO_2 solid solutions. It implies that the incorporation of La_2O_3 to ZrO_2 had made it feasible to achieve the active metal at some lower temperature accompanied by the improved dispersion of metallic species. Furthermore, upon ZrO_2 modification using lanthana, the peaks corresponding to NiO–ZrO_2 solid solutions seemed to be completely suppressed. It implies that La_2O_3 modification also prevents the formation of NiO–ZrO_2 solid solutions. On the other hand, La_2O_3 modification also contributed to enhancing the metal–support interaction, which is evident from the drop in H_2 consumption from Ni/Zr to the Ni/La-Zr catalyst by 22% and from Ni-Ce/Zr to the Ni-Ce/La-Zr catalyst by 17% (Table 1). It is likely that the enhancement in the metal–support interaction is accompanied by the improvement in active phase dispersion. Likewise, Rotgerink et al. [28] also established that the incorporation of La_2O_3 to the Ni/Zr catalyst upgrades the reducibility and Ni dispersion. Some other studies have also found that the improvement in the reducibility of NiO is the outcome of the ability of La_2O_3 to disperse Ni metallic species on the ZrO_2 support [29,30]. After doping by CeO_2, the reduction patterns of catalysts had not significantly changed. Furthermore, in the case of the Ni-Ce/La-Zr catalyst, a peak centred at 225 °C, corresponding to surface oxygen species, became prominent, which may be the outcome of the synergistic effect of CeO_2 and La_2O_3. Actually, the La_2O_3 lattice is recognized for surface and bulk oxygen vacancies, and their density is associated with the La_2O_3 concentration [31].

The quantitative data of H_2-TPD measurements are listed in Table 2.

Table 2. Data of H_2-TPD measurements.

SAMPLE	H_2 uptake ($\mu mol/g_{cat}$)	MSA [a] (m^2/g_{cat})	D_{Ni} [b] (%)	d_{Ni} [c] (nm)
Ni/Zr	107.3	8.4	25.2	4.0
Ni/La-Zr	253.3	19.8	59.6	1.7
Ni-Ce/Zr	173.5	13.6	40.8	2.5
Ni-Ce/La-Zr	182.5	14.3	42.9	2.4

[a] Metal surface area. [b] Nickel dispersion. [c] Average Ni particle diameter.

Given the same concentration of Ni in the catalysts, it was observed how the various modifiers of the ZrO_2 support, as well as the promotion of the active phase with ceria, affect the surface characteristics of the prepared samples. In fact, the "unpromoted" Ni/Zr sample exhibited the lowest H_2 uptake (107.3 mol/g_{cat}), accounting for a minor metal surface area (8.4 m^2/g_{cat}) and dispersion (25.2%), as the result of particles not larger than 4 nm generated during preparation. Instead, promotion of carrier oxide by lanthanum oxide positively influenced the Ni/Zr system, favouring higher metal surface area and dispersion (16.3–19.8 m^2/g and 49.2–59.6%, respectively), as the result of particles with a smaller diameter (1.7–2.1 nm). Regarding the effect of ceria, a promoting effect was markedly evident only on the Ni/Zr system, allowing an enhancement both in terms of metal surface area (13.6 m^2/g) and nickel dispersion (40.8%). In contrast, it is clearly visible that the Ni-Ce/La-Zr sample showed that the surface properties dropped with respect to the sample without Ce (Ni/La-Zr). This suggests that ceria promotion has some inhibiting effect on the Ni/La-Zr catalyst.

The basic properties of Ni/x-Zr (x = 0, La_2O_3) catalysts were estimated by CO_2 desorption measurements. Figure 4 illustrates the TPD profiles of the catalysts. It is recognized that the strength of basic sites is determined by the adsorption and desorption peak of CO_2 in the corresponding temperature range: weak basic sites (50–200 °C), medium/Lewis basic sites (200–400 °C) and strong basic sites (400–650 °C).

The TPD profiles depicted the weak Lewis and moderate Lewis basicity of the reference Ni/Zr and La_2O_3-modified samples. It is clearly visible that the Ni/Zr catalyst experienced no significant variation in the basic character after CeO_2 promotion except a rise in basicity in La_2O_3 modified catalysts.

XPS analyses were conducted to determine and interpret the chemical state of elements present in the catalysts and the correspondent spectra are illustrated in Figure 5.

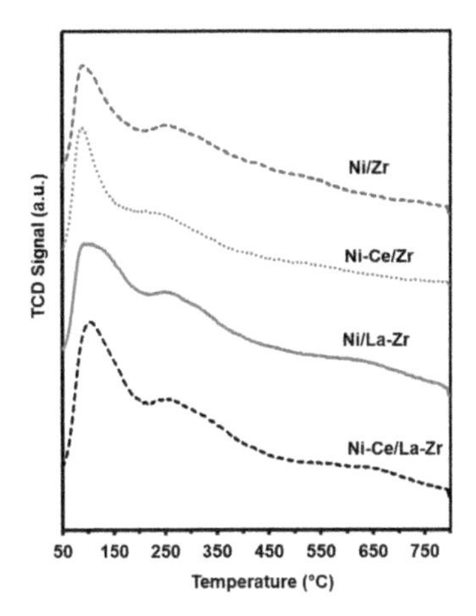

Figure 4. Patterns of fresh Ni/x-Zr and Ni-Ce/x-Zr (x = 0, La_2O_3) catalysts.

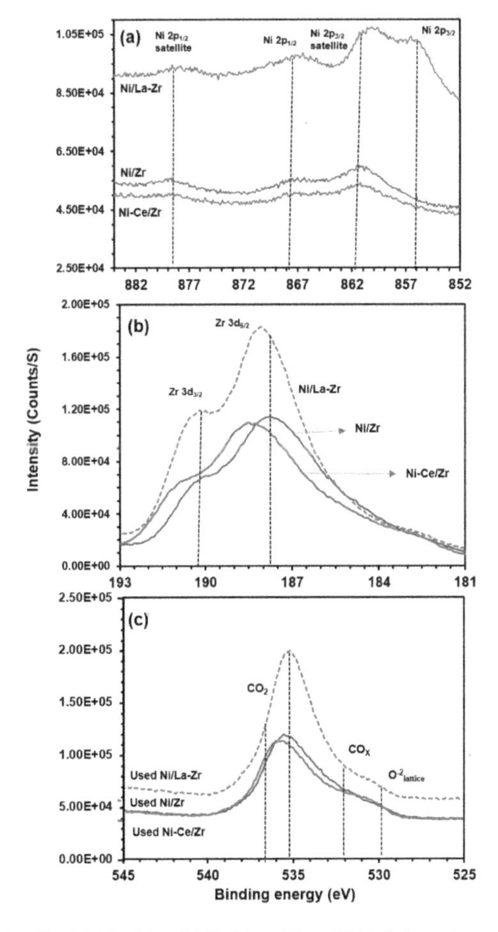

Figure 5. Spectra in the "fresh" (**a**) Ni 2p, (**b**) Zr 3d and "used" (**c**) O 1s regions of Ni/La-Zr, Ni/Zr and Ni-Ce/Zr catalysts.

Ni 2p peaks of all the catalysts Ni/Zr, Ni-Ce/Zr and Ni/La-Zr are presented in Figure 5a. The binding energy of the main Ni 2p peak may be found at 855.6 eV, while the rest of the peaks and satellites are visible at other binding energies. It is apparent from the shape of the main peak (Ni $2p_{3/2}$) that it is the combination of Ni^0 and Ni^{2+}/Ni^{3+} oxides [32,33]. The Ni 2p spectra of Ni/La-Zr and Ni-Ce/Zr catalysts becomes complicated due to the overlapping of La 3d, Ce 3d, and La MNN Auger peaks. As a result, it gives rise to uncertainty in the quantification of the XPS spectra. Generally, the largest uncertainty is confronted as an outcome of the overlapping of Ni $2p_{3/2}$ and La $3d_{3/2}$. However, in our case, the Ni/La-Zr catalyst was found to have the highest Ni concentration. The Zr 3d peaks of Ni/Zr, Ni/La-Zr and Ni-Ce/Zr catalysts are shown in Figure 5b. Zr $3d_{3/2}$ and Zr $3d_{5/2}$ can be found at 190.1 eV and 188.2 eV, respectively, for the Ni/Zr catalyst. Upon support modification using La_2O_3, the intensity of the Zr 3d peak significantly increased and it shifted to a higher binding energy. Likewise, CeO_2 promoted peaks also shifted toward a higher binding energy. The shifting of binding energy to higher values may be attributed to the enhancement of oxygen vacancy as a result of an oxygen deficient state. The phenomenon of shifting of Zr 3d binding energy is related to the large quantity of lattice defects (oxygen vacancies) [34]. The O 1s data for Ni/Zr, Ni-Ce/Zr, Ni/La-Zr, used Ni/Zr and Ni/La-Zr catalysts are illustrated in Figure 5c. The binding energy of O 1s located at 529.6 eV, corresponding to low energy primary oxygen, is consistent for all the catalysts. This peak may be attributed to lattice oxygen ($O^{2-}_{lattice}$). The O 1s XPS data also displayed a small amount of CO_x species, which implies that CO_x species are present as reaction intermediates on the catalyst surface. Consequently, significant variation in chemical states of the catalyst after La incorporation into Ni/Zr implies that support modification using La_2O_3 has substantial influence.

2.2. Catalytic Activity and Stability

The catalytic behavior of the prepared samples was investigated under dry reforming conditions, in terms of carbon dioxide and methane conversion as a function of time on stream for Ni/x-Zr and Ni-Ce/x-Zr (x = 0, La_2O_3) catalysts (Figure 6). The effects of ZrO_2 modification, as well as Ni doping with cerium oxide on the Ni/Zr system, were analyzed. A maximum reaction temperature of 700 °C was chosen, considering that at higher temperature, the formation of encapsulated carbon leads to a rapid deactivation of the system [35,36].

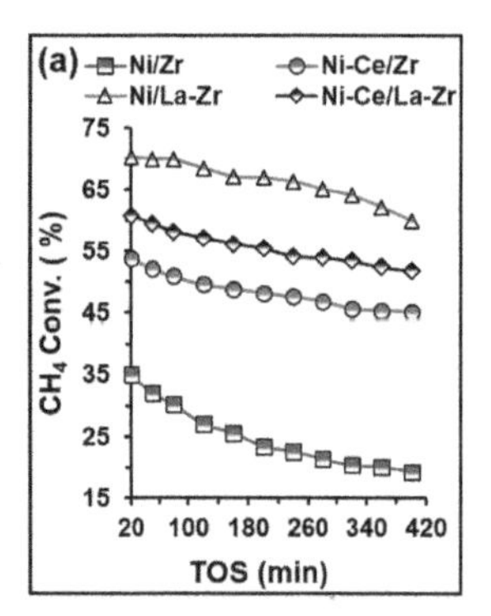

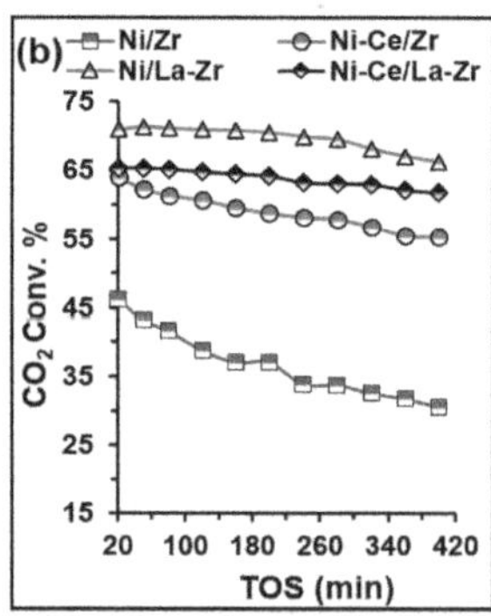

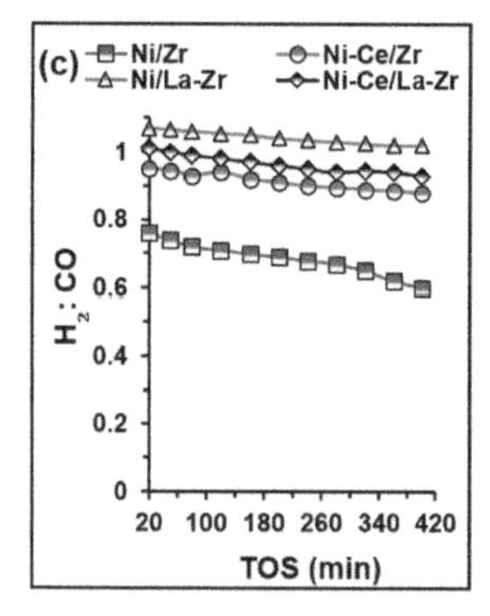

Figure 6. (**a**) CH_4 and (**b**) CO_2 conversion and (**c**) H_2:CO ratio versus time on stream (TOS) for Ni/x-Zr (x = 0, La_2O_3) catalysts at T_R = 700 °C; F/W = 133.33 mL/min.g_{cat}). Total flow rate = 40 mL/min (CH_4 = 17 mL/min, CO_2 = 17mL/min, N_2 = 6 mL/min).

It can be seen that not only La modification affects Ni/Zr activity, but also ceria significantly influences the catalytic behavior of the investigated samples. The "reference" Ni/Zr sample exhibited the worst performance, with an initial CH_4 and CO_2 conversion of 35 and 46%, respectively, which decreased to 20–30% after only 400 min. After the incorporation of La, the conversion rate of the catalyst samples was significantly enhanced. Interestingly, the best catalytic performance was obtained by using the La-modified sample, with similar initial values of CH_4 and CO_2 conversion (ca. 70%) and final conversion of 61 (CH_4) and 67% (CO_2), accounting for a slightly higher stability in respect to the other samples. Regarding the influence of the CeO_2 promoter, from Figure 6 a net improvement in the activity of the Ni-Ce/Zr sample can be observed, not only in terms of initial conversion values of CH_4 and CO_2 (54 and 64% respectively), but also in terms of stability, considering that after 400 min the CH_4 conversion decreased from 54 to 45%, and CO_2 conversion decreased from 64 to 55%, which were lower with respect to the values recorded in the correspondent catalyst without ceria. Instead, 1% CeO_2 promoter on the Ni phase showed an inhibiting effect on the Ni/La-Zr catalyst, since the activity of the Ni-Ce/La-Zr sample dropped to lower conversion values with respect to the correspondent unpromoted catalyst, by exhibiting a more pronounced decay trend of CH_4 conversion from 61 to 52%. In terms of the H_2/CO ratio (Figure 6c), a decreasing trend over time on the stream is visible for Ni/Zr, progressively favouring a bit greater formation of CO (Equation (3)) with a final H_2/CO value of 0.6. However, upon La_2O_3 modification, an equimolar formation of hydrogen and carbon monoxide with a $H_2 / CO \geq 1$ was achieved. On the whole, looking at the properties of the investigated samples, an almost straight-line relationship was found between the initial CO_2 conversion rate and the nickel surface area (see Figure 7). It seems evident that a controlled modification of the carrier structure determines a better surface exposure of the active phase (Ni), and therefore inducing higher CO_2 activation.

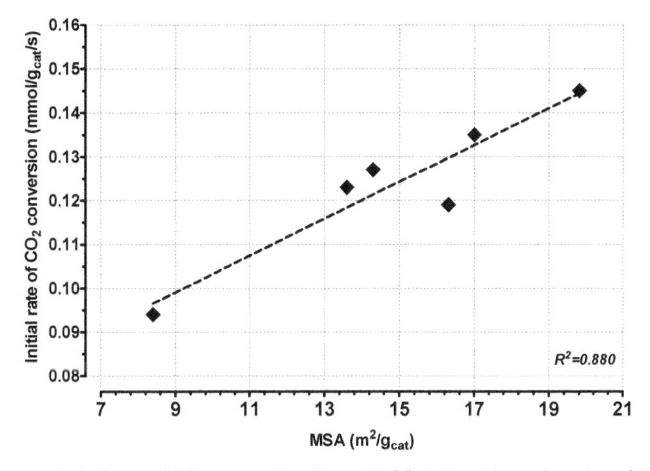

Figure 7. Initial rate of CO_2 conversion (T_R = 700 °C @ 20 min) as a function of MSA.

The catalyst activity must also be put in direct relation to the catalyst structure, considering that, depending on modification by oxides, ZrO_2 can become a prominently more stable phase. In general, a substantially higher activity is observed when ZrO_2 exists as a tetragonal phase, resulting in appreciably suppressed *m*-ZrO_2 phase even in the used catalysts, like the Ni/La-Zr and Ni-Ce/La-Zr samples (see Figure 1a,b). On the other hand, initially, Ni/Zr and Ni-Ce/Zr had a prominent *t*-ZrO_2 phase, however, upon thermal treatment, tetragonal-to-monoclinic phase transformation occurred.

To better analyze the obtained results, a deactivation factor (DF) was also deduced for all the samples as the ratio between the loss of activity calculated between the initial (20 min on stream) and final (400 min on stream) time and the initial conversion value of CH_4. From Figure 8, it is evident that the incorporation of La_2O_3 considerably improves the catalyst stability of the Ni/Zr system, as the deactivation factor decreased from 44% to 22%, with a lower net drop of the activity in respect of the initial time under the same experimental conditions. Moreover, the addition of ceria leads to a further improvement of the catalytic stability for both the undoped Ni/Zr and Ni/La-Zr samples, yielding the smallest loss of activity during time (< 12%) in the Ni-Ce/La-Zr sample.

To investigate the phase analysis following the catalytic treatment, post reaction XRD of the catalysts was performed (Figure 1). It was revealed that the thermal treatment had a profound influence on the phase transformation. For instance, used Ni/Zr catalyst found to have severe lattice distortion and *t*-ZrO_2 had significantly diminished and transformed into *m*-ZrO_2. Likewise, even CeO_2 promotion to Ni/Zr catalyst was not able to completely preserve the *t*-ZrO_2. However, upon support modification using La_2O_3, the *t*-ZrO_2 phase was perfectly retained with no appearance of *m*-ZrO_2. It implies that La_2O_3 incorporation has a significant contribution in stabilizing the *t*-ZrO_2 and achieving the 100% rise in catalytic activity. Similarly, used Ni-Ce/La-Zr also prevented the tetragonal-to-monoclinic transformation and pure *t*-ZrO_2 was preserved.

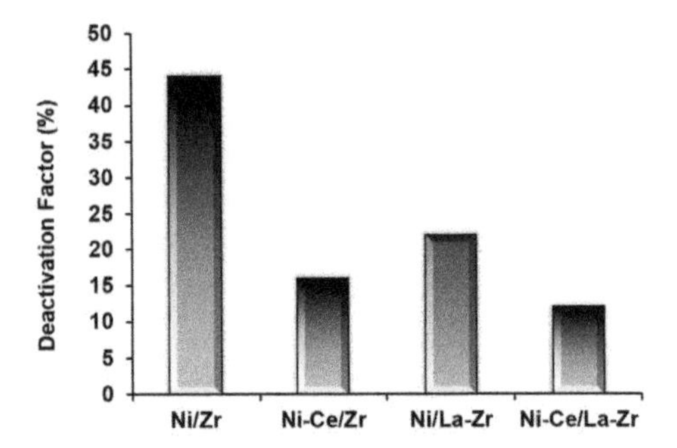

Figure 8. Deactivation factor (DF) of Ni/x-Zr (x = 0, La_2O_3) catalysts, calculated as: [(Initial CH_4 conversion − final CH_4 conversion)/initial conversion of CH_4] (Initial = 20 min; Final = 400 min at T_R = 700 °C).

As a main possible reason for the decay of activity during time, it was obviously necessary to consider the coke formation over the different samples, as it is known that the amount and typology of coke is also influenced by the catalyst properties. On this account, temperature programmed oxidation (TPO) ceria promoted and unpromoted Ni/x-ZrO_2 (0, La_2O_3) catalysts were conducted to determine the type of the deposited carbon during the CO_2 reforming reactions. The TPO peaks are illustrated in Figure 9a.

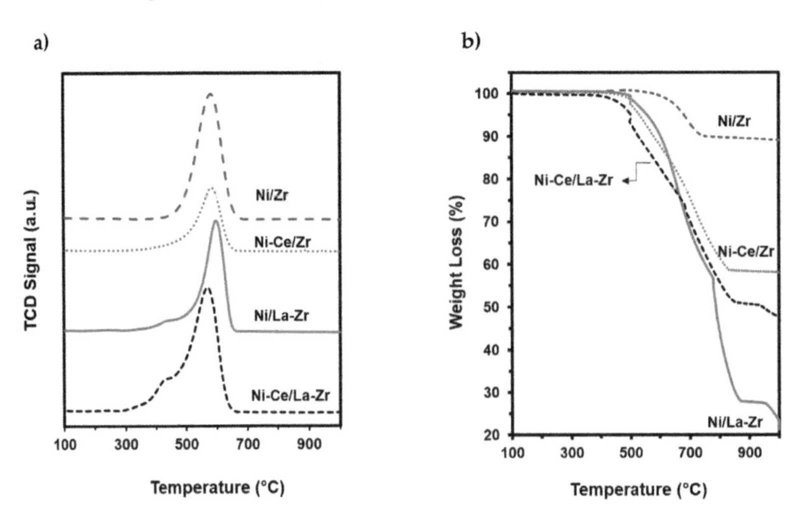

Figure 9. (**a**) TPO profiles of "spent" Ni/x-Zr and Ni-Ce/x-Zr (x = 0, La_2O_3) catalysts after reaction at T_R = 700 °C and (**b**) TGA profiles of "spent" Ni/x-Zr and Ni-Ce/x-Zr (x = 0, La_2O_3) catalysts after reaction at T_R = 700 °C.

The major peak in each profile is positioned around 585 °C, suggesting the presence of carbon nanotubes (CNTs). TGA analysis of spent catalysts was performed on reference and promoted Ni/Zr catalysts (Figure 9b) to quantify the amount of deposited carbon. It was found that no significant weight loss occurs in the case of a pure catalyst, indicating that a low amount of carbon was deposited during the reaction. On the contrary, La-modified catalysts presented a higher weight loss. After CeO_2 promotion, TGA profiles illustrate that the amount of accumulated carbon on the catalyst surface had considerably reduced. Moreover, the corresponding peaks negatively shifted, implying that the type of carbon material formed, is less stable, and burnt at a relatively lower temperature. It implies that carbon nanotubes (CNTs) mainly formed after CeO_2 promotion, and are easily burnt and affect catalyst deactivation less [7,37–39].

Figures 10 and 11 show the TEM images of ceria promoted and unpromoted Ni/x-ZrO_2 (0, La_2O_3) catalysts.

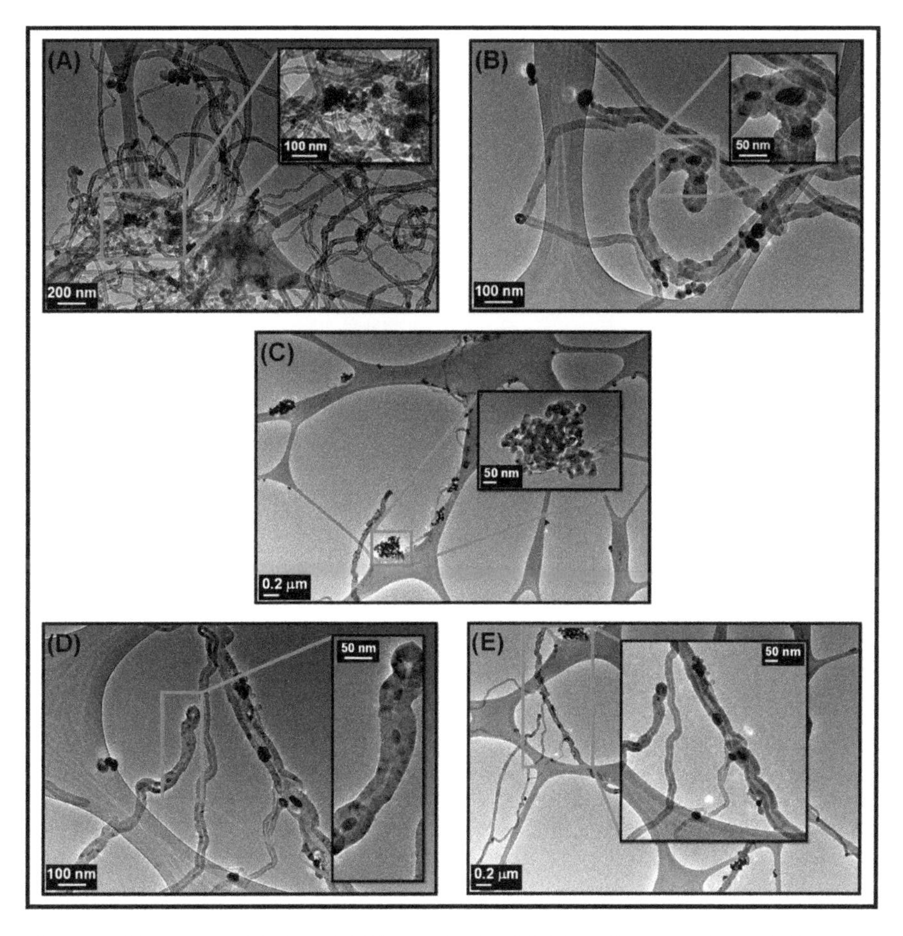

Figure 10. Micrographs of the "spent" Ni/Zr sample (T_R = 700 °C; reaction time = 400 min): (**A**) cluster of CNTs; (**B**) metal sintering; (**C**) encapsulation of metal species; (**D**) and (**E**) metal particles captured inside CNTs.

It can be seen that the majority of the deposited carbon on the Ni/ZrO_2 surface, after the catalytic reaction, is of a filamentous nature, and a tangle of carbon nanotubes (CNTs) is clearly visible (Figure 10A), while the rest of the carbon is of the encapsulating type. It is clear that several particles were encapsulated due to the carbon formation on Ni surface (Figure 10B). Ni/ZrO_2 catalysts also undergo sintering, therefore accumulation of metallic species can be seen (Figure 10C). Conspicuously, metal particles varying in size in the range 2–9 nm were entrapped inside the CNTs (Figure 10D,E), which may be another reason for the lower activity of the pristine catalyst.

When the pristine catalyst Ni/Zr was modified with La_2O_3, morphological investigation revealed the formation of multiwall carbon nanotubes (MWCNTs). The average external diameter of MWCNTs was found to be 28 nm, whereas the metallic species were positioned together in the interior as well as on the exterior of MWCNTs (Figure 11A). It seems that the incorporation of La_2O_3 had no influence in controlling the sintering of metallic particles (Figure 11B). The Ni particles at the tip of the CNTs clearly demonstrated that Ni particles not strongly anchored to the catalyst support formed filaments with a mechanism that included the diffusion of elementary C through the Ni and precipitation on the back side of the particle with consequent formation of the filament. In this case, Ni continues to be active for dry reforming [40–42]. Furthermore, when the Ni/ZrO_2 catalyst was promoted using CeO_2, the situation escalated because the rate of sintering was significantly enhanced upon CeO_2 promotion (Figure 11C), which may justify the deteriorating activity of the Ni-Ce/La-Zr catalyst.

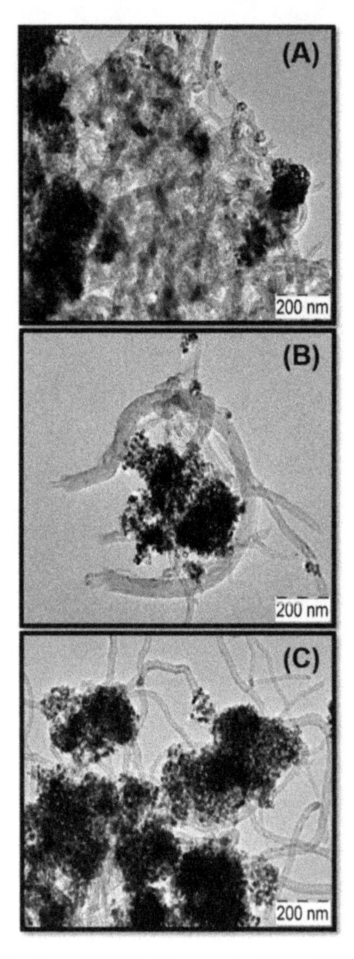

Figure 11. Micrographs of the "spent" Ni/La-Zr sample (T_R = 700 °C; reaction time = 400 min): (**A**) CNTs over catalyst surface; (**B**) metal particles sintering; (**C**) influence of Ce addition.

3. Materials and Methods

3.1. Catalyst Preparation

In this experimental work an incipient wet-impregnation technique was used to prepare the desired catalysts for the dry reforming of methane. The supports (zirconium oxide and modified zirconium oxide with La_2O_3) along with the active metal obtained from the precursor of Ni nitrate [$Ni(NO_3)_2.6H_2O$; 99.9% purity] were dissolved in distilled water. The amount of Ni used in the catalyst was 5 wt%. The solution was kept heated at 90 °C under stirring for 3 h. Later, impregnated catalysts were heated up to 120 °C overnight for drying and then subjected to calcination at 600 °C for 3 h. The $Ce(NO_3)_3.6H_2O$ promoted catalysts with 1 wt% were prepared by co-impregnation of the nitrate salts of the promoter and active metal with support using the same procedure mentioned above. The solution was kept heated at 90 °C under stirring for 3 h. Later, impregnated catalysts were heated up to 120 °C overnight for drying and then subjected to calcination at 600 °C for 3 h. The commercial samples of the support were obtained from KIGAKU DAIICHI KOGYO CO.; LTD, Osaka, Japan. We are grateful to the company for providing us with free samples.

3.2. Catalytic Testing

Dry reforming of CH_4 experiments over Ni/ZrO_2 catalysts were performed at 700 °C and at atmospheric pressure in a vertical stainless steel fixed-bed tubular (i.d., 9.1 mm; length, 0.3 m) micro-reactor (PID Eng& Tech micro activity reference, Madrid, Spain). Activity tests were performed using 0.1 g of a catalyst placed in a quartz reactor between two quartz wool beds. A K-type stainless steel sheathed thermocouple, placed axially at the center of the catalyst bed, measured the temperature during the reaction. Prior to each test, the catalysts were activated under a continuous flow of H_2 (20 mL/min) for 1 h at 600 °C. Experiments were carried out using a feed gas mixture (CH_4, CO_2 and N_2) at the ratio 6/6/1 and overall gas flow rate of 65 mL/min (space velocity: 39,000 mL/)h.g_{cat})). The outlet gas composition was analyzed by on-line gas chromatography (Shimadzu 2014, Kyoto, Japan) fitted out with a thermal conductivity detector (TCD). The CH_4 conversion and hydrogen yield were determined using the following formulae:

$$CH_4\ \text{conversion}(\%) = \frac{(CH_{4,in} - CH_{4,out})}{CH_{4,in}} * 100 \quad (5)$$

$$H_2\ \text{yield}(\%) = \frac{(H_{2,out})}{2 * CH_{4,in}} * 100 \quad (6)$$

3.3. Catalyst Characterization

3.3.1. XRD Characterization

Rigaku (Miniflex) diffractometer (Rigaku Corporation, Dover, DE, USA), with a Cu Kα X-ray radiation working at 40 kV and 40 mA, was used to investigate the structure of the catalysts before and after the reaction. The scanning 2θ range and steps were 10–85° and 0.02° respectively. The raw data file of the instrument was evaluated by X'Pert high score plus software (version 2.1, Panalytical, Malvern, UK). Different phases with their scores were corresponded with the JCPDS data bank.

3.3.2. N_2 Physisorption

The specific surface area and distribution of pore size of the catalysts were obtained using Micromeritics Tristar II 3020 surface area and porosity analyzer. The pore size distribution was calculated by BJH method.

3.3.3. Temperature Programmed Reduction (TPR)

The TPR measurements were performed using Micromeritics Auto Chem II apparatus (Micromeritics, Atlanta, GA, USA). The sample (70 mg) was charged in the TPR cell and flushed with argon at 150 °C for 30 min. Then the sample temperature was reduced to 25 °C. Finally, the furnace temperature was heated to 1000 °C at 10 °C/min ramp at a 40 mL/min flow rate with a H_2/Ar mixture (10:90 vol.%). A thermal conductivity detector (TCD) was used to follow the signals of H_2 consumption.

3.3.4. H_2 Temperature-Programmed Desorption

To evaluate metal surface area (MSA), and surface average nickel particle size (d_{Ni}) of the investigated catalysts, H_2-TPD measurements in the range 20–900 °C (β = 10 °C/min) were carried out at atmospheric pressure by using a linear quartz micro-reactor (i.d., 4 mm) flowing Ar as carrier gas at 50 stp cm^3/min. Before measurements, a catalyst sample (50–100 mg) was reduced for half-an hour in flowing H_2 (25 stp cm^3/min) at 600 °C. Thereafter, the sample was cooled in flowing H_2 to room temperature and then hydrogen was shut off and the sample was purged by the carrier stream until baseline stabilization (≈20 min). Assuming a chemisorption stoichiometry $H_2:Ni_{surf}$ = 1:2 and a spherical shape of the metal particle, the following equations were applied:

$$\text{MSA } (m^2/g_{cat}) = 2 \cdot X_{H_2} \cdot N_{Av}/\sigma_{Ni} \quad (7)$$

$$d_{Ni} \text{ (nm)} = 6000 \cdot (C_{Ni}/100)/(\rho_{Ni} \cdot \text{MSA}) \quad (8)$$

where "X_{H_2}" is the H_2 uptake (mol/g_{cat}), "N_{Av}" is Avogadro's number, "σ_{Ni}" represents the concentration of surface atoms (1.54×10^{19} at/m^2 for Ni), "C_{Ni}" is the metal loading (wt%), while "ρ_{Ni}" is the metal density (8.9 g/cm^3 for Ni).

3.3.5. CO_2 Temperature-Programmed Desorption (CO_2-TPD)

The CO_2 temperature-programmed desorption (TPD) was carried out using a Chemisorption Analyzer (Micromeritics Autochem II apparatus, Micromeritics, Atlanta, GA, USA). The catalyst (50 mg) was reduced at 600 °C for 1 h under He flow (30 mL/min) and then cooled to 50 °C. The CO_2 flow was kept for 60 min, and the sample was then flushed with He to take away any physisorbed CO_2. The desorption profile of the catalysts were recorded by ramping the temperature at a rate of 10 °C/min, while temperature was then linearly increased up to 800 °C. The CO_2 concentration in the output stream was measured with a thermal conductivity detector, and the areas under the peaks were used to determine the amount of desorbed CO_2 during TPD.

3.3.6. Thermo-Gravimetric Analysis (TGA)

The analysis of carbonaceous material deposition post reaction over the catalyst's surface was quantitatively performed using the TGA-15 SHIMADZU analyzer (Shimadzu, Kyoto, Japan) under an air atmosphere. The spent catalyst (10–15 mg) was heated from room temperature to 1000 °C at a heating rate of 20 °C/min and the loss of weight was measured.

3.3.7. TEM Characterization

TEM measurements of the used samples were accomplished on a 120 kV JEOL JEM-2100F transmission electron microscope (JEOL, Peabody, MA, USA).

3.3.8. XPS Analysis

X-ray photoelectron spectroscopy (XPS) spectra were recorded on an Omicron Nanotechnology (ELS5000) spectrometer with a monochromatic Al source. X-ray was employed using a flood gun with a spot size of radius ~ 400 µm. Survey scanning was acquired from −10 to 1350 eV with energy steps of 1 eV, while employing a pass energy of 200 eV. The number of scans was four with a dwell time of 10 ms. Similarly, a slow scan was conducted for the element using a pass energy of a 50 eV and the number of scans was four with a dwell time of 10 ms. The charging effects were corrected by adjusting the binding energy of the C 1s peak from adventitious carbon to 285 eV. The characterization experiments were carried out for fresh (before reaction) and used (after reaction) samples.

4. Conclusions

A high surface area ZrO_2 supported Ni system was prepared using an incipient wet-impregnation method and then modified by heteroatom oxides, like La_2O_3, and the influence of ceria doping in dry reforming of CH_4 at 700 °C was also assessed. It is worthwhile highlighting that the reduction properties of the catalysts were significantly enhanced upon La_2O_3 modification. La_2O_3 modification suppressed the formation of $NiO–ZrO_2$ solid solutions, with a complete reduction of the active phase at temperatures below 750 °C. Furthermore, La_2O_3 modification also contributed to enhancing the metal-support interaction and active phase dispersion. However, the weak-medium basicity of all modified catalyst samples were substantially unchanged after Ce-promotion. Phase analysis revealed that the *t*-ZrO_2 phase of the Ni/Zr catalyst had almost completely transformed into the *m*-ZrO_2 phase after thermal treatment. However, when La_2O_3 incorporated into the ZrO_2 support, the *t*-ZrO_2 phase was protected and prominently stabilized, leading to a superior catalytic performance. Interestingly, CH_4 conversion increased to 2x, and CO_2 conversion to 1.5x, that of pristine Ni/ZrO_2. However, when the modified catalysts were promoted by CeO_2, a decline in catalytic activity was observed in the case of the Ni-Ce/La-Zr catalyst. Eventually, it was revealed that the high activity of La_2O_3 modified catalysts was achieved by protecting the stability of the *t*-ZrO_2 phase.

Author Contributions: A.S.-F., Y.A., A.A.I., S.O.K. and A.H.F. synthesized the catalysts, performed all the experiments and characterization tests and wrote the manuscript. F.F., A.A. and G.B. performed the characterization tests for CO_2, H_2-TPD, XPS and TEM, and proofread the manuscript. A.E.A. contributed to the analysis of the data and proofread the manuscript.

Funding: The authors would like to express their sincere appreciation to the Deanship of Scientific Research at King Saud University for funding this research project (No. RGP-1435-078).

Acknowledgments: The authors would like to extend their sincere appreciation to the Deanship of Scientific Research at King Saud University for its funding this research group No. (RGP-1435-078).

Conflicts of Interest: The authors declare no conflict of interest.

References

1. Pan, X.; Fan, Z.; Chen, W.; Ding, Y.; Luo, H.; Bao, X. Enhanced ethanol production inside carbon-nanotube reactors containing catalytic particles. *Nat. Mat.* **2007**, *6*, 507–511. [CrossRef] [PubMed]
2. Hamza Fakeeha, A.; Arafat, Y.; Aidid Ibrahim, A.; Shaikh, H.; Atia, H.; Elhag Abasaeed, A.; Armbruster, U.; Sadeq Al-Fatesh, A. Highly Selective Syngas/H_2 Production via Partial Oxidation of CH_4 Using (Ni, Co and Ni–Co)/ZrO_2–Al_2O_3 Catalysts: Influence of Calcination Temperature. *Processes* **2019**, *7*, 141. [CrossRef]
3. Gao, X.Y.; Hidajat, K.; Kawi, S. Facile synthesis of Ni/SiO_2 catalyst by sequential hydrogen/air treatment: A superior anti-coking catalyst for dry reforming of methane. *J. CO2 Util.* **2016**, *15*, 146–153. [CrossRef]
4. Xu, Y.; Lin, Q.; Liu, B.; Jiang, F.; Xu, Y.; Liu, X. A Facile Fabrication of Supported Ni/SiO_2 Catalysts for Dry Reforming of Methane with Remarkably Enhanced Catalytic Performance. *Catalysts* **2019**, *9*, 183. [CrossRef]
5. Karam, L.; Casale, S.; El Zakhem, H.; El Hassan, N. Tuning the properties of nickel nanoparticles inside SBA-15 mesopores for enhanced stability in methane reforming. *J. CO2 Util.* **2017**, *17*, 119–124. [CrossRef]
6. El Hassan, N.; Kaydouh, M.; Geagea, H.; El Zein, H.; Jabbour, K.; Casale, S.; El Zakhem, H.; Massiani, P. Low temperature dry reforming of methane on rhodium and cobalt based catalysts: Active phase stabilization by confinement in mesoporous SBA-15. *Appl. Catal. A Gen.* **2016**, *520*, 114–121. [CrossRef]
7. Al-Fatesh, A.S.; Arafat, Y.; Atia, H.; Ibrahim, A.A.; Ha, Q.L.M.; Schneider, M.; M-Pohl, M.; Fakeeha, A.H. CO_2-reforming of methane to produce syngas over Co-Ni/SBA-15 catalyst: Effect of support modifiers (Mg, La and Sc) on catalytic stability. *J. CO2 Util.* **2017**, *21*, 395–404. [CrossRef]
8. Moradi, G.R.; Rahmanzadeh, M.; Khosravian, F. The effects of partial substitution of Ni by Zn in $LaNiO_3$ perovskite catalyst form ethane dry reforming. *J. CO2 Util.* **2014**, *6*, 7–11. [CrossRef]

9. Kambolis, A.; Matralis, H.; Trovarelli, A.; Papadopoulou, C. Ni/CeO_2-ZrO_2 catalysts for the dry reforming of methane. *Appl. Catal. A* **2010**, *377*, 16–26. [CrossRef]
10. Lemonidou, A.A.; Vasalos, I.A. Carbon dioxide reforming of methane over 5 wt.% Ni/CaO-Al_2O_3 catalyst. *Appl. Catal. A* **2002**, *228*, 227–235. [CrossRef]
11. Han, J.W.; Park, J.S.; Choi, M.S.; Lee, H. Uncoupling the size and support effects of Ni catalysts for dry reforming of methane. *Appl. Catal. B* **2017**, *203*, 625–632. [CrossRef]
12. Świrk, K.; Gálvez, M.E.; Motak, M.; Grzybek, T.; Rønning, M.; Da Costa, P. Dry reforming of methane over Zr- and Y-modified Ni/Mg/Al double-layered hydroxides. *Catal. Commun.* **2018**, *117*, 26–32.
13. Sarkar, D.; Adak, S.; Chu, M.; Cho, S.; Mitra, N. Influence of ZrO_2 on the thermo-mechanical response of nano-ZTA. *Ceram. Int.* **2007**, *33*, 255–261. [CrossRef]
14. Lercher, J.; Bitter, J.; Hally, W.; Niessen, W.; Seshan, K. Design of stable catalysts for methane-carbon dioxide reforming. *Studies Surf. Sci. Catal.* **1996**, *101*, 463–472.
15. Li, X.; Chang, J.; Park, S. Carbon as an intermediate during the carbon dioxide reforming of methane over zirconia-supported high nickel loading catalysts. *Chem. Lett.* **1999**, *10*, 1099–1100. [CrossRef]
16. Chuah, G.K.; Liu, S.H.; Jaenicke, S.; Li, J. High surface area zirconia by digestion of zirconium propoxide at different pH. *Microporous Mesoporous Mat.* **2000**, *39*, 381–392. [CrossRef]
17. Jaenicke, S.; Chuah, G.K.; Raju, V.; Nie, Y.T. Structural and Morphological Control in the Preparation of High Surface Area Zirconia. *Catal. Surv. Asia* **2008**, *12*, 153–169. [CrossRef]
18. Fan, M.-S.; Abdullah, A.Z.; Bhatia, S. Utilization of greenhouse gases through carbon dioxide reforming of methane over Ni–Co/MgO–ZrO_2: preparation, characterization and activity studies. *Appl. Catal. B* **2010**, *100*, 365–377. [CrossRef]
19. Yamasaki, M.; Habazaki, H.; Asami, K.; Izumiya, K.; Hashimoto, K. Effect of tetragonal ZrO_2 on the catalytic activity of Ni/ZrO_2 catalyst prepared from amorphous Ni–Zr alloys. *Catal. Commun.* **2006**, *7*, 24–28. [CrossRef]
20. Rezaei, M.; Alavi, S.M.; Sahebdelfar, S.; Yan, Z.-F. A highly stable catalyst in methane reforming with carbon dioxide. *Scripta Materialia* **2009**, *61*, 173–176. [CrossRef]
21. Choque, V.; de la Piscina, P.R.; Molyneux, D.; Homs, N. Ruthenium supported on new TiO_2–ZrO_2 systems as catalysts for the partial oxidation of methane. *Catal. Today* **2010**, *149*, 248–253. [CrossRef]
22. Verykios, X.E. Catalytic dry reforming of natural gas for the production of chemicals and hydrogen. *Int. J. Hydrogen Energy* **2003**, *28*, 1045–1063. [CrossRef]
23. Slagtern, A.; Schuurman, Y.; Leclercq, C.; Verykios, X.; Mirodatos, C. Specific features concerning the mechanism of methane reforming by carbon dioxide over Ni/La_2O_3 catalyst. *J. Catal.* **1997**, *172*, 118–126. [CrossRef]
24. Titus, J.; Roussiere, T.; Wasserschaff, G.; Schunk, S.; Milanov, A.; Schwab, E.; Wagner, G.; Oeckler, O.; Gläser, R. Dry reforming of methane with carbon dioxide over NiO–MgO–ZrO_2. *Catal. Today* **2016**, *270*, 68–75. [CrossRef].
25. Ay, H.; Üner, D. Dry reforming of methane over CeO_2 supported Ni, Co and Ni–Co catalysts. *Appl. Catal. B* **2015**, *179*, 128–138. [CrossRef]
26. Chen, J.; Zhang, X.; Arandiyan, H.; Peng, Y.; Chang, H.; Li, J. Low temperature complete combustion of methane over cobalt chromium oxides catalysts. *Catal. Today* **2013**, *201*, 12–18. [CrossRef]
27. Goula, M.A.; Charisiou, N.D.; Siakavelas, G.; Tzounis, L.; Tsiaoussis, I.; Panagiotopoulou, P.; Goula, G.; Yentekakis, I.V. Syngas production via the biogas dry reforming reaction over Ni supported on zirconia modified with CeO_2 or La_2O_3 catalysts. *Int. J. Hydrogen Energy* **2017**, *42*, 13724–13740.
28. Rotgerink, H.L.; Paalman, R.; Van Ommen, J.; Ross, J. Studies on the promotion of nickel—alumina coprecipitated catalysts: II. Lanthanum oxide. *Appl. Catal.* **1988**, *45*, 257–280. [CrossRef]
29. Wang, S.; Lu, G.; Millar, G.J. Carbon dioxide reforming of methane to produce synthesis gas over metal-supported catalysts: state of the art. *Energy Fuels* **1996**, *10*, 896–904. [CrossRef]
30. Kumar, P.; Sun, Y.; Idem, R.O. Comparative Study of Ni-based Mixed Oxide Catalyst for Carbon Dioxide Reforming of Methane. *Energy Fuels* **2008**, *22*, 3575–3582. [CrossRef]
31. Huang, S.J.; Walters, A.B.; Vannice, M.A. TPD, TPR and DRIFTS studies of adsorption and reduction of NO on La_2O_3 dispersed on Al_2O_3. *Appl. Catal. B* **2000**, *26*, 101–118. [CrossRef]
32. Grosvenor, A.P.; Biesinger, M.C.; Smart, R.S.C.; McIntyre, N.S. New interpretations of XPS spectra of nickel metal and oxides. *Surf.Sci.* **2006**, *600*, 1771–1779. [CrossRef]

33. Charisiou, N.D.; Siakavelas, G.; Tzounis, L.; Sebastian, V.; Monzon, A.; Baker, M.A.; Hinder, S.J.; Polychronopoulou, K.; Yentekakis, I.V.; Goula, M.A. An in depth investigation of deactivation through carbon formation during the biogas dry reforming reaction for Ni supported on modified with CeO_2 and La_2O_3 zirconia catalysts. *Int. J. Hydrogen Energy* **2018**, *43*, 18955–18976. [CrossRef]
34. Liu, G.; Liu, A.; Meng, Y.; Shan, F.; Shin, B.; Lee, W.; Cho, C. Annealing dependence of solution-processed ultra-thin ZrOx films for gate dielectric applications. *J. Nanosci. Nanotechnol.* **2015**, *15*, 2185–2191. [CrossRef]
35. Italiano, G.; Delia, A.; Espro, C.; Bonura, G.; Frusteri, F. Methane decomposition over Co thin layer supported catalysts to produce hydrogen for fuel cell. *Int. J. Hydrogen Energy* **2010**, *35*, 11568–11575. [CrossRef]
36. Frusteri, F.; Italiano, G.; Espro, C.; Cannilla, C.; Bonura, G. H_2 production by methane decomposition: Catalytic and technological aspects. *Int. J. Hydrogen Energy* **2012**, *37*, 16367–16374. [CrossRef]
37. Frusteri, F.; Frontera, P.; Modafferi, V.; Bonura, G.; Bottari, M.; Siracusano, S.; Antonucci, P.L. Catalytic features of Ni/Ba–Ce0. 9–Y0. 1 catalyst to produce hydrogen for PCFCs by methane reforming. *Int. J. Hydrogen Energy* **2010**, *35*, 11661–11668.
38. Frontera, P.; Macario, A.; Monforte, G.; Bonura, G.; Ferraro, M.; Dispenza, G.; Antonucci, V.; Aricò, A.S.; Antonucci, P.L. The role of Gadolinia Doped Ceria support on the promotion of CO_2methanation over Ni and NiFe catalysts. *Int. J. Hydrogen Energy* **2017**, *42*, 26828–26842. [CrossRef]
39. Al-Fatesh, A.S.; Arafat, Y.; Ibrahim, A.A.; Atia, H.; Fakeeha, A.H.; Armbruster, U.; Abasaeed, A.E.; Frusteri, F. Evaluation of Co-Ni/Sc-SBA–15 as a novel coke resistant catalyst for syngas production via CO_2 reforming of methane. *Appl. Catal. A* **2018**, *567*, 102–111. [CrossRef]
40. Bonura, G.; Di Blasi, O.; Spadaro, L.; Arena, F.; Frusteri, F. A basic assessment of the reactivity of Ni catalysts in the decomposition of methane for the production of "CO_x-free" hydrogen for fuel cells application. *Catal. Today* **2006**, *116*, 298–303. [CrossRef]
41. Frusteri, F.; Spadaro, L.; Arena, F.; Chuvilin, A. TEM evidence for factors affecting the genesis of carbon species on bare and K-promoted Ni/MgO catalysts during the dry reforming of methane. *Carbon* **2002**, *40*, 1063–1070. [CrossRef]
42. Branca, C.; Frusteri, F.; Magazzù, V.; Mangione, A. Characterization of carbon nanotubes by TEM and infrared spectroscopy. *J. Phys. Chem. B* **2004**, *108*, 3469–3473. [CrossRef]

Article

Performance and Stability of Metal (Co, Mn, Cu)-Promoted $La_2O_2SO_4$ Oxygen Carrier for Chemical Looping Combustion of Methane

Stefano Cimino *, Gabriella Mancino and Luciana Lisi *

Istituto Ricerche sulla Combustione – CNR, 80125 Napoli, Italy; gabriellamancino85@libero.it
* Correspondence: stefano.cimino@cnr.it (S.C.); luciana.lisi@cnr.it (L.L.)

Received: 18 December 2018; Accepted: 23 January 2019; Published: 2 February 2019

Abstract: Oxygen carrier materials based on $La_2O_2SO_4$ and promoted by small amounts (1% wt.) of transition metals, namely Co, Mn and Cu, have been synthesized and characterized by means of X-ray diffraction (XRD), Brunauer–Emmett–Teller (BET), Temperature-programmed reduction/oxidation (TPR/TPO) and thermogravimetry-mass-Fourier transform infrared spectrometry (TG-MS-FTIR) experiments under alternating feeds in order to investigate their potential use for the Chemical Looping Combustion process using either hydrogen or methane as the fuel. The chemical looping reactivity is based on the reversible redox cycle of sulfur from S^{6+} in $La_2O_2SO_4$ to S^{2-} in La_2O_2S and entails a large oxygen storage capacity, but it generally requires high temperatures to proceed, challenging material stability and durability. Herein we demonstrate a remarkable improvement of lattice oxygen availability and activity during the reduction step obtained by cost-effective metal doping in the order Co > Mn > Cu. Notably, the addition of Co or Mn has shown a significant beneficial effect to prevent the decomposition of the oxysulfate releasing SO_2, which is identified as the main cause of progressive deactivation for the unpromoted $La_2O_2SO_4$.

Keywords: oxygen storage capacity; thermal stability; cyclic operation; deactivation; oxysulfate; oxysulfide

1. Introduction

Chemical Looping Combustion (CLC) represents one of the most investigated approaches for clean combustion of fossil fuels because it provides an easy CO_2 capture [1–3] by splitting the process into two cyclic half-steps. The same two-step technology has been also proposed for Chemical Looping Reforming [1–3].

Large oxygen storage capacity (OSC), fast oxidation and reduction kinetics and high stability of the oxygen carrier (OC) are crucial parameters for process development [4]. A large variety of OCs have been proposed, most of them based on transition metals (Cu, Ni, Fe, Mn, Co) as bulk or supported simple oxides or as more complex oxides such as perovskite-like materials [3,5,6]. Perovskites show a better thermal stability and adjustable properties related to the partial or total substitution of A and B metal cations in their ABO_3 structure [6].

Recently, Miccio et al. [7,8] have dispersed iron and iron/manganese oxides into a geopolymer matrix to provide low density materials with good resistance to mechanical stresses that can be suitable for fluidized bed reactors.

Additionally, $CaSO_4$ has been investigated due to its potentially higher oxygen transport capacity compared to transition metal oxides, deriving from the redox cycle of sulfur [3,9,10]. Nevertheless, a significant sulfur release due to decomposition at high temperature represents an unsolved issue for $CaSO_4$, in addition to agglomeration and sintering that are typical of metal oxides (MeOx). An attempt to face the loss of efficiency has been proposed by adding elemental iron [9].

Machida and co-workers [11–16] have deeply investigated rare earths based oxysulfates ($Ln_2O_2SO_4$) as stable, durable, and high performing oxygen storage materials in three ways catalysts for automotive emission control. Sulfur can undergo a complete reduction from S^{6+} in the oxysulfate to S^{2-} in the oxysulfide structure by reaction with a reducing agent and it can be reversibly re-oxidized to S^{6+} by reaction with molecular oxygen, provided that a suitable temperature is chosen for the two processes. Indeed, both the $Ln_2O_2SO_4$ and its corresponding reduced compound Ln_2O_2S crystallize in a similar structure consisting of alternating layers of $Ln_2O_2{}^{2+}$ and $SO_4{}^{2-}$ or S^{2-} anions. The phase transformation is therefore regarded as the removal or insertion of O^{2-} surrounding sulfur [13]. The thermal stability of $Ln_2O_2SO_4$ decreases with increasing the atomic number of Ln [11,17]. Accordingly, it has been reported that $Lu_2O_2SO_4$ cannot be reduced to the corresponding oxysulfide, since it transforms directly into Lu_2O_3 and SO_2 under reducing atmosphere [16]. Furthermore, Machida et al. [12–14] found that among other Ln-oxysulfates, $Pr_2O_2SO_4$ is able to operate the redox cycles at the lowest temperatures, due to the beneficial effect of the existence of Pr^{3+}/Pr^{4+} surface species on the redox of sulfur. They also reported that sulfate reduction becomes easier with increasing distortion of the tetragonal SO_4 unit in the oxysulfate, as evidenced in $Pr_2O_2SO_4$ structure [13]. However, due to the high cost and scarce availability of praseodymium, substitution of the more common lanthanum is difficult to pursue [13]. In conclusion, the oxygen storage properties of $Ln_2O_2SO_4$ can be enhanced (i) by introducing redox species and/or (ii) by increasing distortion of the SO_4 unit as in the case of the partial substitution of Ce for La in $(La_{1-x}Ce_x)_2O_2SO_4$ [14] and/or (iii) by the addition of noble metals on the surface (1% wt. Pd or Pt), likely promoting the redox cycle of $Ln_2O_2SO_4$ by hydrogen/oxygen spillover effect [11–14].

The cost-effective addition of small quantities of Cu [10] or Na and Fe [18] to $La_2O_2SO_4$, as well as Ni to $Pr_2O_2SO_4$ [19] has been recently reported to enhance oxygen storage capability (OSC) and storage/release reaction rates, in agreement with the distortion of the SO_4 unit [13,15], thus allowing operation at lower temperatures than with parent oxysulfates. It can be argued that cations with different dimensions can alter the lattice to a different extent, thus affecting the OSC and stability of materials.

In this paper, we set out to investigate the effect of the addition of a low amount (1% wt.) of three different transition metals (Cu, Mn, Co) on the OSC as well as on stability/durability of a $La_2O_2SO_4$ oxygen carrier. Chemical looping reaction experiments have been performed under alternating feed conditions at a fixed temperature in the range 700–900 °C, using either hydrogen or methane as the fuel (i.e., reducing agent). The emission of sulfur compounds under repeated redox cycles has been investigated to provide a detailed analysis of conditions promoting the undesired decomposition of the oxygen carrier.

2. Results

In Table 1 the list of metal-promoted OCs is reported with the corresponding values of Brunauer–Emmett–Teller (BET) surface area. The original surface area of the as-prepared $La_2O_2SO_4$ is rather low (2.7 m^2/g), in agreement with previous results reported for the simple preparation method based on the thermal decomposition of bulk lanthanum sulfate at 1027 °C under inert flow [20,21]. The further addition of 1% wt. of transition metals followed by heat treatment at 900 °C does not significantly alter the specific surface area of the materials.

In Figure 1a X-ray diffraction (XRD) patterns of fresh $La_2O_2SO_4$ and transition metal-promoted $La_2O_2SO_4$ are reported. All spectra show the characteristic lines of monoclinic lanthanum oxysulfate (JCPDS 085-1535). The presence of the transition metals causes a decrease of crystallinity, as shown by the lower intensity of XRD peaks, which is particularly evident for the case of Co-$La_2O_2SO_4$. As mentioned in the Introduction, the structure of $La_2O_2SO_4$ consists of alternating stacks of $La_2O_2{}^{2+}$ layers and layers of anions $SO_4{}^{2-}$ [12,15], therefore the possible inclusion of the transition metal into the lattice can generate a distortion since the radius of the transition metal ion differs from that of La^{3+} (1.03 Å). Notably, for the same oxidation state (including the metallic state), cobalt shows the smallest dimensions. In particular, the ionic radius of Me^{3+}, supposed to substitute La^{3+} in the lattice, is 0.60 Å for cobalt and 0.70 Å for manganese. Copper, which is generally in the +2 oxidation state, has a ionic radius as Cu^{2+} similar to Mn^{3+} (0.73Å). That could explain the lowest crystallinity of the Co-$La_2O_2SO_4$ material.

Table 1. List of oxygen carriers and corresponding values of Brunauer–Emmett–Teller (BET) surface area.

Oxygen Carrier	BET Area (m^2/g)
$La_2O_2SO_4$	2.7
Cu-$La_2O_2SO_4$	2.7
Mn-$La_2O_2SO_4$	2.4
Co-$La_2O_2SO_4$	2.4

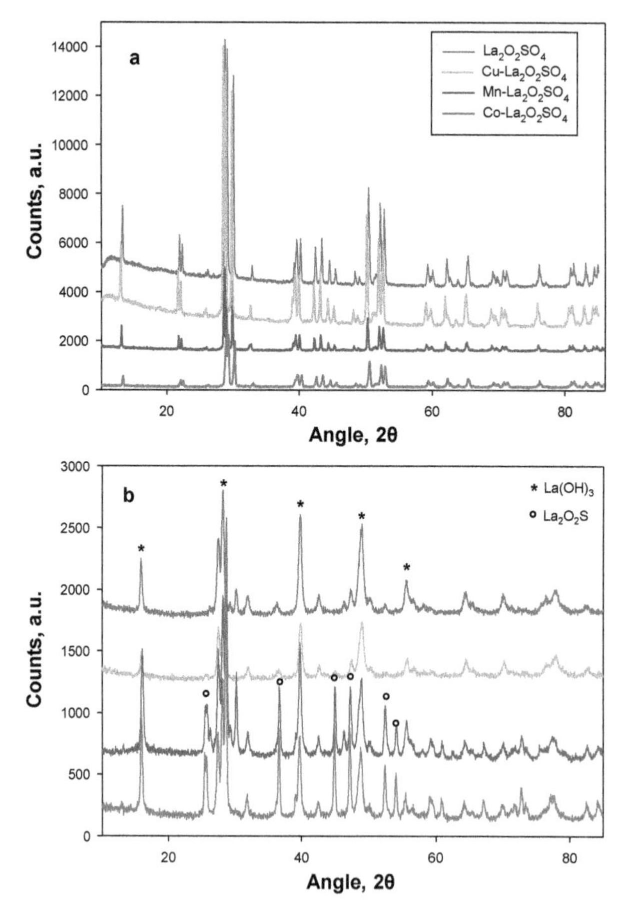

Figure 1. X-ray diffraction (XRD) patterns of (**a**) as prepared and (**b**) used (15 cycles at 900°C) $La_2O_2SO_4$ and metal (Cu, Mn, Co) promoted $La_2O_2SO_4$.

The effect of metal doping on the thermal stability of $La_2O_2SO_4$ has been studied by thermogravimetry-mass spectrometry (TG-MS) experiments ramping up to 1300 °C at 10 °C/min under N_2 flow: the corresponding derivative weight loss plots (Figure 2a) are superimposed and flat for all materials, indicating an almost constant weight up to ca. 1050 °C. Above this temperature, which corresponds well to the decomposition temperature previously reported for La oxysulfate [21], all materials experience a significant weight loss, at similar increasing rates up to 1300 °C, where the process is not yet completed (during 10 min under isothermal conditions). Accordingly, the MS traces for SO_2 (m/z = 64), O_2 (m/z = 32) and CO_2 (m/z = 48) in the evolved gas display similar trends for undoped $La_2O_2SO_4$ (Figure 2b) as well as for metal- doped materials (exemplified in Figure 2c for the case of Mn-$La_2O_2SO_4$). In particular, a limited CO_2 release can be observed in the temperature range of 250–500 °C corresponding to the decomposition of same superficial La- carbonates. A rather small SO_2 peak in the range of 800–950 °C suggests the decomposition of some residual La sulfates. Eventually, a large SO_2 release begins at temperatures around 1050 °C, which is accompanied by the evolution of O_2, thus suggesting the occurrence of La oxysulfate decomposition to La_2O_3. Poston et al. [21] detected (by XRD analysis) the simultaneous formation of La_2O_3 and La_2O_2S phases during thermal treatment of $La_2O_2SO_4$ at T $\geq$ 1050 °C in air. According to our results, the oxygen release never occurs independently from SO_2. Therefore, it can be concluded that under N_2 flow none of the doped or undoped La oxysulfates is able to release oxygen and turns spontaneously into the corresponding oxysulfide without decomposing.

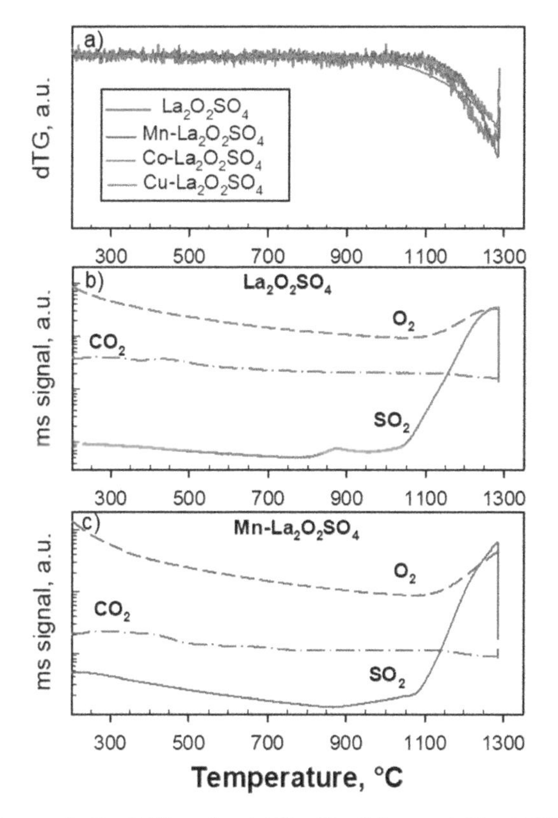

Figure 2. (**a**) dTG curves for $La_2O_2SO_4$ and metal (Cu, Mn, Co) promoted $La_2O_2SO_4$ during heating under a N_2 flow. Heating rate: 10 °C min^{-1}. Corresponding MS temporal profiles for O_2 (m/z = 32), CO_2 (m/z = 44), and SO_2 (m/z = 64) concentration in the evolved gas relevant to (**b**) $La_2O_2SO_4$ and (c) Mn-$La_2O_2SO_4$.

Sequential H_2-TPR/O_2-TPO analysis have been performed to compare the redox behavior of the carriers and results are reported in Figure 3a,b.

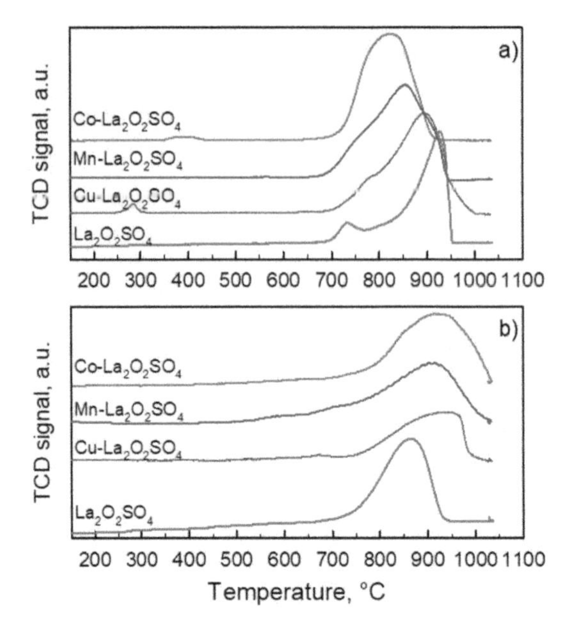

Figure 3. (**a**) H_2-TPR profiles of $La_2O_2SO_4$ and metal (Cu, Mn, Co) promoted $La_2O_2SO_4$; (**b**) O_2-TPO profiles recorded during the subsequent re-oxidation of OC materials. Heating 10°C/min from room temperature up to 1030 °C.

The main H_2 consumption event starts for all materials (promoted or not) at ca. 630 °C, and the total H_2 uptake accounts for the complete transformation of $La_2O_2SO_4$ into the La_2O_2S [10]. However, Co-$La_2O_2SO_4$ is most easily reduced among others, as testified by the clear shift towards lower temperatures of its H_2 consumption peak showing a maximum at ca. 810 °C. Notably, Cu- and Mn-promoted $La_2O_2SO_4$ samples show their peaks respectively at 855 °C and 890 °C, whereas the undoped La oxysulfate requires as much as 925 °C. Furthermore, Cu- and Co-promoted OCs display additional small peaks in the low temperature region that can be assigned respectively to the reduction of some segregated/supported CuO and Co_3O_4 [22]. In the case of Mn-$La_2O_2SO_4$ a very small peak at ca. 560 °C could be related to the reduction of some Mn_2O_3 [23], whereas the eventual presence of Mn_3O_4 could not be detected since its reduction requires higher temperatures [23] and would, therefore, be masked by the onset of reduction of $La_2O_2SO_4$.

The same trend is confirmed by the results of the TG-MS experiments carried out in the thermo-balance under a 2%H_2/N_2 flow: Figure 4a and Table 2 show all samples experience a similar weight loss (ca. 15.2%), which approaches the theoretical value (15.8%) estimated from the reduction reaction:

$$La_2O_2SO_4 + 4H_2 \rightarrow La_2O_2S + 4H_2O \quad (1)$$

Table 2. Weight loss with onset, inflection, and final temperatures for the reduction of $La_2O_2SO_4$ and metal (Cu, Mn, Co) promoted $La_2O_2SO_4$ under a flow of 2% H_2/N_2; amount of SO_2 emitted below 950 °C.

	Δ m % wt.	T_{in}	T_{infl} °C	T_f	SO_2 Emitted % wt.
$La_2O_2SO_4$	15.2	635	874	972	0.060
Cu-$La_2O_2SO_4$	15.1	625	817	910	0.065
Mn-$La_2O_2SO_4$	15.2	625	784	862	0.009
Co-$La_2O_2SO_4$	15.1	600	690	770	0.009

Small differences are due to some impurities, such as carbonates, which are present in the original samples (Figure 2b,c). In contrast, the complete transformation of $La_2O_2SO_4$ into La_2O_3 would account for a weight loss as large as 19.8%. The onset temperature of weight loss associated with the reduction of $La_2O_2SO_4$ (ca.630 °C) is not affected by doping with Cu or Mn, whereas Co addition lowers it by 30 °C (Table 2). However, all transition metal (oxides) dopants act as a catalyst to facilitate the surface reaction with the reducing gas, confirming the order of activity is Cu < Mn < Co.

In particular, as shown in Table 2, Co-doping lowers the temperature required to complete the transformation of $La_2O_2SO_4$ into La_2O_2S by as much as 200 °C. In the case of Pd- and Pt-promoted $La_2O_2SO_4$ or $Pr_2O_2SO_4$ [11,14] it was inferred that the easier reducibility at lower temperatures is caused by the higher reactivity of atomic hydrogen produced via H_2 spillover from noble metal nanoparticles with respect to molecular H_2. On the contrary, Me-promoted $La_2O_2SO_4$ materials show similar onset temperatures but, thereafter, reduction proceeds at a significantly faster rate, suggesting this effect could be rather a consequence of distortion of the oxysulfate lattice.

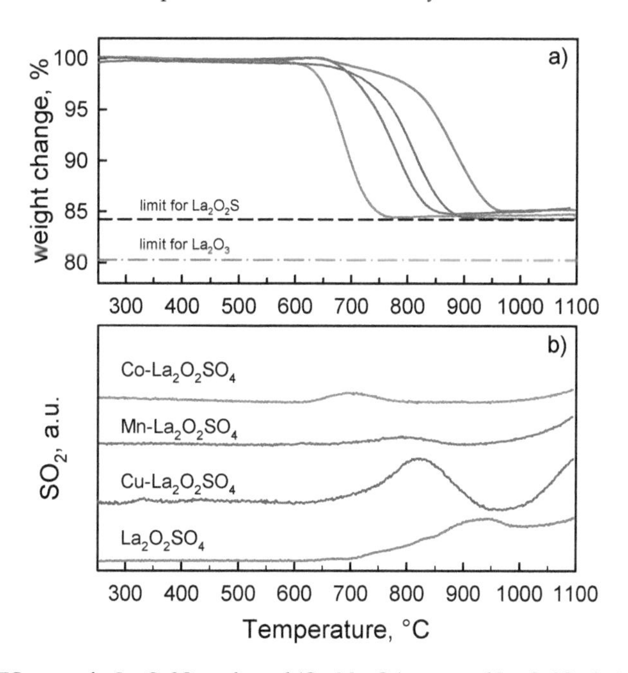

Figure 4. (**a**) TG curves for $La_2O_2SO_4$ and metal (Cu, Mn, Co) promoted $La_2O_2SO_4$ during reduction under a flow of 2% H_2/N_2. Heating rate: 10 °C min^{-1}; (**b**) Corresponding MS profiles for SO_2 (m/z = 64) in the evolved gas.

A serious issue for the durability of La oxysulfates comes from their irreversible decomposition [14] while reacting under a reducing atmosphere. Figure 4b presents the MS profiles for SO_2 emission recorded during TG experiments under H_2/N_2 flow. In fact, besides the H_2 consumption and the corresponding emission of water deriving from its oxidation (not shown), a small SO_2 emission is observed for each material, peaking at the same temperature of the inflection point of the corresponding TG curve (Table 2). Osseni et al. [17] observed by XRD the bulk formation of Lu_2O_3 during $Lu_2O_2SO_4$ reduction with H_2 at 650, 690 and 800 °C via reaction (2), which possibly occurs to a limited extent also when Ln = La:

$$Ln_2O_2SO_4 + H_2 \rightarrow Ln_2O_3 + SO_2 + H_2O \quad (2)$$

In the present case the amount of SO_2 emitted, estimated by peak area after specific calibration, is always rather small: it accounts for ca. 0.06% of the original weight for the case of the undoped and Cu-doped $La_2O_2SO_4$, whereas it drops down to 0.009% for Co- and Mn-doped materials, possibly due to their faster oxygen mobility and lower temperature level required to complete reduction.

Moreover, a second SO_2 emission starts from 925–945 °C for all OCs. In fact, the two SO_2 emission peaks are partly overlapped for the unpromoted $La_2O_2SO_4$. It can be argued that the second SO_2 emission corresponds to the decomposition of some residual lanthanum oxysulfate [19]; however, the reaction of La-oxysulfide with residual water [19] present in the chamber cannot be ruled out

$$La_2O_2S + 2H_2O \rightarrow La_2O_3 + SO_2 + 2H_2 \quad (3)$$

Therefore, present results suggest that a faster oxygen mobility in the OC enables operation under conditions far from those leading to a significant and irreversible decomposition.

As shown in Figure 3b, during TPO the undoped La_2O_2S starts to be reoxidized at ca. 700 °C displaying a broad peak centered at 850 °C, with an estimated oxygen consumption corresponding to the theoretical value needed for the restoration of $La_2O_2SO_4$ [10]. All metal promoted OCs in their oxysulfide form display similar onset temperatures for reoxidation, although the associated O_2 consumption peak is somehow broader than for the undoped material and its maximum occurs at around 900 °C. Unresolved signals also appear in the low-mid temperature region of the TPO profiles of metal-doped OC materials that are related to the reoxidation of Cu or MeOx phases [10].

The oxygen storage performance has been next evaluated at constant temperature under alternating feed stream conditions, where reducing (fuel) and oxidizing (air) gases are cycled. Figure 5 shows the weight changes recorded for doped and undoped OC materials at 800 °C during three cycles obtained alternating 5% H_2/N_2 and air flows. During the first reduction step, all metal-promoted materials undergo a weight loss of ca. 15%, which is compatible with an almost complete reduction of $La_2O_2SO_4$ to La_2O_2S. However, the unpromoted $La_2O_2SO_4$ material shows a larger weight loss that slightly exceeds the value expected for the transformation into the corresponding oxysulfide. Moreover, $La_2O_2SO_4$ needs ca. 40 min to complete its weight loss, whereas all metal promoted samples require a shorter time: the rate of reaction increases in the order Cu < Mn < Co.

Re-oxidation to lanthanum oxysulfate occurs immediately after the exposure to air flow for the three metal promoted materials (showing superimposed TG traces in Figure 5), whilst it takes about 20 min for the unpromoted sample. In fact, the larger weight loss displayed by $La_2O_2SO_4$ during its first reduction step is irreversible, as confirmed by the incomplete recovery of its original weight after reoxidation. The second and the third cycles show similar qualitative features, Co-$La_2O_2SO_4$ quickly turning into Co-La_2O_2S and Cu- and Mn-promoted carriers taking few more minutes. On the contrary, the reduction of pure $La_2O_2SO_4$ appears even harder than in the first cycle, since the weight loss is not completed within the time allowed for this step. Slowing down of the reduction rate of $La_2O_2SO_4$ after the first reduction/reoxidation cycle was already observed [10] during cyclic H_2-TPR/O_2-TPO experiments with maximum temperatures of 1027 °C. Therefore, a progressive loss of the initial redox properties of $La_2O_2SO_4$ occurs also when the chemical looping process is operated at a fixed and relatively moderate temperature level.

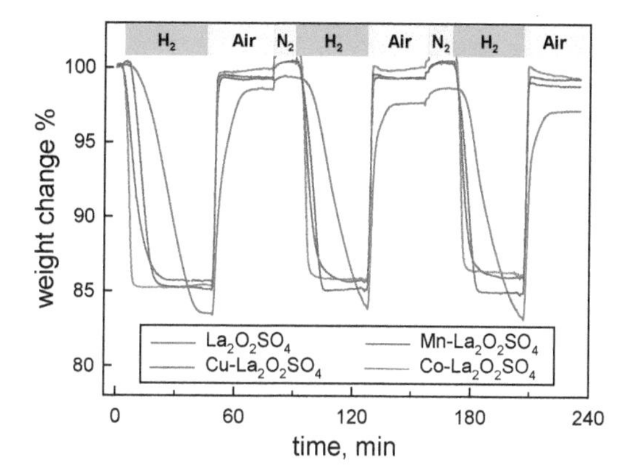

Figure 5. Weight changes during chemical looping reactions under cycled feed stream of 5% CH_4/N_2 and air at 800 °C over $La_2O_2SO_4$ and metal (Cu, Mn, Co) promoted $La_2O_2SO_4$.

Lowering the reaction temperature mainly affects the time required to complete the reduction step with H_2, which indeed represents the limiting step, whereas reoxidation in air results always very fast. In fact, due to its intrinsically easier reducibility, only Co-$La_2O_2SO_4$ is able to complete full redox cycles at 700 °C (not shown) within the given time (45 min).

In order to extend this investigation to a more common fuel, chemical looping combustion in the flow microbalance has been performed using methane (5% vol. in N_2) as the reducing agent. Due to its much lower reactivity with respect to H_2, the operating temperature has been preliminarily set at 900 °C. The oxygen storage performance of the carriers has been verified during 15 cycles (Figure 6), which also allows testing their thermo-chemical stability.

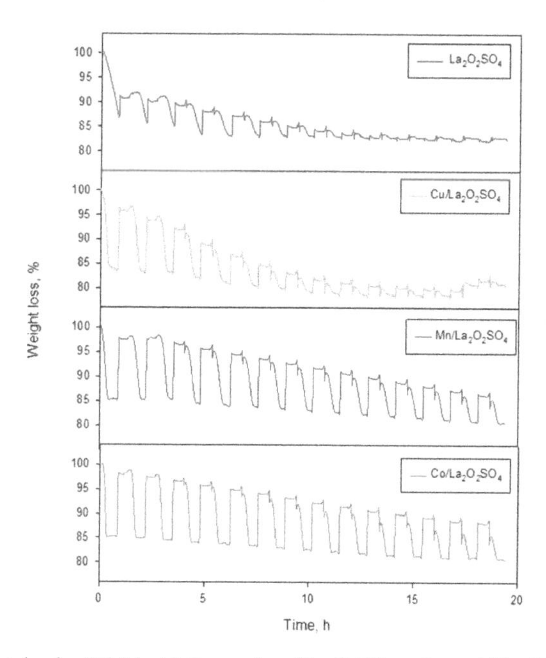

Figure 6. Reducing (under CH_4)/oxidation cycles of $La_2O_2SO_4$ and metal (Cu, Mn, Co) promoted $La_2O_2SO_4$ carried out at 900 °C (15 cycles).

All OCs, except the unpromoted one, initially display the typical weight loss of ca. 15%, suggesting the transformation of $La_2O_2SO_4$ into La_2O_2S can occur also by reaction with methane. Judging by the slope of the weight loss profile, the reaction rate increases in the order Cu < Mn < Co, thus following the same trend of reactivity found with H_2 as fuel. On the contrary, the weight loss of $La_2O_2SO_4$ is not yet completed after 45 min under CH_4 flow. Thereafter, reoxidation by air does not completely restore the original weight of samples, particularly for those less reactive materials. In fact, the OSC of the undoped $La_2O_2SO_4$ is progressively reduced in the subsequent cycles and vanishes after the 10th cycle.

Doping by copper has a modest beneficial effect on the residual OSC, but the weight of the sample progressively approaches a stable level compatible with the irreversible transformation into La_2O_3. On the other hand, the promoting effect of manganese and particularly that of cobalt is evident both in terms of reaction rates as well as residual oxygen storage capacity, which is still ca. 45% of its initial value for Co-$La_2O_2SO_4$ during the last cycle.

Figure 7 shows the infrared (IR) spectra of evolved gas collected at three times (initial, middle, end) during the 7th reducing step for Co-$La_2O_2SO_4$. The band around 2340 cm^{-1} is assigned to CO_2 formed by the reaction between CH_4 and lattice oxygen from the material (Equation (4)). CO_2 is the only product detected during the initial phase of reaction over Co- and Mn-doped materials, which promote the deep oxidation of methane with 100% selectivity.

$$La_2O_2SO_4 + CH_4 \rightarrow La_2O_sS + CO_2 + 2H_2O \quad (4)$$

$$La_2O_2SO_4 + 2CH_4 \rightarrow La_2O_sS + 2CO + 4H_2O \quad (5)$$

Thereafter, the intensity of the CO_2 signal declines, as a consequence of a reduction in the lattice oxygen mobility/availability and, therefore, in the overall reaction rate. Notably, the appearance of a weak doublet band at 2170–2115 cm^{-1} (insert of Figure 7) indicates the formation of only small amounts of CO in addition to CO_2 via the partial oxidation of methane (Equation (5)), while the characteristic band centered at 3020 cm^{-1} is associated to increasing amounts of unconverted methane. As expected, a comparison of IR spectra of all OCs (not shown) highlights that the order of deep oxidation rate of methane follows the same trend of lattice oxygen mobility/availability in the OC materials.

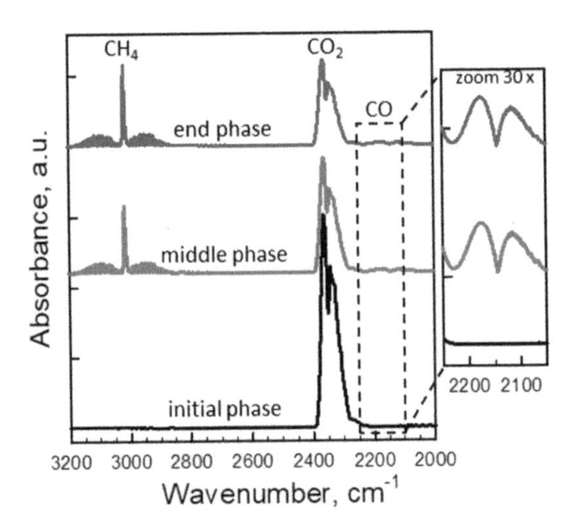

Figure 7. Fourier transform infrared (FTIR) spectra of the evolved gas leaving the TG at three times during the reduction step (under CH_4 flow) of Co-$La_2O_2SO_4$ at 900 °C (7th cycle of Figure 6). The inset shows a magnification (30x) of the region of typical CO absorption bands.

XRD patterns of $La_2O_2SO_4$ and metal-promoted $La_2O_2SO_4$ at the end of 15 cycles are presented in Figure 1b. The low intensity of signals compared to the corresponding ones in Figure 1a is due to the small amount of each material (12–16 mg) available after the experiments in the flow microbalance. All samples show quite complex patterns suggesting the presence of two or more crystalline phases. In particular they all show a well detectable peak at ca. 13° that can be assigned to lanthanum hydroxide (JCPDS 36–1481) produced by hydration of La_2O_3. This confirms the loss of oxygen storage capacity is caused by the irreversible decomposition of $La_2O_2SO_4$ with formation of lanthanum oxide. While most of the signals in the XRD patters of $La_2O_2SO_4$ and Cu-$La_2O_2SO_4$ can be attributed to $La(OH)_3$, Mn- and Co-promoted carriers still show well detectable peaks at 25°, 28°, 36.9°, 44.9° characteristic of $La_2O_2S_2$ phase. Therefore, the addition of Mn and Co can prevent to some extent the decomposition of lanthanum oxysulfate/oxysulfide. This represents an interesting result considering that the enhancement of oxygen mobility is probably associated with a higher distortion of SO_4 units caused by inclusion of cobalt or manganese ions into the oxysulfate/oxysulfide structure. Therefore, the better crystallized, un-promoted $La_2O_2SO_4$ is more prone to loose sulfate units than the more amorphous Co-$La_2O_2SO_4$.

In order to get further insights into the possible reactions occurring during cycled anaerobic oxidation of methane over the best performing Co-$La_2O_2SO_4$, further TG experiments have been carried out analyzing evolved gases with the mass spectrometer. Figure 8 shows the weight change and the corresponding mass signals for the main gaseous products (m/z = 2, 18, 44, 64) as recorded at 900 °C by alternating reducing and oxidizing flows (40min), separated by a N_2 purge (30 min) in between. As soon as exposed to methane, Co-$La_2O_2SO_4$ is quickly and completely reduced according to equation (4), giving H_2O and CO_2 as primary products, as confirmed by the sharp and simultaneous MS peaks recorded in the effluent gas. After the expected weight loss (ca. 15.5%) is completed, a small weight increase is observed, that is probably related to some coke deposition on the sample deriving from CH_4 decomposition catalyzed by Co [24]. Accordingly, a sudden emission of H_2 shows up after the production of CO_2 and water has stopped, and it continues during the purge phase, as long as some methane is still available in the reacting chamber. While the lattice oxygen from the OC material is consumed to oxidize methane, an evident peak of SO_2 emission appears that ends when the OC is fully reduced. Notably, the maximum SO_2 concentration is ca. two orders of magnitude lower than CO_2. This phenomenon resembles what observed during H_2-TPR experiments (Figure 4), and indicates a limited decomposition of $La_2O_2SO_4$ can occur also by reaction with methane leading to SO_2 and the products of methane partial oxidation (Equation (6)):

$$La_2O_2SO_4 + CH_4 \rightarrow La_2O_3 + SO_2 + 2\,H_2 + CO \tag{6}$$

However, the formation of CO (m/z = 28) cannot be easily detected because of the use of N_2 as the carrier gas.

Due to the formation of some La_2O_3, the OC material cannot fully recover its original weight during the fast oxidation occurring as soon as it is exposed to air flow. During this phase, a small CO_2 peak appears, caused by the oxidation of coke deposits previously formed on the material. The stepwise increase in the signal of H_2O is assigned to humidity contained in the air flow. Moreover, a new but rather small emission of SO_2 is detected during the reoxidation, therefore the occurrence of reaction (7) should be also considered:

$$La_2O_2S + 3/2\,O_2 \rightarrow La_2O_3 + SO_2 \tag{7}$$

During the subsequent cycles, the temporal profiles of the gaseous species are repeatable.

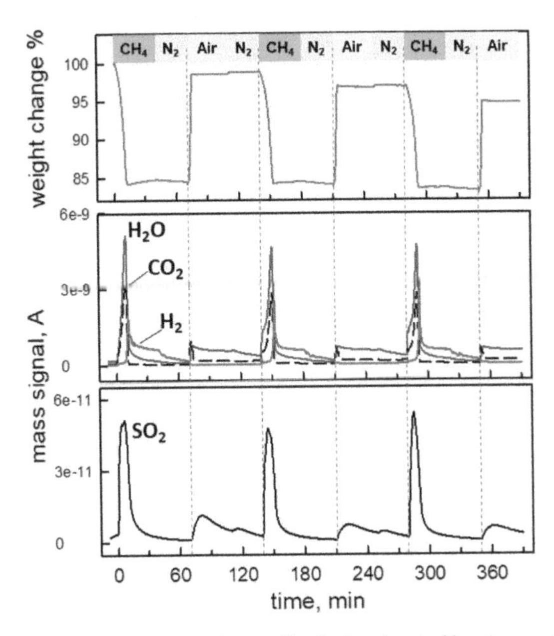

Figure 8. Weight change and gas concentration profiles during chemical looping reactions under cycled feed stream of 5% CH_4/N_2 and air at 900 °C over Co-$La_2O_2SO_4$.

Considering the high oxygen mobility in Co-$La_2O_2SO_4$, the cyclic anaerobic oxidation of methane has been tested also at 800 °C and 750 °C in order to verify if the adverse impact of decomposition reactions (Equations (6) and (7)) could be limited. Results are shown in Figures 9 and 10, respectively. As expected, at 800 °C the reduction of the oxysulfate takes place at a slower rate but the corresponding weight loss is completed within ca. 30 min under CH_4 flow. Accordingly, water and CO_2 are produced during the whole reduction phase, resulting in wider emission peaks than at 900 °C. Some H_2 appears only at the end of this step, but is not accompanied by any detectable weight increase. It is inferred that CH_4 decomposition does not significantly proceed at 800 °C, whereas some methane partial oxidation can occur when lattice oxygen from the OC is almost completely exploited. The SO_2 emission during the reduction step at 800 °C is at least one order of magnitude lower than at 900 °C, confirming the hypothesis on the beneficial effect of a lower reaction temperature to limit the decomposition of the oxysulfate into the corresponding oxide. Re-oxidation under air at 800 °C is still very rapid, and the weight recovery is almost complete. However, the second (and third) reduction step is slower than the first one so that the transformation into the oxysulfide phase ends only during the purging step. As a consequence, CO_2 and H_2O profiles show broader peaks with similar area but lower maxima with respect to the first cycle and shifted towards the end of the period. The SO_2 emission during reoxidation remains quite low. The results of the third cycle are repeatable.

As shown on the top of Figure 10, when operating at 750 °C the duration of the reduction step has been prolonged up to 90 min. The slower rate of reduction of the OC material limits its final weight loss to 15.3%, indicating that the transformation into the oxysulfide phase is not yet completed and ca. 96.5% of the total OSC has been consumed. The corresponding production of water and CO_2 by methane oxidation with lattice oxygen is characterized by the presence of two emission peaks: the first one is sharp and located at the beginning of the exposure to CH_4, whereas the second peak is broad and centered at ca. 45 min. It can be argued that the initial formation of an outer shell of La_2O_2S may slow down the reaction due to the increase of diffusional limitations (in the solid state). H_2 formation is initially low and increases towards the end of the period, testifying an increase of process selectivity towards the products of partial oxidation of methane due to the slow oxygen availability from the

OC material. However, no SO_2 emission can be detected (sub-ppm level) during operation at 750 °C. The subsequent reoxidation by air is still fast. As already observed at 800 °C, the second reduction step is slower than the first one: based on the final weight of the OC, ca. 80% of its total OSC has been utilized. Moreover, the fraction of La_2O_2S displays a lower attitude to be fully re-oxidized, since the sample does not completely recover its previous weight. The results of the third cycle are repeatable, confirming that the partial reduction of the OS performance is not related to material stability issues (decomposition) but rather to an incomplete transformation between $La_2O_2SO_4$ and La_2O_2S phases due to the reduced oxygen mobility.

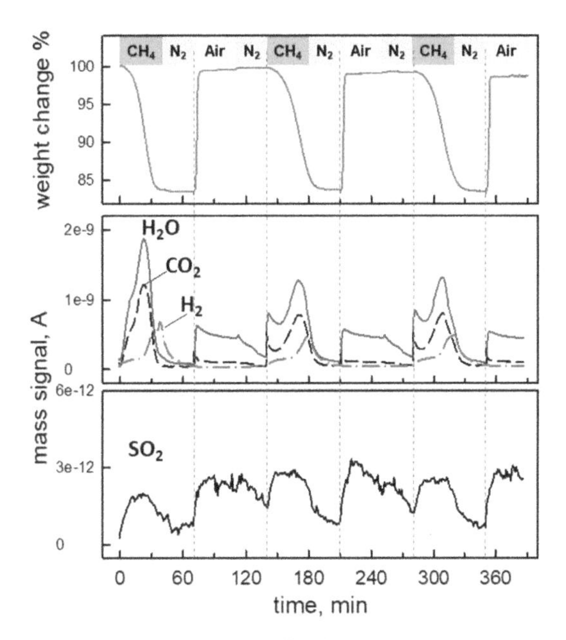

Figure 9. Weight change and gas concentration profiles during chemical looping reactions under cycled feed stream of 5% CH_4/N_2 and air at 800 °C over Co-$La_2O_2SO_4$.

Overall the results indicate that the Co-promoted $La_2O_2SO_4/La_2O_2S$ couple can be favorably operated at 800 °C in the chemical looping combustion of methane: it shows a high oxygen mobility that promotes complete oxidation of methane and full exploitation of its outstanding OS capacity, and a remarkable stability/durability during cyclic operation. Notably, the stability of the $La_2O_2SO_4/La_2O_2S$ system could be strongly enhanced in the presence of even small contents of S-bearing compounds [14], commonly found in the fuel stream. On the other hand, further studies should address the possibility to enhance lattice oxygen mobility and overall reactivity at lower temperatures, by increasing the specific surface area of the OC [18], and by varying the content of transition metal (oxide) dopant which could effectively work as an intermediate oxygen gateway [16] apart from its role as structural promoter.

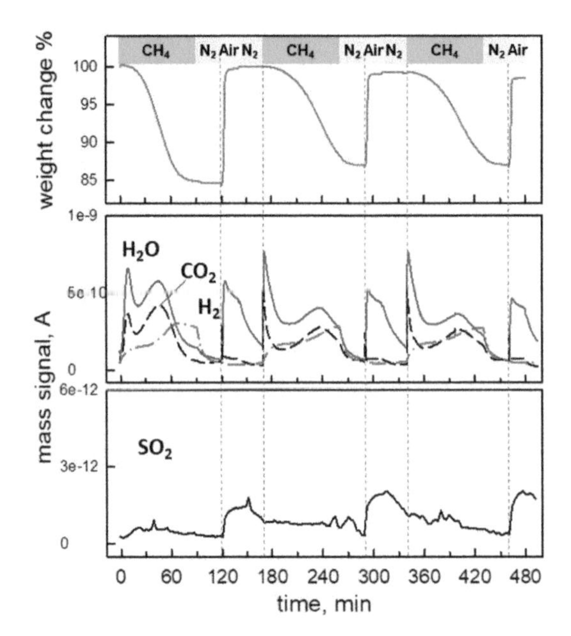

Figure 10. Weight change and gas concentration profiles during chemical looping reactions under cycled feed stream of 5% CH_4/N_2 and air at 750 °C over Co-$La_2O_2SO_4$.

3. Materials and Methods

3.1. Preparation of Oxygen Carriers

Lanthanum oxysulfate ($La_2O_2SO_4$) was obtained starting from $La_2(SO_4)_3$ (Sigma-Aldrich, 99.99% purity, St. Louis, MO, USA) by heating it at 1027 °C for 5 h under helium flow [10,20]. Different fractions of $La_2O_2SO_4$ powder were impregnated with water solutions of $Cu(NO_3)_2{\cdot}2.5H_2O$ (Sigma-Aldrich, purity ≥ 98%, St. Louis, MO, USA), $Mn(NO_3)_2{\cdot}4H_2O$ (Sigma-Aldrich, purity ≥ 97%, St. Louis, MO, USA) and $Co(NO_3)_2{\cdot}6H_2O$ (Sigma-Aldrich, purity ≥ 98%, St. Louis, MO, USA) to obtain samples with a nominal 1wt% metal loading. After impregnation, samples were dried for 12 h at 120 °C in a stove and eventually heat treated at 900 °C for 2 h under He.

3.2. Physical and Chemical Characterization

The surface area was evaluated according to the BET method by N_2 adsorption at 77K with a Quantachrome Autosorb 1-C analyzer (Quantachrome Instruments, Boynton Beach, FL, USA).

Powder X-ray diffraction (XRD) analysis was performed with a Bruker D2 Phaser diffractometer (Bruker, Billerica, MA, USA) operated at diffraction angles ranging between 10 and 80° 2θ with a scan rate of 0.02° 2θ s^{-1}.

Temperature Programmed Reduction (TPR) and Oxidation (TPO) experiments were carried out with a Micromeritics Autochem II TPD/TPR instrument (Micromeritics, Norcross, GA, USA) equipped with a TC detector. The samples (20–30 mg) were first reduced under a flow of 2% H_2 in Ar (50 Ncm^3/min) by heating up to 1027 °C at 10 °C/min; thereafter, they were reoxidized under a flow of 0.5% O_2 in He (50 Ncm^3/min) using the same temperature ramp.

TG analysis were performed in a Setaram Labsys Evo TGA-DTA-DSC 1600 (Setaram Instrumentation, Caluire, France) flow microbalance loading 15–25 mg of sample in an alumina crucible and ramping the temperature at 10 °C/min up to 1300 °C or 1150 °C, respectively under a flow (100 cc/min) of pure N_2 or 2% H_2 in N_2.

3.3. *Chemical Looping Reaction*

Cyclic isothermal reduction/oxidation tests were carried out in the same Setaram Labsys Evo flow microbalance at 700, 750, 800 and 900 °C, by switching alternatively the flow between 5% H_2 (or CH_4) in N_2 and air, at a fixed total flow rate of 100 cc/min. If not otherwise stated, the duration of the reduction phase was 45 min, whilst that of oxidation phase was 30 min, with an intermediate purge by pure N_2 (2–30 min). The evolved gases, leaving the flow microbalance through two independent heated capillaries, were analyzed by a Pfeiffer Thermostar G mass spectrometer and a Perkin Elmer Spectrum GX spectrometer in order to detect all IR-active gaseous species. In this last case, IR spectra were rationed against a background spectrum of pure N_2 and were collected every 1-10 min depending on the duration of the experiment.

4. Conclusions

Oxygen carrier materials based on $La_2O_2SO_4$ promoted by small amounts (1% wt.) of transition metals such as Co, Mn and Cu have been synthesized and characterized by means of XRD, BET, TPR/TPO and TG-MS-FTIR experiments in order to investigate their potential use in the Chemical Looping Combustion process with either hydrogen or methane as fuel. Results demonstrate that doping the parent $La_2O_2SO_4$ with transition metals can promote lattice oxygen mobility in the temperature range 700–900 °C following the order Cu < Mn < Co, catalyzing the complete reduction of $La_2O_2SO_4$ to form La_2O_2S by reaction with either H_2 and CH_4. On the other hand, the reoxidation of La_2O_2S by molecular oxygen to restore the $La_2O_2SO_4$ phase requires temperatures ≥700 °C to proceed quickly and is poorly affected by the addition of those transition metals.

Notably, the addition of Co or Mn has shown a marked beneficial effect to limit or inhibit the decomposition of the oxysulfate that releases some SO_2. Such undesired side reaction proceeds to some extent together with the main reduction of the $La_2O_2SO_4$ by either H_2 or CH_4 and it causes a progressive and irreversible loss of the original oxygen storage capacity through the formation of the stable La_2O_3 phase. Eventually, metal doping does not affect the thermal decomposition of $La_2O_2SO_4$ occurring above ca. 1030 °C under inert flow, nor it promotes any spontaneous reduction of the lanthanum oxysulfate to oxysulfide.

The enhanced performance of the Co-promoted carrier can be assigned to H_2/CH_4 activation on metal sites and to the higher distortion of SO_4 units in the oxysulfate lattice caused by the partial inclusion of Co^{3+} ion, much smaller than La^{3+}, which enhances oxygen mobility.

TG-MS-FTIR experiments of CLC with alternating feeds have shown that Co-$La_2O_2SO_4$ is able to oxidize CH_4 at 800 °C producing CO_2 and water with reasonable rates and high selectivity, avoiding the deterioration of performance related to a progressive decomposition. At higher temperatures, the oxygen storage performance is progressively lost due to the irreversible formation of some La_2O_3, that occurs as a side reaction during the reducing step of the OC material by the fuel and is accompanied by the release of SO_2. At lower temperatures, the mobility of lattice oxygen becomes the limiting factor that can preclude the complete exploitation of the OSC within a reasonable contact time with the fuel stream.

Author Contributions: Data curation, G.M.; Investigation, S.C., G.M. and L.L.; Supervision, S.C. and L.L.; Writing—original draft, S.C. and L.L.; Writing—review & editing, S.C.

Funding: This research received no external funding.

Acknowledgments: Luciano Cortese is kindly acknowledged for XRD analysis.

Conflicts of Interest: The authors declare no conflict of interest.

References

1. Tang, M.; Xu, L.; Fan, M. Progress in oxygen carrier development of methane-based chemical-looping reforming: A review. *Appl. Energy* **2015**, *151*, 143–156. [CrossRef]

2. Bhavsar, S.; Najera, M.; Solunke, R.; Veser, G. Chemical looping: To combustion and beyond. *Catal. Today* **2014**, *228*, 96–105. [CrossRef]
3. Adanez, J.; Abad, A.; Garcia-Labiano, F.; Gayan, P.; de Diego, L.F. Progress in Chemical-Looping Combustion and Reforming technologies. *Prog. Energy Combust. Sci.* **2012**, *38*, 215–282. [CrossRef]
4. Zheng, X.; Che, L.; Hao, Y.; Su, Q. Cycle performance of Cu-based oxygen carrier based on a chemical-looping combustion process. *J. Energy Chem.* **2016**, *25*, 101–109. [CrossRef]
5. Imanieh, M.H.; Rad, M.H.; Nadarajah, A.; González-Platas, J.; Rivera-López, F.; Martín, I.R. Novel perovskite ceramics for chemical looping combustion application. *J. CO2 Util.* **2016**, *13*, 95–104. [CrossRef]
6. Liu, L.; Taylor, D.D.; Rodriguez, E.E.; Zachariah, M.R. Influence of transition metal electronegativity on the oxygen storage capacity of perovskite oxides. *Chem. Commun.* **2016**, *52*, 10369–10372. [CrossRef] [PubMed]
7. Miccio, F.; Bendoni, R.; Piancastelli, A.; Medri, V.; Landi, E. Geopolymer composites for chemical looping combustion. *Fuel* **2018**, *225*, 436–442. [CrossRef]
8. Bendoni, R.; Miccio, F.; Medri, V.; Landi, E. Chemical looping combustion using geopolymer-based oxygen carriers. *Chem. Eng. J.* **2018**, *341*, 187–197. [CrossRef]
9. Bi, W.; Chen, T.; Zhao, R.; Wang, Z.; Wu, J.; Wu, J. Characteristics of a $CaSO_4$ oxygen carrier for chemical-looping combustion: Reaction with polyvinylchloride pyrolysis gases in a two-stage reactor. *RSC Adv.* **2015**, *5*, 34913–34920. [CrossRef]
10. Lisi, L.; Mancino, G.; Cimino, S. Chemical looping oxygen transfer properties of Cu-doped lanthanum oxysulphate. *Int. J. Hydrogen Energy* **2015**, *40*, 2047–2054. [CrossRef]
11. Machida, M.; Kawamura, K.; Ito, K. Novel oxygen storage mechanism based on redox of sulfur in lanthanum oxysulfate/oxysulfide. *Chem. Commun. (Camb.)* **2004**, *2*, 662–663. [CrossRef] [PubMed]
12. Machida, M.; Kawamura, K.; Ito, K.; Ikeue, K. Large-Capacity Oxygen Storage by Lanthanide Oxysulfate/Oxysulfide Systems. *Chem. Mater.* **2005**, *17*, 1487–1492. [CrossRef]
13. Machida, M.; Kawano, T.; Eto, M.; Zhang, D.; Ikeue, K. Ln Dependence of the Large-Capacity Oxygen Storage/Release Property of Ln Oxysulfate/Oxysulfide Systems. *Chem. Mater.* **2007**, *19*, 954–960. [CrossRef]
14. Ikeue, K.; Eto, M.; Zhang, D.J.; Kawano, T.; Machida, M. Large-capacity oxygen storage of Pd-loaded $Pr_2O_2SO_4$ applied to anaerobic catalytic CO oxidation. *J. Catal.* **2007**, *248*, 46–52. [CrossRef]
15. Zhang, D.; Yoshioka, F.; Ikeue, K.; Machida, M. Synthesis and Oxygen Release/Storage Properties of Ce-Substituted La-Oxysulfates, $(La_{1-x}Ce_x)_2O_2SO_4$. *Chem. Mater.* **2008**, *20*, 6697–6703. [CrossRef]
16. Zhang, D.; Kawada, T.; Yoshioka, F.; Machida, M. Oxygen Gateway Effect of $CeO_2/La_2O_2SO_4$ Composite Oxygen Storage Materials. *ACS Omega* **2016**, *1*, 789–798. [CrossRef]
17. Osseni, S.A.; Denisenko, Y.G.; Fatombi, J.K.; Sal'nikova, E.I.; Andreev, O.V. Synthesis and characterization of $Ln_2O_2SO_4$ (Ln = Gd, Ho, Dy and Lu) nanoparticles obtained by coprecipitation method and study of their reduction reaction under H_2 flow. *J. Nanostruct. Chem.* **2017**, *7*, 337–343. [CrossRef]
18. Zhang, W.; Arends, I.W.C.E.; Djanashvili, K. Nanoparticles of lanthanide oxysulfate/oxysulfide for improved oxygen storage/release. *Dalt. Trans.* **2016**, *45*, 14019–14022. [CrossRef]
19. Tan, S.; Li, D. Enhancing Oxygen Storage Capability and Catalytic Activity of Lanthanum Oxysulfide (La_2O_2S) Nanocatalysts by Sodium and Iron/Sodium Doping. *ChemCatChem* **2018**, *10*, 550–558. [CrossRef]
20. Sal'nikova, E.I.; Kaliev, D.I.; Andreev, P.O. Kinetics of phase formation upon the treatment of $La_2(SO_4)_3$ and $La_2O_2SO_4$ in a hydrogen flow. *Russ. J. Phys. Chem. A* **2011**, *85*, 2121–2125. [CrossRef]
21. Poston, J.A.; Siriwardane, R.V.; Fisher, E.P.; Miltz, A.L. Thermal decomposition of the rare earth sulfates of cerium(III), cerium(IV), lanthanum(III) and samarium(III). *Appl. Surf. Sci.* **2003**, *214*, 83–102. [CrossRef]
22. Arnone, S.; Bagnasco, G.; Busca, G.; Lisi, L.; Russo, G.; Turco, M. Catalytic combustion of methane over transition metal oxides. *Stud. Surf. Sci. Catal.* **1998**, *119*, 65–70.
23. Cimino, S.; Lisi, L.; Tortorelli, M. Low temperature SCR on supported MnOx catalysts for marine exhaust gas cleaning: Effect of KCl poisoning. *Chem. Eng. J.* **2016**, *283*, 223–230. [CrossRef]
24. Zhang, Y.; Smith, K.J. CH_4 decomposition on Co catalysts: Effect of temperature, dispersion, and the presence of H_2 or CO in the feed. *Catal. Today* **2002**, *77*, 257–268. [CrossRef]

Article

Deactivation of Commercial, High-Load o-Xylene Feed VO_x/TiO_2 Phthalic Anhydride Catalyst by Unusual Over-Reduction

Oliver Richter and Gerhard Mestl *

Clariant AG, 83052 Bruckmühl, Germany; Oliver.Richter@clariant.com
* Correspondence: Gerhard.Mestl@clariant.com; Tel.: +49-806-14903-825

Received: 13 April 2019; Accepted: 7 May 2019; Published: 9 May 2019

Abstract: An unusual temporal behavior of the by-product spectrum, as well as the temperature profiles of a commercial phthalic anhydride reactor, indicated a non-typical change of the incumbent catalyst. In order to understand these observations, catalyst samples were taken from this reactor and analyzed by standard physico-chemical methods. Catalyst samples from another commercial reference reactor with most similar operating conditions and catalyst lifetime were also taken for comparison. The detailed physical analysis did not indicate unusual thermal stress leading to catalyst deactivation by rutilisation or sintering of the titania phase. The chemical analysis did not reveal significant amounts of any of the known catalyst poisons, which would also contribute to an untypical catalyst deactivation/behavior. Quantitative X-ray diffraction measurements on the other hand revealed an unusually high degree of reduction of the vanadium species in the final polishing catalyst layer. Such an abnormal degree of catalyst reduction, and hence, irreversible damaging, was concluded to likely originate from a unit shutdown without sufficient air purging of the catalyst bed. Combustion analysis of the deactivated catalyst confirmed unusually high carbon contents in the finishing catalyst bed (L4) accompanied with a significant loss in the specific surface area by plugging the catalyst pores with high-molecular carbon deposits. According to the well-known Mars–van-Krevelen-mechanism, o-xylene and reaction intermediates remain adsorbed on the catalyst surface in case of a shutdown without air purging and will continue to consume lattice oxygen, accordingly reducing the catalytic species. This systematic investigation of used catalyst samples demonstrated the importance of sufficient air purging during and after a unit shutdown to avoid abnormal, irreversible damage and thus negative impact to catalyst performance.

Keywords: phthalic anhydride; vanadia-titania catalyst; unusual deactivation; physico-chemical characterization; over-reduction; vanadia species; coke deposition

---

1. Introduction

Phthalic anhydride (PA) has been produced commercially since the late 1960s by passing a mixed gas containing o-xylene (oX) and air at elevated temperatures through salt-bath cooled multi-tubular reactors packed with vanadia/titania catalysts, which enable the selective catalytic oxidation of oX to PA [1]. Mixtures of oX and naphthalene, as well as pure naphthalene, are also used as feedstocks on a commercial scale. Beside the nature of the feedstock, PA production processes are discriminated by different reactor lengths from 2.5 to 3.7 m, different tube inner diameters from 21 to 25 mm, different air rates from 2.2 to 4.2 $Nm^3/h/tube$, and the o-xylene air mixing ratio from 42 to 100 g/Nm^3. Obviously, catalyst suppliers have to fine-tune process catalysts to the respective environment of a customer plant to allow maximum performance. With a world production of about 4 million metric tons PA in 2017 [2], this reaction is not only commercially interesting but also scientifically, as it exhibits a rather complex reaction network (Figure 1) that is still not fully understood [3].

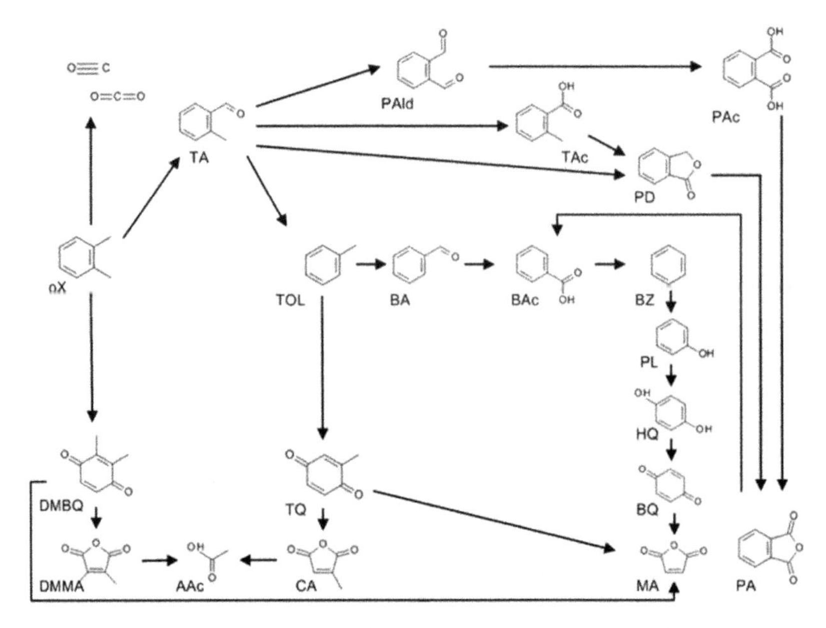

Figure 1. Reaction scheme of o-xylene oxidation (adapted from [3]).

Due to high reaction rates, such vanadia/titania catalysts are nowadays produced as so-called egg-shell catalysts, where the catalytically active material is coated as a thin shell on an inert steatite carrier ring. To ensure a homogenous shell thickness for optimum catalyst performance, Clariant PA catalysts are produced by an innovative coating process [4]. Multi-layer catalyst systems (Figure 2) are currently used on commercial scale to cope best with the high exothermicity of the reaction [5].

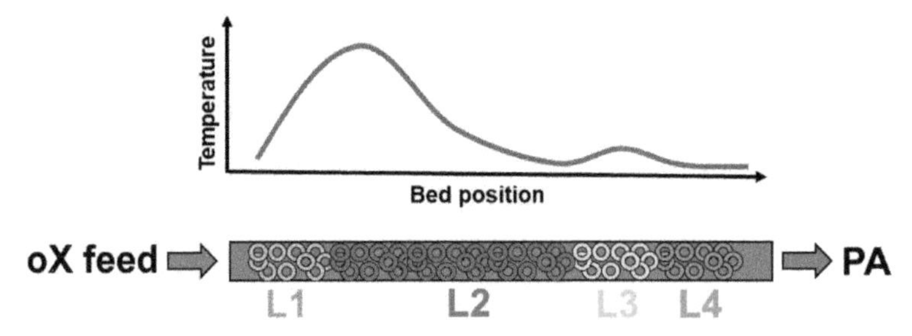

Figure 2. Typical four layer-system and axial temperature profile with Clariant's commercial phthalic anhydride (PA) catalyst.

Different catalyst suppliers have different layer strategies to achieve optimum PA yields and optimum PA quality especially at high feed loadings up to 100 g/Nm^3. Clariant has developed a specific catalyst layer management during the last few years that makes optimized hot spot control possible [6,7]. The different layers of the catalyst show significant differences in catalytic activity and selectivity, matching the gas phase composition changing along the reactor. The chemistry of the different catalyst layers in the reactor is hence tailored to their respective tasks by adding promoters like antimony, cesium or phosphorous to the main catalyst components vanadia and titania [8,9].

Modern, commercial PA oX feed catalysts [8,9] are four-layer systems (Figure 2) when based on proprietary Clariant knowhow, in which the first layer (L1) has the task to start up the oX partial oxidation heating up the incoming fresh gas to reaction temperature as fast as possible. This catalyst layer is highly active, but rather short to avoid undesired over-oxidation and excessive hotspot temperatures, especially during the beginning of the catalyst lifetime. This so called "starter-layer" also assures an early maximum o-xylene conversion guaranteeing long service life. The second layer (L2) is the main and most selective working layer, i.e., 90–95 mol % of the oX feed to the reactor is converted in this catalyst bed. This layer is designed for maximum PA selectivity by optimizing mass as well as heat transport properties within the active mass shell. It has to be adjusted to the respective customer process conditions, i.e., oX concentration in the feed, air rates and reactor dimensions. The third layer catalyst (L3) has a slightly lower selectivity to PA as the second while simultaneously higher activity. This catalyst layer ensures almost complete oX conversion over the entire catalyst lifetime compensating for the normal ageing of the first layers, which always occurs with time on stream. The last layer (L4), close to the reactor exit, is designed for optimum PA product quality; hence its task is to selectively convert all remaining under-oxidation by-products as completely as possible to PA and to combust all high boiling impurities as well as over-oxidation by-products to CO_x.

It has to mentioned here that PA process catalysts need a certain time on stream before they reach their maximum performance. Prior to the start-up of a commercial reactor, the loaded catalyst is calcined in-situ to combust the organic binder in the active mass. In addition, catalyst precursors, like vanadia, start to form the active species on the anatase support. The initial start-up of the reactor after in-situ calcination usually takes about 3 days and the catalytically active species further develop resulting in a well-established hotspot in the upper catalyst layer L2 (Figure 2). Subsequently, the ramp-up phase to the design o-xylene load starts, which can take up to three months depending on the plant operation conditions. The ramp-up is followed by the so-called break-in period taking another 3 months during which the optimum selectivity develops leading to minimum by-product formation. The above described ramp-up and break-in phases are already accompanied by beginning, very slow deactivation processes especially taking place in the hotspot zone [10].

Afterwards, stable operation is achieved for a service life of up to 5 years, again depending on the plant conditions. The very slow catalyst deactivation with time on stream is typically compensated by an accordingly slow increase of the reactor coolant temperature of about 2–5 °C per year [11].

The reactor off-gas needs to be regularly sampled and analyzed for monitoring the reactor performance and accordingly adjusting the reactor coolant temperature. On an industrial scale, the analysis is often limited to the main over-oxidation by-products CO_x, maleic anhydride, citraconic anhydride, and benzoic acid as well as to the most important under-oxidation intermediate phthalide, and unconverted o-xylene. For practical relevance, the mass-based PA yield is calculated by using a simplified, empirical correlation:

$$\text{PA yield} = 139.52\ \text{wt}\ \% - \left(800\ \frac{\text{g}\cdot\text{wt}\ \%}{\text{Vol}\ \%\cdot\text{Nm}^3}\cdot\frac{\text{CO}+\text{CO}_2}{\text{load}}\right) + (100\ \text{wt}\ \% - \text{purity}) - (1.25\cdot \text{MA}) - (1.1\cdot \text{oX}), \quad (1)$$

wherein,

PA yield = mass-based phthalic anhydride yield [wt %]
CO = CO content in reactor off-gas [Vol %]
CO_2 = CO_2 content in reactor off-gas [Vol %]
load = o-xylene loading in feed stream [g/Nm^3]
purity = o-xylene feed purity [wt %]
oX = o-xylene slip in reactor off-gas [wt %]
MA = maleic anhydride content in reactor off-gas [wt %].

Such a modern, commercial catalyst system reaches molecular selectivity up to 83–84 mol % PA at essentially full oX conversions, corresponding to 115–117 wt % mass-based PA yield, with lifetimes of 3–5 years on stream depending on the economic strategy of the plant management [5].

Due to the technical importance of these types of vanadia/titania catalysts, there is a wealth of literature available on their chemistries, molecular structures and catalyst deactivation mechanisms.

In brief, PA catalysts are often seen as being monolayer-type catalysts [12], although their vanadia loading is often above the theoretical monolayer capacity. Accordingly, it is reported that such catalysts hardly contain any crystalline V_2O_5 [13,14]. However, often minute amounts of crystalline V_2O_5 can be detected by X-ray diffraction (XRD) due to the fact that the vanadia loadings are above monolayer capacity.

After calcination, activation and formation, the high V_2O_5 loadings form an amorphous overlayer of polymeric vanadates being one to a few atomic layers thick on top of the TiO_2 anatase phase [15]. Furthermore, it is known that V^{5+} species being in direct contact with the TiO_2 support are reduced to V^{4+} ions which are being incorporated into the TiO_2 surface [13]. Catalytic tests after removing the polymeric V^{5+} species but keeping the V^{4+} surface species intact revealed inferior catalytic performance. Hence, the polymeric V^{5+} species covering the reduced vanadia surface on the TiO_2 support seem to be related to high catalytic activity and selectivity. Moreover, it is also known that free uncovered TiO_2 surface with its acidic OH groups lead to o-xylene cracking and subsequent total oxidation of the cracking products [13]. Many studies concerning permanent or reversible catalyst deactivation have been also published [10,11,16–29]. Different deactivation phenomena are described in detail:

- over-reduction of the active surface vanadium oxide species,
- coke formation by adsorbed reaction intermediates,
- deposition of catalyst poisons, like alkali salts,
- changing selectivity due to loss of promoters from the surface,
- fouling by deposits like dust plugging pores,
- sintering and accompanied loss of surface area,
- transformation of the TiO_2 anatase phase into the catalytically inactive but thermodynamically more stable rutile phase.

While there are many studies available in the literature investigating different root causes for deactivation, most of them deal with model catalyst compositions being very different from modern, commercially proven, multi-layer catalyst systems after typical run times of 3–5 years. For this reason, we have made it to our task to investigate catalyst deactivation on a commercial scale. In this manuscript, we will report about the deactivation of a high-load oX feed PA catalyst installed in a commercial PA reactor (design air rate 4.0 Nm^3/h/tube, design oX load 95 g/Nm^3) and we will compare the observed analytical data with another commercial reference high-load oX feed catalyst which had not shown any unusual behavior during its service life.

2. Results

2.1. Development of Process Parameters, Reactor Performance and Catalyst Bed Temperatures

The commercial PA reactor was started up and ramped up to design conditions without any issues. Figure 3a–d displays the most relevant operation conditions and reveals that the coolant temperature had to be slowly decreased while increasing accordingly the o-xylene load. There were no drastic changes in the air rate, and hence all parameters indicated a smooth, standard ramp-up. Unsurprisingly, the catalyst revealed the expected high and stable performance in terms of low by-product formation and high PA yields within the first months, despite several plant shutdowns (Figure 3b–d). However, the under-oxidation level unexpectedly rose after about 252 days time-on-stream (TOS) despite counteracting by increasing the coolant temperature (Figure 3b). At the same time, the over-oxidation by-product formation and the phthalic anhydride yields remained stable (Figure 3c,d). This observation already hints at a rather unusual phenomenon occurring during operation.

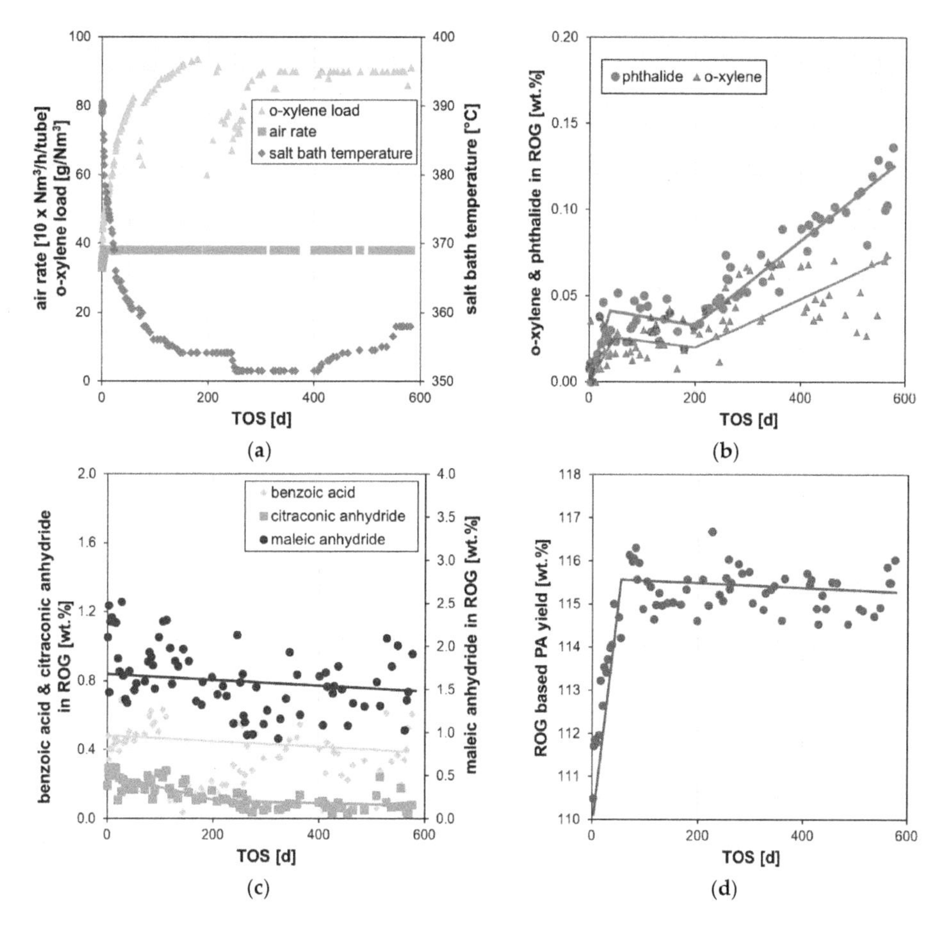

Figure 3. Evolution vs. time-on-stream (TOS) of: (**a**) process parameters (o-xylene (oX) load, air rate, salt bath temperature); (**b**) under-oxidation by-products in the reactor off-gas (phthalide, o-xylene); (**c**) over-oxidation by-products in the reactor off-gas (maleic anhydride, benzoic acid, citraconic anhydride); (**d**) reactor off-gas based PA yield.

To follow the temperature development across the entire PA reactor, 15 single point thermocouples (Ø 1 mm) distributed over several reactor tubes and vertical bed positions were installed during catalyst loading (Section 4.1). A detailed analysis of all available temperature readings in the catalyst bed was performed to determine potential root causes for this unusual, very fast increase of the under-oxidation by-products.

As indicated by the thermocouples, the hotspot temperature declined unusually fast after about 90 d TOS despite the load increase (Figure 4a). When restarting the unit after about 117 d TOS, the hotspot temperature slightly rose as expected (Figure 4a) due to the well-known temporary reactivation of the catalyst after a proper unit shutdown. However, it is important to note that the L2 bed temperatures remained at the unusually low level after the restart, and the hotspot stayed in L3 despite increasing the coolant temperatures (Figure 4a). When comparing the temperature profile evolution of this PA reactor with that of a reference reactor (compare Figure 4a,b), the above described, fast decline of the hotspot temperatures in L2 is very unusual as it was not observed in the reference and is normally not observed in other PA plants as well. Hence, it is reasonable to conclude that the unusual behavior of this PA reactor indicates potential catalyst damaging e.g., by poisoning or unusual thermal stress. In order to investigate this assumption, used catalyst from several tubes was sampled and analyzed. Details of the used catalyst sampling as well as the physico-chemical analyses are reported in the following sections.

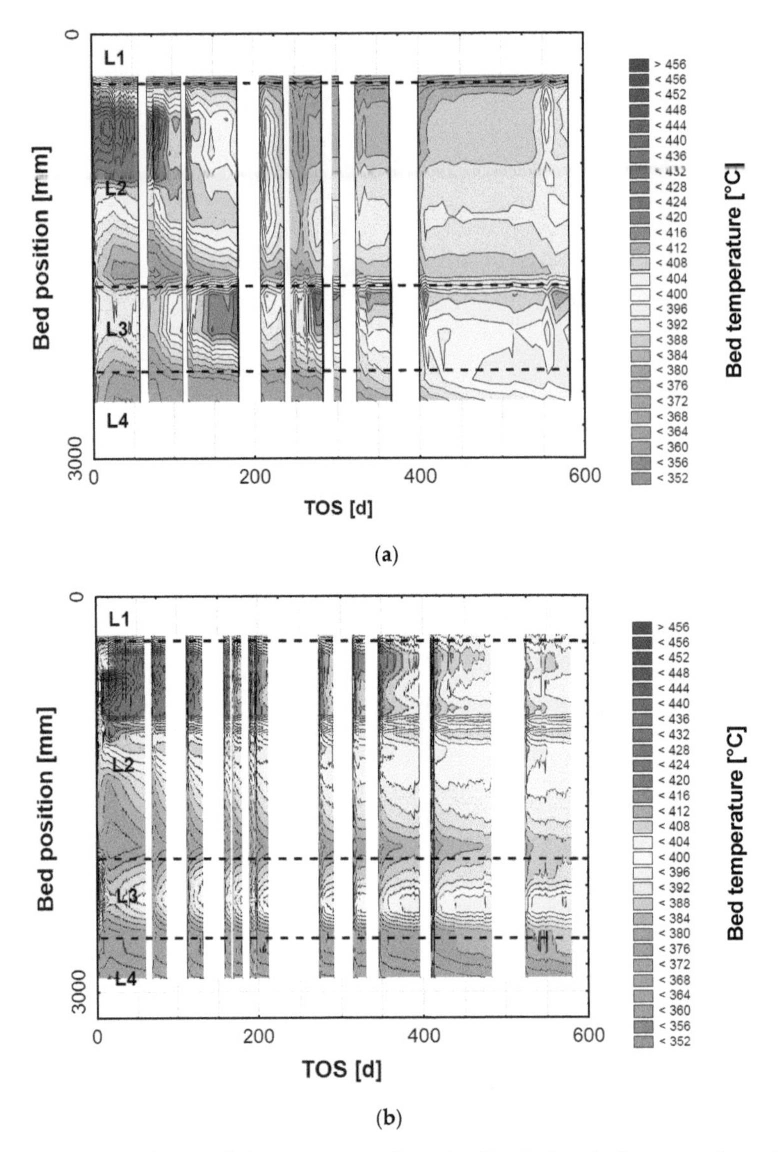

Figure 4. Evolution of catalyst bed temperature as determined by single point thermocouples vs. TOS: (**a**) Investigated reactor; (**b**) commercial reference reactor. The reactor inlet is at position 0 mm and the catalyst layers are labelled according to their loading position. White, vertical areas in the contour plot indicate plant shutdowns.

2.2. *Analysis of Used Catalyst Samples*

2.2.1. Crystalline Phase Composition

A detailed bulk phase analysis by X-ray diffraction (XRD, see Section 4.2) measurements revealed several crystalline phases. Unknown phases/peaks were not detected. Rietveld refinements were calculated to quantify these detected crystalline phases. Figure 5 shows exemplarily the diffractogram together with its Rietveld refinement of a sample of the L4 catalyst.

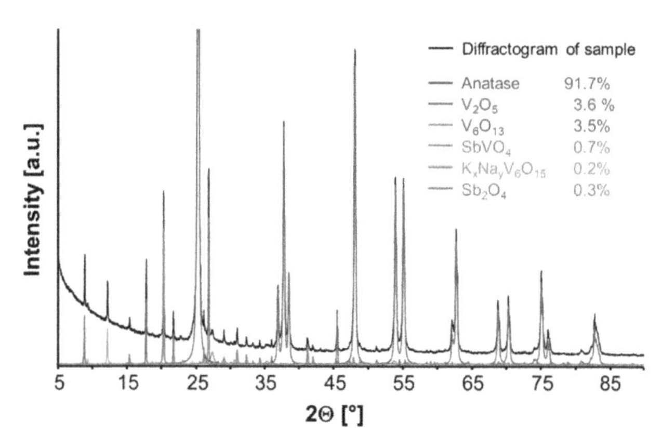

Figure 5. Exemplary X-ray diffractogram of sample from bottom (L4) of one respective reactor tube, combined with its Rietveld refinement.

In summary, the main crystalline bulk phase was titania (TiO_2) in its anatase modification as expected. A minor phase transition of anatase to the thermodynamically more stable, but catalytically inactive rutile was detected in almost all layers with the exception of L4 (Figure 6a). Especially, the samples of the upper L2 catalyst bed, where the hotspot is typically positioned, revealed the highest degree of rutile formation due to the highest thermal stress there as expected. No, or only a minor degree of, rutilisation is observed in PA catalysts after their service lives as typically confirmed by the analysis of used catalyst samples from other commercial reference plants. Comparing the detected minor rutilisation here with the commercial reference (Figure 6a), the rutile levels of only up to about 4 wt % in L2 are within the expected range and hence will not lead to the observed unusual catalyst deactivation. Most of the crystalline vanadium pentoxide (Figure 6b) and all the crystalline antimony trioxide (Sb_2O_3) have disappeared during operation due to their expected spreading over the anatase surface leading to the catalytically active species. Only a very minor solid-state reaction was observed forming crystalline antimony vanadate (Figure 6c) in-situ during operation. A very small amount of volatile V species was obviously formed as well (compare Section 2.2.3) leading to a certain, minor loss in the vanadium content. As can be seen from the comparison with the commercial reference (Figure 6b,c), the formation of crystalline antimony vanadate and the change of the crystalline vanadium pentoxide in the upper layers (L1 to L3) are also within the expected ranges.

However, it is important to note that the detected change of the crystalline vanadium pentoxide in the lowest layer L4 is significantly enhanced as compared to the commercial reference (Figure 6b). High amounts of reduced vanadium oxides (V_6O_{13}) as well as traces of an antimony oxide spinel phase (Sb_2O_4) were detected as well (Figures 6d and 7a). Comparing to used catalyst samples of the commercial reference reactor (Figure 6d), the detected traces of the antimony spinel are on the regular level and hence will not contribute to unusual catalyst deactivation. However, the significantly increased contents of reduced vanadium oxides (Figure 7a) clearly indicate an abnormal degree of reduction of the catalytically active species and as consequence an irreversible damaging of the catalyst. In addition, the above-mentioned enhanced loss of crystalline vanadium pentoxide from the L4 catalyst obviously originated from an increased catalyst reduction. Such an abnormal catalyst reduction, and hence, irreversible damaging could arise from a unit shutdown without sufficient purging of the catalyst bed with air. According to the well-known Mars–van-Krevelen-mechanism, o-xylene and reaction intermediates remain adsorbed on the catalyst surface and are oxidized by lattice oxygen, accordingly reducing the catalytic species [30]. Due to insufficient purging with air, the oxygen content in the gas phase is reduced to unusual levels and the catalyst lattice as well as surface are not sufficiently re-oxidized. Finally, catalytically inactive, reduced phases e.g., like V_6O_{13} are increasingly formed. In addition, potassium-sodium-vanadium bronze ($K_xNa_yV_6O_{15}$) also containing reduced V species was detected in all samples (Figure 7b). This phase indicates, in addition, a certain sodium/potassium contamination of the active mass. Comparing to the commercial reference (Figure 7b), the content of the crystalline potassium-sodium-vanadium bronze is within the expected range and hence does not indicate an unusual catalyst poisoning by sodium and potassium (see Section 2.2.3).

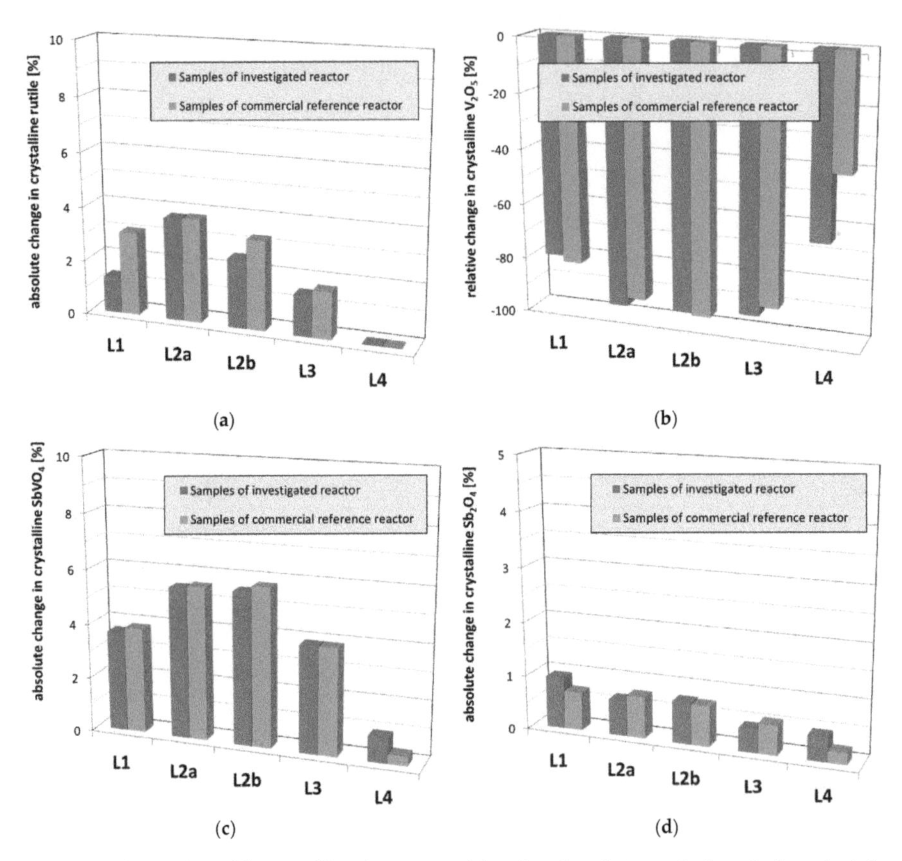

Figure 6. Comparison of the crystalline phase composition of used catalyst samples from the investigated reactor and the commercial reference reactor: (**a**) absolute changes in crystalline rutile; (**b**) relative changes in crystalline V_2O_5; (**c**) absolute changes in crystalline $SbVO_4$; (**d**) absolute changes in crystalline Sb_2O_4. The samples of each layer were averaged.

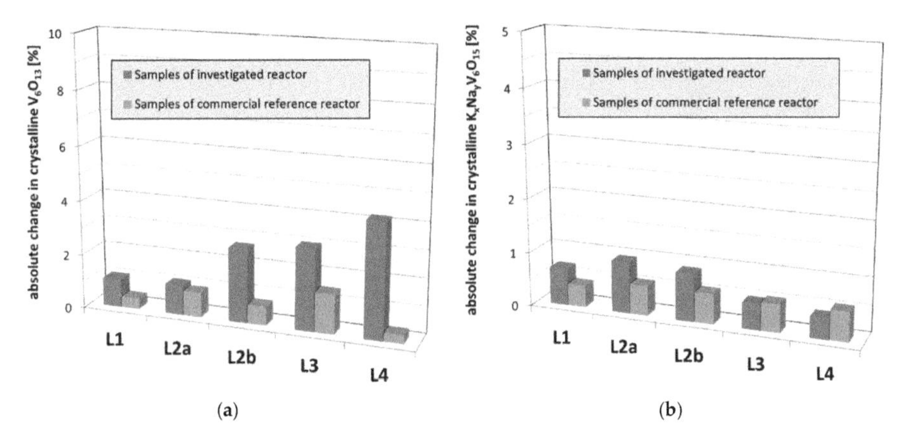

Figure 7. Comparison of the crystalline phase composition of used catalyst samples from the investigated reactor and the commercial reference reactor: (**a**) absolute changes in crystalline V_6O_{13}; (**b**) absolute changes in crystalline $K_xNa_yV_6O_{15}$. The samples of each layer were averaged.

2.2.2. Specific Surface Area

The specific surface area of each layer of the used catalyst samples were measured by nitrogen adsorption (see Section 4.2). As depicted in Figure 8, an unusual loss of the specific surface area was not detected for the upper catalyst layers (L1 to L3). Compared to used catalyst samples from the commercial reference, a total loss of up to 7% is expected, especially for the upper L2 in the region of the typical hotspot position. A loss in the surface area can arise from deposits plugging catalyst pores and/or sintering of titania. As shown in detail below (compare Section 2.2.3), an unusually high level of carbon was detected in L4. Hence, the observed loss in the specific surface area here is mainly due to plugging the catalyst pores by high-molecular carbon deposits reducing the catalyst activity.

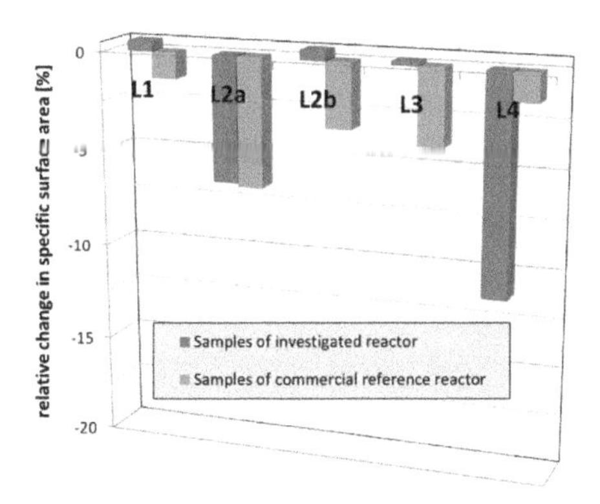

Figure 8. Comparison of the relative changes in specific surface area of used catalyst samples from the investigated reactor and the commercial reference reactor. The samples of each layer were averaged.

2.2.3. Chemical Composition

The analysis of the chemical bulk composition of the samples was done by atomic absorption spectroscopy and combustion analysis (see Section 4.2). The contents of the catalyst compounds antimony and cesium do not show unexpected deviations between the used catalyst samples and the unused, fresh catalysts. The plots of the other catalyst compounds vanadium (calculated as V_2O_5, Figure 9a) and phosphorous (P, Figure 9b) reveal some certain loss along the catalyst bed. It is known from literature that vanadium oxide [31] as well as phosphorous [32] can form volatile species under reaction conditions. Losing about 20% vanadium (calculated as V_2O_5) and about 70% phosphorous is within the expected range for a catalyst of about 2.3 years lifetime as confirmed by analysis results of used catalyst samples from the other commercial reference (Figure 9a,b), and hence do not contribute to the unusual catalyst deactivation. The amount of iron (Fe) detected on the catalyst is also within the expected range, as shown by analysis results of the reference used catalyst samples (Figure 9c). Iron typically arises from rust from the up-stream section which is blown into the catalyst bed by process air. Significant amounts of iron, e.g., rust, will increase the total oxidation to CO/CO_2 limiting phthalic anhydride yields. As shown in Figure 9d, the detected amounts of sodium (Na) and potassium (K), known catalyst poisons damping the catalyst activity, are in a normal range for all catalyst samples and hence do not contribute to unusual catalyst deactivation. Sodium and potassium poisons in the respective concentration ratios of the salt bath coolant ($NaNO_2$ + KNO_3) are likely caused by micro-cracks in the reactor tubes and/or tube sheets contaminating the catalyst. Additionally, sodium contaminations are caused by salt-water aerosols (NaCl) blown into the catalyst bed by the air blower.

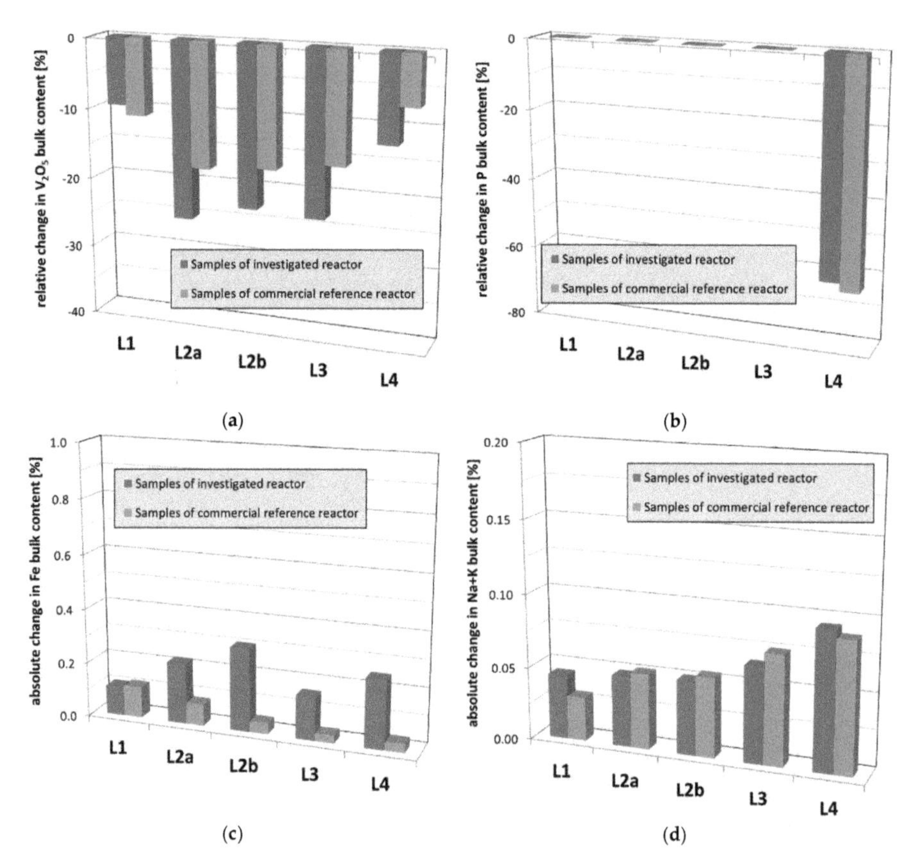

Figure 9. Comparison of the chemical bulk composition of used catalyst samples from the investigated reactor and the commercial reference reactor: (**a**) relative change in vanadium bulk content (calculated as V_2O_5); (**b**) relative change in phosphorous (P) bulk content; (**c**) absolute changes in iron (Fe) bulk content; (**d**) absolute changes in sodium and potassium (Na + K) bulk content. The samples of each layer were averaged.

The carbon contents of the upper three catalyst layers are within the expected range and confirmed by the reference (Figure 10). Due to the well-known enhanced adsorption of molecular reaction intermediates in the initial section of the catalyst bed (L1), the carbon content is typically increased here but it does not reduce the specific surface area or block pores (compare Section 4.2). Hence, it does not lead to the unusual catalyst deactivation. However, the carbon contents in the polishing L4 catalyst bed are unusually high and occur together with the simultaneous reduction of the specific surface area. This, hence, does result in catalyst activity reduction (compare Section 4.2). This loss of specific surface area of L4 is obviously caused by plugging the catalyst pores with high-molecular carbon deposits.

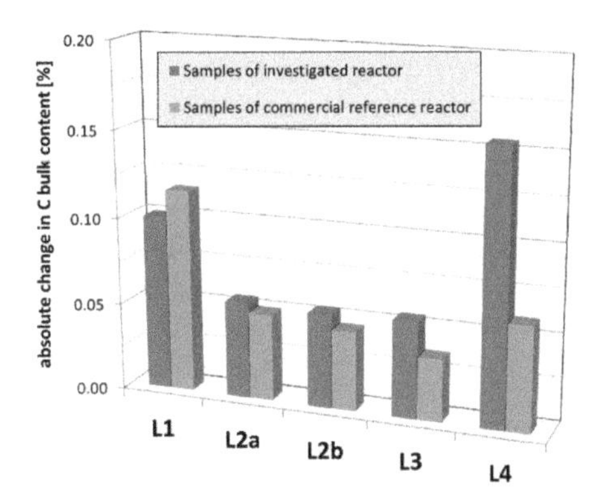

Figure 10. Comparison of absolute changes in carbon (C) bulk contents of used catalyst samples from the investigated reactor and the commercial reference reactor. The samples of each layer were averaged.

3. Discussion

All used catalyst samples of the investigated reactor were analyzed by standard physico-chemical methods and compared to another commercial reference reactor with most similar operating conditions and catalyst lifetime not showing this unusual deactivation behavior. The detailed physical analysis does not give indications for unusual thermal stress or insufficient salt-bath cooling which would cause an unusual catalyst deactivation by rutilisation or sintering of the titania phase. In addition, chemical analysis of the used catalyst samples did not reveal significant amounts of any of the known catalyst poisons sodium, potassium or iron, which would also contribute to an untypical catalyst deactivation/behavior. Quantitative Rietveld refinements of X-ray diffraction measurements indicate however an unusually high degree of reduction of the vanadium species in most of the catalyst bed, which would result in irreversible catalyst deactivation. Such an abnormal degree of catalyst reduction, and hence, irreversible damaging, was concluded to likely originate from a unit shutdown without sufficient purging of the catalyst bed by air. According to the well-known Mars–van-Krevelen-mechanism, o-xylene and reaction intermediates remain adsorbed on the catalyst surface in such a case and will be oxidized by lattice oxygen, accordingly reducing the catalytic species. Due to insufficient purging by air, the oxygen content in the process air depletes to unusual levels due to its continuing consumption. The crystal lattice as well as catalyst surface are not sufficiently re-oxidized under such circumstances. Finally, catalytically inactive, reduced phases like V_6O_{13} are enhanced formed. Combustion analysis of the deactivated catalyst revealed too unusually high carbon contents in the lower catalyst bed (L4) accompanied with a significant loss in the specific surface area and thus catalyst activity. In summary, the loss in surface area is caused by plugging the catalyst pores with high-molecular carbon deposits. Such unusual formation of high-molecular deposits in the lower catalyst bed is caused by the enhanced degree of catalyst reduction, respectively catalyst deactivation in most of the catalyst bed. Hence, it is concluded that the detected unusually fast decline of the hotspot temperatures as well as the fast-increasing under-oxidation by-products observed in this reactor originated from:

- abnormal, irreversible catalyst deactivation in most of the catalyst bed due to an enhanced degree of catalyst reduction,
- abnormal loss in the specific surface area/catalyst activity of the polishing layer catalyst bed due to enhanced formation of high-molecular carbon deposits plugging pores,
- both abnormal catalyst surface processes were induced by an irregular reactor shutdown during which the air purge was insufficient to remove the organic species from the reactor and the catalyst surface.

This systematic investigation of used catalyst samples by Clariant demonstrated the importance of sufficient air purging after disruption of the feed stream to VO_x/TiO_2 phthalic anhydride catalyst systems to avoid abnormal, irreversible damage and thus negative impact to catalyst performance.

4. Materials and Methods

4.1. Sampling of Used Catalyst

Several samples of used catalyst were collected in order to investigate potential catalyst damaging (e.g., by poising or unusual thermal stress). Six reactor tubes positioned close to the installed single point thermocouple tubes were discharged by vacuum suction (Figure 11a). The catalysts of all tubes were fractioned in five parts each related to their relative bed positions (Figure 11b).

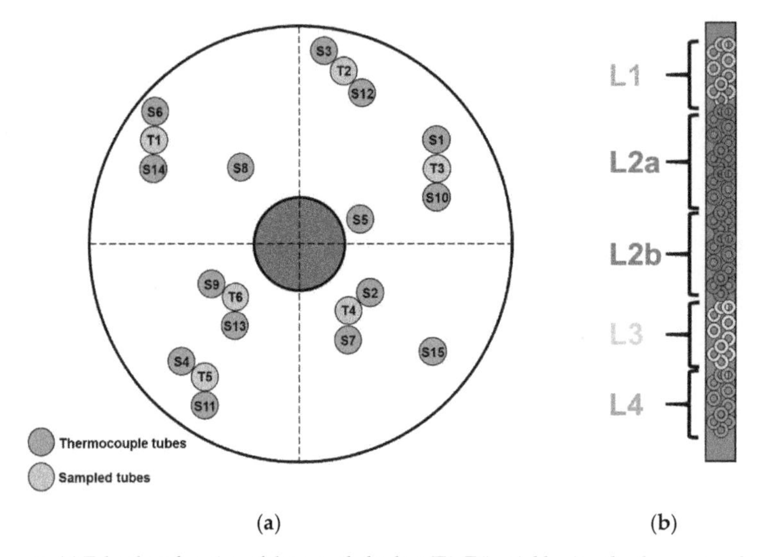

Figure 11. (**a**) Tube sheet location of the sampled tubes (T1–T6) neighboring the thermocouple tubes (S1–S15); (**b**) scheme of the five different samples taken along the catalyst bed.

As described in detail in the section below (Section 4.2), the sieved active mass powder (without steatite carrier rings) of all used catalyst samples were analyzed at Clariant´s R&D site Heufeld/Germany with regards to

- crystalline bulk phase compositions by X-ray diffraction measurements combined with Rietveld refinements,
- specific surface area measurements by nitrogen adsorption,
- chemical bulk contents of V, Sb, P, Na, K and Fe by atomic absorption spectroscopy,
- chemical bulk contents of C by combustion analysis.

It has to be kept in mind when evaluating the analytical data that a certain error will occur in the measured values because of an enhanced intermixing/cross-contamination of the active masses of the different layers when discharging the tubes by vacuum suction.

4.2. Physico-Chemical Analysis of Used Catalyst

For X-ray diffraction analysis, samples were pressed into disks and mounted onto the sampling stage of a diffractometer (D4 Endavour, BRUKER) equipped with an energy-dispersive one-dimensional detector (Lynxeye, BRUKER). Diffractograms were recorded in Bragg-Brentano geometry with Cu $K\alpha$ radiation in the 2Θ range of 5° to 90°. The incoming as well as the diffracted beams were directed through 0.3° fixed divergence slits. Rietveld refinements were calculated with the DIFFRAC^{plus} TOPAS software (version 4.2, BRUKER Corp., Billerica, MA, USA) using crystal data files (CIF type) from the Pearson´s crystal database (release October 2009, CRYSTAL IMPACT, Bonn, Germany).

The specific surface areas were measured using nitrogen adsorption (Gemini, MICROMERITICS GmbH, 85716 Unterschleißheim, Germany). Five points in the linear range of the adsorption isotherm (p/p_0 = 0.10–0.30) measured at 77 K were used to calculate the surface area according to the BET method. Before adsorption, 0.2 g of the sample was degassed by vacuuming to 0.1 mbar at 350 °C.

For determining the bulk content of V, Sb, P and Cs, 250 mg of each active mass powder sample was dissolved in 2 mL bi-distillated water, 2 mL hydrofluoric acid (40% pA), and 5 mL nitric acid (65% pA). The closed Teflon vessel was then mixed and heated up by a microwave oven (Multiwave Go, PARR Instrument, Moline, IL 61265-1770, USA) within 15 min to 180 °C and held for additional 15 min. After cooling down to room temperature, the clear solution was mixed with 2.5 mL potassium chloride solution (20 wt.-%) and 0.5 mL scandium standard (1000 mg/L). Analyzing the content of Na, K and Fe, 1.0 g of each sample was dissolved in 2 mL bi-distillated water, 5 mL hydrochloric acid (37% pA) and 0.5 mL cesium chloride solution (20 wt %). The closed plastic vessel was then heated up by a heating block to 85 °C and held for 30 min.

For final elemental analysis of V, Sb, P and Cs, an inductively coupled plasma atomic emission spectrometer (ICP-AES) (Spectro Arcos, AMETEK, Newark, DE 19702, USA) was used with a plasma power of 1400 W and a spray gas flow rate of 0.85 L/min. In the case of elemental analysis of Na, K and Fe, an atomic absorption spectrometer (AAS) (iCE 3000, THERMO FISCHER SCIENTIFIC, Waltham, MA 02451, USA) was used.

The total carbon content of the active mass samples was determined by combustion analysis (CS-200, LECO Corp., Saint Joseph, MI 49085, USA). About 200 mg of the fine pestled sample powder was weighted into a ceramic vessel and heated up above 1000 °C. To accelerate the combustion process, iron and tungsten chips were added. The CO_2 evolution of the combustion off-gas was analyzed by an infrared detector and used for recalculating the original carbon content of the sample.

Author Contributions: O.R. conducted all the measurements and data analysis. G.M. as the principal investigator was responsible for the research strategy and the scientific discussions leading to the understanding of the observed phenomenon.

Funding: This research received no external funding.

Conflicts of Interest: The authors declare no conflict of interest.

References

1. Suter, H. *Phthalsäureanhydrid und seine Verwendung*, 1st ed.; Dr. Dietrich Steinkopff Verlag: Darmstadt, Germany, 1972; pp. 1–6.
2. The Global Phthalic Anhydride Market. 2018. Available online: https://www.researchandmarkets.com/reports/4515077/global-phthalic-anhydride-market-segmented-by (accessed on 26 November 2018).
3. Marx, R.; Wölk, H.-J.; Mestl, G.; Turek, T. Reaction scheme of o-xylene oxidation on vanadia catalyst. *Appl. Catal.* **2011**, *398*, 37–43. [CrossRef]
4. Mestl, G.; Gückel, C.; Estenfelder, M.; Käding, B. Method for Applying a Wash Coat Suspension to a Carrier Strucutre. Patent EP 2160241, 31 May 2007.

5. Richter, O.; Mestl, G. *Selective Oxidation of o-Xylene to Phthalic Anhydride: Still Room for Improvement in an Established Catalyst System*; Jahrestreffen Reaktionstechnik: Würzburg, Germany, 22–27 May 2017.
6. Gückel, C.; Dialer, H.; Estenfelder, M.; Pitschi, W. Use of a Multi-Layer Catalyst for Producing Phthalic Anhydride. Patent WO 2006092304, 2 March 2005.
7. Gückel, C.; Dialer, H.; Estenfelder, M.; Pitschi, W. Method for Producing a Multi-Layer Catalyst for Obtaining Phthalic Anhydride. Patent WO 2006092305, 2 March 2005.
8. Richter, O.; Mestl, G.; Lesser, D.; Marx, R.; Fromm, N.; Schulz, F.; Schinke, P.; Pitschi, W. Catalyst Arrangment with Optimized Void Fraction for the Production of Phthalic Acid Anhydride. Patent WO 2015162227, 24 April 2004.
9. Richter, O.; Mestl, G.; Schulz, F.; Pitschi, W.; Fromm, N.; Schinke, P. Catalytic Converter Arrangement with Optimized Surface for Producing Phthalic Anhydride. Patent WO 2015162230, 24 April 2014.
10. Nikolov, V.A.; Anastasov, A.I. Pretreatment of a Vanadia-Titania Catalyst for Partial Oxidation of o-Xylene under Industrial Conditions. *Ind. Eng. Res.* **1992**, *31*, 80–88. [CrossRef]
11. Anastasov, A.I. Deactivation of industrial V_2O_5-TiO_2 catalyst for oxidation of o-xylene into phthalic anhydride. *Chem. Eng. Proc.* **2003**, *42*, 449–460. [CrossRef]
12. Saleh, R.Y.; Wachs, I.E.; Chan, S.S.; Chersich, C.C. The interaction of V_2O_5 with TiO_2 (anatase): Catalyst evolution with calcination temperature and o-xylene oxidation. *J. Catal.* **1986**, *98*, 102–114. [CrossRef]
13. Dias, C.R.; Portela, M.F.; Bond, G.C. Synthesis of Phthalic Anhydride: Catalysts, Kinetics, and Reactor Modeling. *Catal. Rev.* **1997**, *39*, 169–207. [CrossRef]
14. Nikolov, V.A.; Klissurski, D.G.; Anastasov, A. I Phthalic Anhydride from o-Xylene Catalysis: Science and Engineering. *Catal.s Rev.* **1991**, *33*, 319–374. [CrossRef]
15. Gasior, M.; Haber, J.; Machej, T. Evolution of V_2O_5-TiO_2 catalysts in the course of the catalytic reaction. *Appl. Catal.* **1987**, *33*, 1–14. [CrossRef]
16. Argyle, M.D.; Bartholomew, C.H. Heterogenous Catalyst Deactivation and Regeneration: A Review. *Catalyst* **2015**, *5*, 145–269. [CrossRef]
17. Pernicone, N. Methods for Laboratory-Scale Evaluation of Catalyst Life in Industrial Plants. *Appl. Catal.* **1985**, *15*, 17–31. [CrossRef]
18. Georgieva, A.T.; Anastasov, A.I.; Nikolov, V.A. Deactivation Properties of a High-Productive Vanadia-Titania Catalyst for Oxidation of o-Xylene to Phthalic Anhydride. Braz. *J. Chem. Eng.* **2008**, *25*, 351–364.
19. Krajewski, W.; Galantowicz, M. Effect of catalyst deactivation on the process of oxidation of o-xylene to phthalic anhydride in an industrial multitubular reactor. *Catal. Deactiv.* **1999**, 447–452.
20. Castillo-Araiza, C.O.; López-Isunza, F. The role of catalyst activity on the steady state and transient behavior of an industrial-scale fixed bed catalytic reactor for the partial oxidation of o-xylene on V_2O_5/TiO_2 catalysts. *Chem. Eng. J.* **2011**, *176*, 26–32. [CrossRef]
21. Calderbank, P.H. Kinetics and Yields in the Catalytic Oxidation of o-Xylene to Phthalic Anhydride with V_2O_5 Catalysts. *Adv. Chem. Ser.* **1974**, *133*, 646–653.
22. Dias, C.R.; Portela, M.F.; Bond, G.C. Oxidation of o-Xylene to Phthalic Anhydride over V_2O_5/TiO_2 Catalyst I. Influence of Catalyst Composition, Preparation Method and Operating Conditions on Conversion and Product Selectivities. *J. Catal.* **1995**, *157*, 344–352. [CrossRef]
23. Dias, C.R.; Portela, M.F.; Bond, G.C. Oxidation of o-Xylene to Phthalic Anhydride over V_2O_5/TiO_2 Catalyst II. Transient Catalytic Behaviour. *J. Catal.* **1995**, *157*, 353–358. [CrossRef]
24. Dias, C.R.; Portela, M.F.; Bond, G.C. Oxidation of o-Xylene to Phthalic Anhydride over V_2O_5/TiO_2 Catalyst III. Study of Organic Residie Formed on the Catalyst Surface. *J. Catal.* **1996**, *162*, 284–294. [CrossRef]
25. Dias, C.R.; Portela, M.F.; Bond, G.C. Oxidation of o-Xylene to Phthalic Anhydride over V_2O_5/TiO_2 Catalyst Part 4. Mathematical Modelling Study and Analysis of the Reaction Network. *J. Catal.* **1996**, *164*, 276–287.
26. Dias, C.R.; Portela, M.F.; Bond, G.C. Deactivation of V_2O_5/TiO_2 catalyst in the oxidation of o-xylene to phthalic anhydride. *Stud. Surf. Sci. Catal.* **1994**, *88*, 475–482.
27. Mongkhonsi, T.; Kershenbaum, L. The effect of deactivation of a V_2O_5/TiO_2 (anatase) industrial catalyst on reactor behavior during the partial oxidation of o-xylene to phthalic anhydride. *Appl. Catal.* **1998**, *170*, 33–48. [CrossRef]
28. Bond, G.C.; König, P. The Vanadium Pentoxide-Titanium Dioxide System. *J. Catal.* **1982**, *77*, 309–322. [CrossRef]

29. Nikolov, V.A.; Klissurski, D.G.; Hadjiivanov, K.I. Deactivation of a V_2O_5-TiO_2 Catalyst for the Oxidation of o-Xylene to Phthalic Anhydride. *Stud. Surf. Sci. Catal.* **1987**, *34*, 173–182.
30. Wainwright, M.S.; Hoffman, T.W. The Oxidation of Ortho-xylene on Vanadium Pentoxide Catalysts. I. Transient Kinetic Measurements. *Can. J. Chem. Eng.* **1977**, *55*, 552–556. [CrossRef]
31. Chapman, D.M. Behavior of titania-supported vanadia and tungsta SCR catalysts at high temperatures in reactant streams: Tungsten and vanadium oxide and hydroxide vapor pressure reduction by surficial stabiblization. *Appl. Catal.* **2011**, *392*, 143–150. [CrossRef]
32. Lesser, D.; Mestl, G.; Turek, T. Transient behavior of vanadyl pyrophosphate catalyst during the partial oxidation of n-butane in industrial-sized, fixed bed reactors. *Appl. Catal.* **2016**, *510*, 1–10. [CrossRef]

Article

Dry Reforming of Methane over NiLa-Based Catalysts: Influence of Synthesis Method and Ba Addition on Catalytic Properties and Stability

Ruan Gomes [1], Denilson Costa [1], Roberto Junior [2], Milena Santos [1], Cristiane Rodella [3], Roger Fréty [1], Alessandra Beretta [2] and Soraia Brandão [1,*]

1 Instituto de Química, Universidade Federal da Bahia, Rua Barão de Jeremoabo, 147, Ondina, CEP: 40170-115 Salvador, BA, Brasil; ruansag@gmail.com (R.G.); denilson_costa19@hotmail.com (D.C.); milenadesantanasantos@gmail.com (M.S.); ro_fre@hotmail.fr (R.F.)

2 Laboratory of Catalysis and Catalytic Process, Dipartimento di Energia, Politecnico di Milano, via La Masa 34, 20156 Milano, Italy; roberto.batistadasilva@polimi.it (R.J.); alessandra.beretta@polimi.it (A.B.)

3 Laboratório Nacional de Luz Síncrotron—LNLS, 13083-100 Campinas (SP), Brasil; cristiane.rodella@lnls.br

* Correspondence: soraia.ufba@gmail.com; Tel.: +55-71-3283-6882

Received: 5 March 2019; Accepted: 25 March 2019; Published: 30 March 2019

Abstract: CO_2 reforming of CH_4 to produce CO and H_2 is a traditional challenge in catalysis. This area is still very active because of the potentials offered by the combined utilization of two green-house gases. The development of active, stable, and economical catalysts remains a key factor for the exploitation of natural gas (NG) with captured CO_2 and biogas to produce chemicals or fuels via syngas. The major issue associated with the dry reforming process is catalyst deactivation by carbon deposition. The development of suitable catalyst formulations is one strategy for the mitigation of coking which becomes especially demanding when noble metal-free catalysts are targeted. In this work NiLa-based catalyst obtained from perovskite precursors $La_{1-x}Ba_xNiO_3$ (x = 0.0; 0.05; 0.1 and 0.2) and NiO/La_2O_3 were synthesized, characterized by in situ and operando XRD and tested in the dry reforming of methane. The characterization results showed that the addition of barium promoted $BaCO_3$ segregation and changes in the catalyst structure. This partly affected the activity; however, the incorporation of Ba improved the catalyst resistance to deactivation process. The Ba-containing and Ba-free NiLa-based catalysts performed significantly better than NiO/La_2O_3 catalysts obtained by wet impregnation.

Keywords: dry reforming of methane; nickel catalysts; barium carbonate; deactivation by coking

1. Introduction

The world energy matrix is essentially based on the use of fossil fuels with an increasing share of natural gas; this factor, together with the improved efficiency of energy conversion systems, has largely contributed in the last 10–15 years to the mitigation of CO_2 emissions in the electric power sector worldwide. In order to impact on the CO_2 footprint of the chemical and transportation sectors, transition strategies have been developed by oil and energy companies which emphasize the crucial role of a growing exploitation of NG reserves for the production of fuels and chemicals through the indirect conversion into synthesis gas and platform intermediates like methanol [1–4].

The steam reforming of methane (SRM) is the most widespread industrial process for syngas production with H_2/CO ratio close to 3, which is suitable for the production of fuels such as hydrogen, methanol, dimethyl ether and important chemicals like ammonia [5–7]. Other alternative processes that yield syngas with different ratios, such as autothermal reforming (ATR), partial oxidation (POX) and dry reforming of methane (DRM) have been studied and they are promising technologies for industrial application [8–13]. The dry reforming of methane has traditionally attracted attention because it yields a lower H_2/CO syngas ratio which is effective in obtaining hydrocarbons and oxygenated by Fischer–Tropsch synthesis, and it also consumes two greenhouse gases CH_4 and CO_2 [14]. Besides, DRM represents an interesting solution for exploitation of bio-gas as raw material for the fuel and chemical sectors, alternatively to the now more commonly practiced energetic use.

The dry reforming process consists of a highly endothermic reaction (Equation (1)):

$$CH_4 + CO_2 \rightleftharpoons 2H_2 + 2CO \qquad \Delta H^0 = 247.4\ kJ\ mol^{-1} \tag{1}$$

It is typically accompanied by simultaneous occurrence of reverse water-gas shift reaction—RWGS (Equation (2)):

$$H_2 + CO_2 \rightleftharpoons CO + H_2O \qquad \Delta H^0 = 41.2\ kJ\ mol^{-1} \tag{2}$$

DRM is susceptible to carbon deposition through methane decomposition (Equation (3)) and/or the Boudouard reaction (Equation (4)) [15]. Equilibrium calculations and data in the literature [14,16,17] show that carbon deposition is favored in conditions of high CH_4/CO_2 ratios:

$$CH_4 \rightleftharpoons 2H_2 + C \qquad \Delta H^0 = 74.9\ kJ\ mol^{-1} \tag{3}$$

$$2CO \rightleftharpoons CO_2 + C \qquad \Delta H^0 = -172.5\ kJ\ mol^{-1} \tag{4}$$

Additionally, high temperatures may induce active phase sintering and irreversible reactions between active phase and support leading to catalyst deactivation.

Many studies report that catalysts belonging to group VIII metals are good options for DRM catalysts. As a result, catalysts based on noble metals (Rh, Ru, Pt, and Pd) and nickel have been developed. Even though noble metals exhibit better catalytic performance and higher coke-resistance when compared to nickel catalysts, their high prices and low availability limit their industrial application [18,19]. Aiming to enhance the stability against coke formation and active phase sintering, nickel-based catalysts have been synthesized by diverse routes and on different supports [4,11,20].

One of the methods for improving catalyst resistance to sintering and carbon deposition is the insertion of transition metals into well-defined structures [8,21]. Perovskite-type oxides with the chemical formula ABO_3, after a reduction process, generate stable and well-dispersed nanoparticles which are suitable for reforming reactions. In addition, it is possible to use different compositions partially replacing the cations in positions A and B, and obtaining materials with different chemical properties [21–23].

Studies have been carried out into the reaction mechanism of DRM. There is a consensus that this reaction follows a bifunctional mechanism where CH_4 is activated and cracked preferentially on metallic sites and CO_2 is activated by the support [24–26]. Since the catalysts applied in this reaction are prone to deactivate, the support plays a fundamental role. Several studies in the literature have proposed that the nature of the support affects the mechanism of carbon species oxidation. Ni-based catalysts supported on CeO_2, La_2O_3, CaO, MgO, and BaO are alternative ways to inhibit the deactivation by carbon deposition, due to the relative increase in global basicity [27]. However, for promoted perovskites, $La_{1-x}Ba_xNiO_3$ (x = 0.05; 0.1 and 0.2), few investigations have been done with the addition of barium in the perovskite structures due to the fact that this element promotes the segregation of barium oxides and barium carbonate [11,24].

Still, the intimate dispersion of Ba-species within the NiLa-based catalyst might be effective in contrasting the coking kinetics. Therefore, the aim of this study was to evaluate such catalysts and the impact of barium on their stability under DRM conditions. The synthesis method (nickel perovskite reduction and wetness impregnation) was also investigated.

2. Results and Discussion

2.1. X-Ray Diffraction and BET Specific Surface Area

Figure 1a shows the diffractograms of the precursors after calcination at 800 °C ($La_{1-x}Ba_xNiO_3$) and at 500 °C (NiO/La_2O_3). All precursors prepared by the citrate method showed an XRD pattern similar to that of pure $LaNiO_3$ perovskite (PDF card 00-010-0341). This result agrees with various literature data [8,28,29]. Nevertheless, segregated NiO (PDF card 00-001-1239) was found in the $La_{0.8}Ba_{0.2}NiO_3$ sample. Moreover, the barium carbonate phase (PDF card 00-005-0378) was detected in the barium-containing catalysts. Thus, the calcination temperature was not enough to promote the decomposition of all the carbonates. The NiO/La_2O_3 synthesized by wet impregnation method showed the diffraction lines characteristic of NiO (PDF card 00-004-0835) and La_2O_3 (PDF card 00-002-0688) phases. Solid broader and less intense diffraction lines were observed probably due to the calcination temperature, 300 °C lower than that used for the preparation of the perovskites.

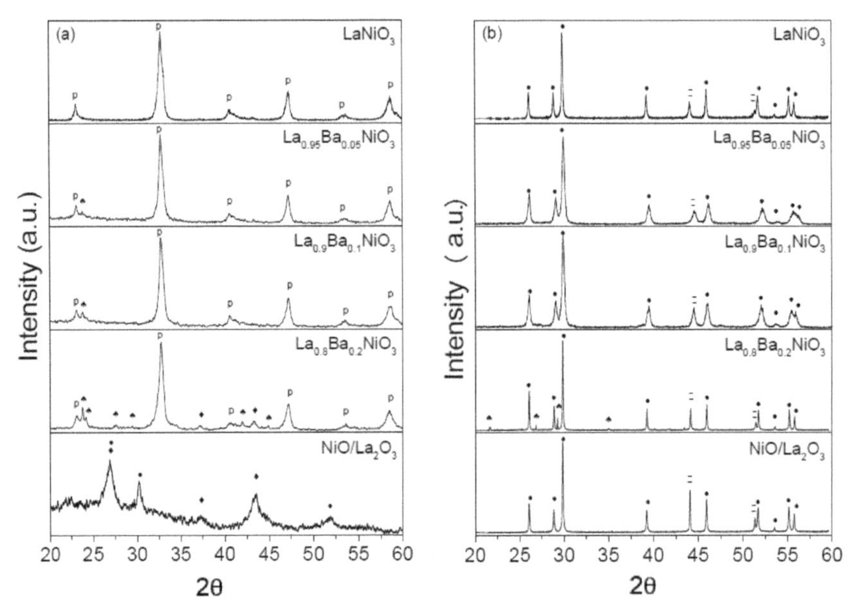

Figure 1. (**a**) X-ray diffractograms of the precursors $La_{1-x}Ba_xNiO_3$ and NiO/La_2O_3 (**b**) X-ray diffractograms for the reduced precursor $La_{1-x}Ba_xNiO_3$ and NiO/La_2O_3. (p = $LaNiO_3$, ♦ NiO, • La_2O_3, □ Ni and ♠ $BaCO_3$).

The diffraction patterns of the reduced catalysts are shown in Figure 1b. After reduction, all the catalysts exhibited diffraction lines characteristic of the Ni^0 and La_2O_3 phases. However, the diffraction pattern for the $La_{0.8}Ba_{0.2}NiO_3$ catalyst confirmed that the $BaCO_3$ phase persisted after reduction, indicating that this catalyst has $BaCO_3$ in its composition. For the catalysts $La_{1-x}Ba_xNiO_3$ (x = 0.05; 0.1 and 0.2) the Ni (111) XRD diffraction line shifted to higher 2-theta values while the XRD Ni (200) diffraction line for the samples $La_{1-x}Ba_xNiO_3$ (x = 0.05 and 0.1) was no longer detected.

The average Ni^0 crystallite size was calculated using the diffractograms presented in Figure 1b, considering the line at $2\theta = 44.4°$ and using the Scherrer equation. The results show that nickel catalysts generated by the reduction of $La_{1-x}Ba_xNiO_3$ (x = 0.05, 0.1, and 0.2) have larger mean Ni^0 crystallite size than the catalyst obtained by reduction of $LaNiO_3$. The values were 19 nm for $LaNiO_3$, 34 nm for $La_{0.95}Ba_{0.05}NiO_3$, 36 nm for both $La_{0.9}Ba_{0.1}NiO_3$ and $La_{0.8}Ba_{0.2}NiO_3$, values lower than the one generated by reduction of NiO/La_2O_3 (40 nm).

The specific surface area for all perovskites are rather low and approximately the same, ranging from 4 m^2/g for $La_{0.8}Ba_{0.2}NiO_3$ up to 6 m^2/g for $LaNiO_3$ and 6 m^2/g for NiO/La_2O_3 sample. From our research group's experience [8], BET surface areas of initial perovskites are low, between 3 to 5 m^2/g, being marginally increased after in situ reduction at 700–800 °C.

2.2. In Situ X-Ray Diffraction under Reducing Atmosphere (XRD-H_2)

The analysis was performed to monitor the crystalline structure changes of the precursor during reduction (Figure 2).

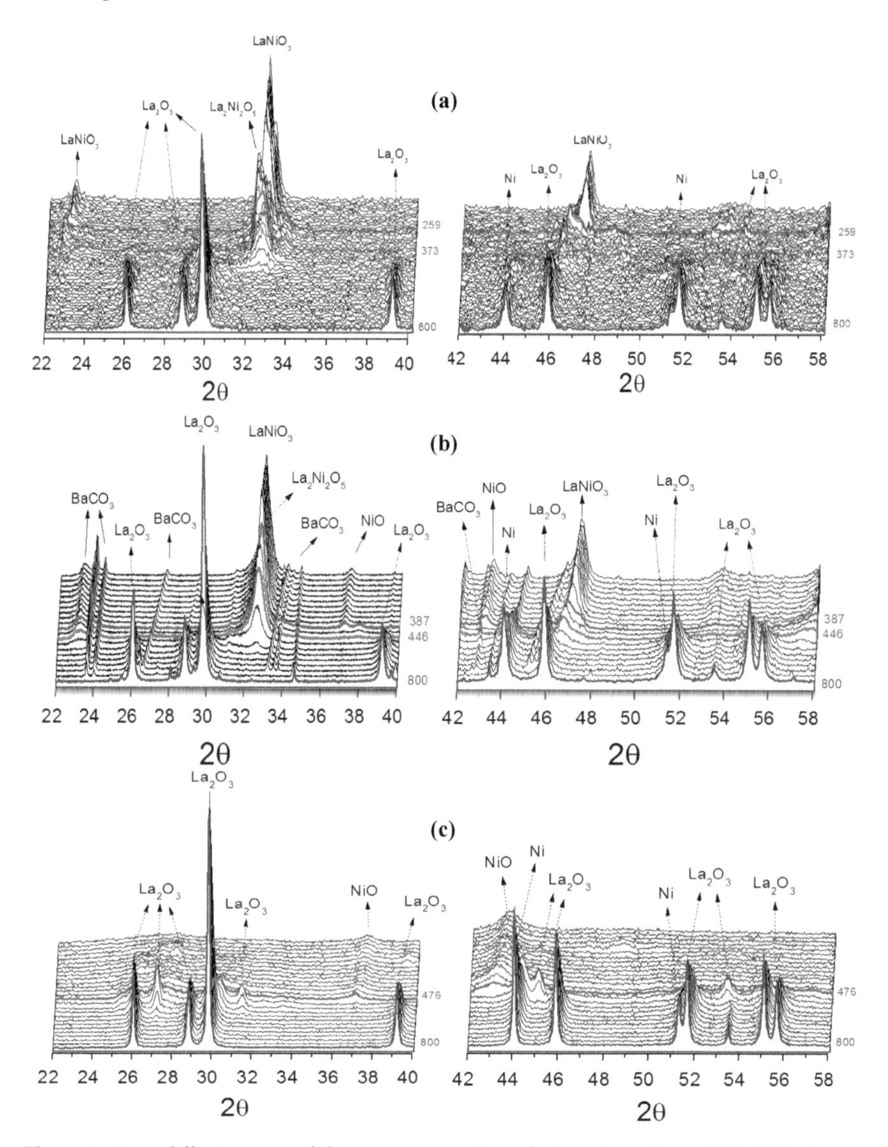

Figure 2. X-ray diffractograms of the precursors under reducing atmosphere. (**a**) $LaNiO_3$ (upper panels); (**b**) $La_{0.8}Ba_{0.2}NiO_3$ (central panels); and (**c**) NiO/La_2O_3 (bottom panels).

Figure 2a displays the XRD patterns of $LaNiO_3$, where at about 259 °C, the diffraction lines characteristic of the phase begin to decrease in intensity and shift to lower 2θ values. These events indicate the transformation of $LaNiO_3$ to $La_2Ni_2O_5$ (PDF card 00-036-1230), which is an oxygen–deficient perovskite. While the diffraction lines associated to $La_2Ni_2O_5$ decrease in intensity (indicating that this phase is being destroyed), the lines attributed to metallic Ni (PDF card 00-001-1258) and La_2O_3 (PDF card 01-074-2430) phases start to appear, suggesting the formation of Ni^0/La_2O_3, the effective catalyst. Thus, based on the diffraction patterns, the main reduction steps are described by Equations (5) and (6):

$$2LaNiO_3(s) + H_2(g) \rightarrow La_2Ni_2O_5(s) + H_2O(g) \quad (5)$$

$$La_2Ni_2O_5(s) + 2H_2(g) \rightarrow 2Ni(s) + La_2O_3(s) + 2H_2O(g) \quad (6)$$

The sample $La_{0.8}Ba_{0.2}NiO_3$ showed a similar reduction process, although the transition events occurred at higher temperatures. Furthermore, no change was observed in the $BaCO_3$ crystalline structure, except its crystallization due to the thermal treatment.

For the NiO/La_2O_3 catalyst, the reducing treatment promoted the formation of the Ni^0 phase at about 476 °C. At 476 °C, diffraction lines at 2θ = 27.10, 31.50, and 45.10 became detectable but with increasing temperature they were no longer observed. These diffraction lines were attributed to hexagonal-to-cubic transition of La_2O_3 (PDF card 00-004-0856), which suggests a mixture of La_2O_3 phases at high temperatures.

2.3. *In Situ X-Ray Diffraction under Reaction Atmosphere (XRD-CH_4/CO_2)*

With the obtained reduced catalysts, their possible evolution under a reaction atmosphere was then studied. Figures 3 and 4 show the phases of evolution when the catalysts were exposed to a CH_4/CO_2 mixture from room temperature to 800 °C.

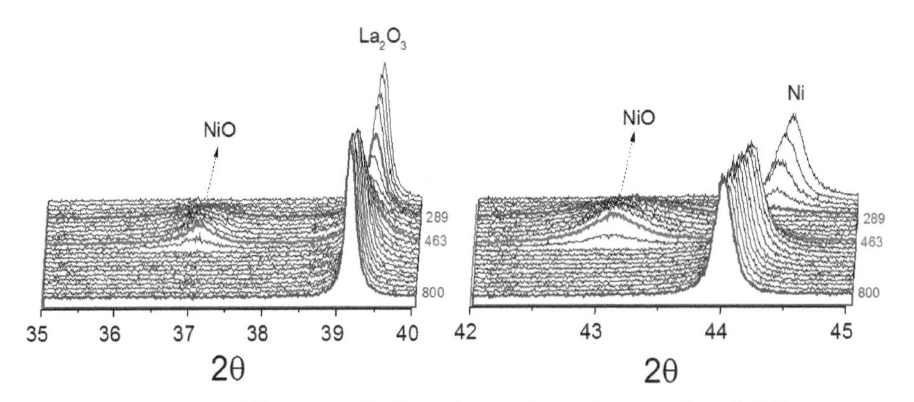

Figure 3. X-ray diffractograms for the catalysts under reaction atmosphere. $LaNiO_3$.

For all catalysts, metallic nickel was oxidized at low temperature to nickel oxide when exposed to the reaction atmosphere, suggesting that CO_2 was responsible for the nickel oxidation. Comparing the catalysts obtained from $LaNiO_3$ and $La_{0.8}Ba_{0.2}NiO_3$, it was found that the metallic nickel present in reduced $LaNiO_3$ catalyst is more susceptible to oxidation. At 289 °C, the diffraction lines associated to the La_2O_3 phase decreased in intensity and the diffraction line (111) attributed to metallic nickel was no longer detected, demonstrating that the metallic nickel was destroyed. At 344 °C, the diffraction lines of NiO, (PDF card 00-036-1230), started to appear, reaching a maximum at 463 °C. At around 463 °C, the line (111) attributed to metallic nickel starts to be detected, suggesting the catalyst was being regenerated. For the catalyst obtained from $La_{0.8}Ba_{0.2}NiO_3$, the oxidation event takes place at higher temperatures, starting at about 432 °C and ending at 610 °C. Furthermore, the nickel sites were not totally oxidized, suggesting only a partial oxidation process, which was characterized mainly by the decrease in the (111) metallic nickel diffraction line intensity. A likely explanation of this effect is the possibility of the partial blockage of nickel sites by $BaCO_3$ during the reduction process, making them less susceptible to oxidation. At temperatures higher than 610 °C orthorhombic-to-hexagonal phase transition of $BaCO_3$ was observed. This was expected since the literature has reported the orthorhombic-hexagonal transition at high temperatures [30,31]. The catalyst obtained from NiO/La_2O_3, similarly to $LaNiO_3$ catalyst, showed the re-oxidation of metallic nickel to NiO by the reaction mixture, before its new reduction. In summary, with increasing temperature, the main diffraction lines of Ni, La_2O_3 and $BaCO_3$ for all catalysts shift in a non-parallel way in 2θ axis.

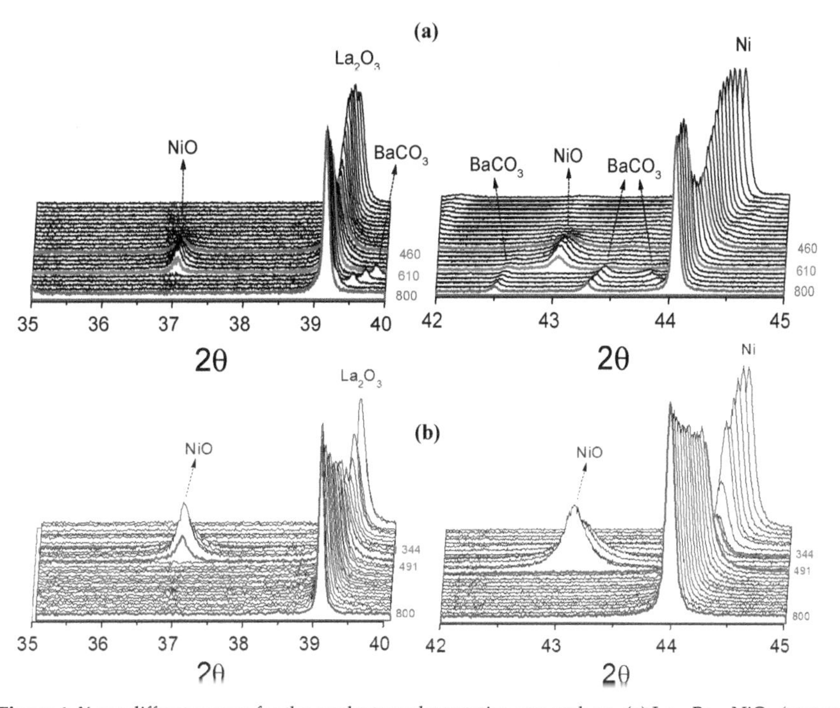

Figure 4. X-ray diffractograms for the catalysts under reaction atmosphere. (**a**) $La_{0.8}Ba_{0.2}NiO_3$ (upper panels) and (**b**) NiO/La_2O_3 (bottom panels).

2.4. Temperature-Programmed Reduction (TPR- H_2)

The TPR-H_2 profiles of the precursors $La_{1-x}Ba_xNiO_3$ (x = 0.0, 0.05, 0.1, and 0.2), Figure 5, showed that these precursors have a similar reduction profile, where two main reduction events were identified. The first one, in the range 250–360 °C, was attributed to the reduction of $LaNiO_3$ to $La_2Ni_2O_5$, Equation (5), and the second one, in the range 478–502 °C, was attributed to the transformation of $La_2Ni_2O_5$ to metallic nickel and La_2O_3, Equation (6). These results are in good agreement with the X-ray results previously discussed and the literature data [32], where the $LaNiO_3$ reduction process also occurred in two main events.

According to the perovskite reduction reaction stoichiometry, Equations (5) and (6), the area under the second peak should ideally be twice that of the first peak. Changes in that ratio implies that reducible species other than $LaNiO_3$ are present in the system. In fact, for all the perovskite precursors, ratios lower than the theoretical ones were found. These decrease to a minimum when Ba-substitution reaches a maximum, suggesting the segregation of other reducible phases (or species). In the range 316–391 °C a reduction event was identified and attributed to NiO reduction. This is in total agreement with the X-ray diffraction where nickel oxide was reduced in that temperature range. In addition, the TPR-H_2 results showed that Ba-substitution, in general, shifts the final reduction temperature to higher values.

In the case of NiO/La_2O_3, the sample shows two sets of reduction events. The low temperature events were attributed to the reduction of bulk NiO and to NiO species weakly bound to La_2O_3. The second set of NiO reduction events occurred at temperatures close to 600 °C. The corresponding NiO species with stronger interaction with support may generate Ni particles which are more stable and less prone to sintering [33–35].

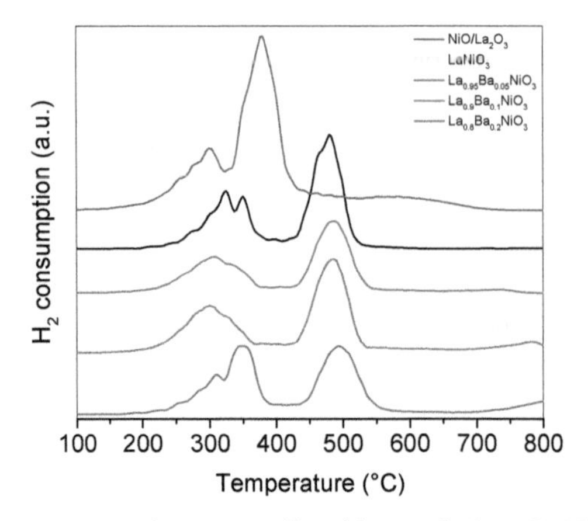

Figure 5. Temperature programmed reaction profiles of the samples $La_{1-x}Ba_xNiO_3$ and NiO/La_2O_3.

3. Stability Tests

The onset of catalytic activity in DRM of the various catalysts was qualitatively measured by temperature-programmed surface reaction tests, which is temperature-ramped experiments under $CH_4/CO_2/Ar$ flow; the experiments were performed after in situ reduction at 800 °C of the precursors. The results are not reported for the sake of brevity. It is herein briefly mentioned that incipient conversion of reactants was observed at about 350–400 °C over the Ba-free and Ba-containing perovskites. Interestingly, the onset of the reaction was accompanied by the unique production of H_2O and CO which grew with increasing temperature; only at temperature higher than 550–600 °C H_2 was progressively produced and H_2O concentration declined in line with the chemical thermodynamics.

The NiO/La_2O_3 catalyst showed similar trends although "delayed" at higher temperatures, thus indicating a lower activity of the catalyst obtained by impregnation than the corresponding catalyst obtained by co-precipitation of perovskite precursor.

These preliminary experiments allowed to identify 700 °C as a suitable temperature for measuring the catalyst stability under reacting conditions. The characterization results also support the assumption that at this temperature Ni is fully reduced in all the formulations.

The catalyst activity was then tested using a feed of $CH_4/CO_2/Ar = 25/25/50$ mL·min^{-1} (GHSV = 2×10^5 NL·h^{-1}·kg^{-1}) along 24 h on stream at 700 °C (after reduction in H_2/He flow at 800 °C). The results of the experiments are reported in Figure 6.

During the first 4 h on stream, the catalyst $LaNiO_3$ showed a continuous increase of the activity followed by a substantial stabilization. H_2 and CO molar fractions showed analogous trends stabilizing around 17 and 35%, respectively. Verykios et al. [26,36] observed similar trends and suggested that during the reaction initial hours (induction period) substantial changes occur on the catalyst; in particular, FTIR results suggested that the increase of activity could be associated with the increase of $La_2O_2CO_3$ and formate species concentration, which were believed intermediate species of the dry reforming process [8,11,26]. The tests on the impregnated NiO/La_2O_3 catalyst showed a rapid decay of conversion and syngas yield.

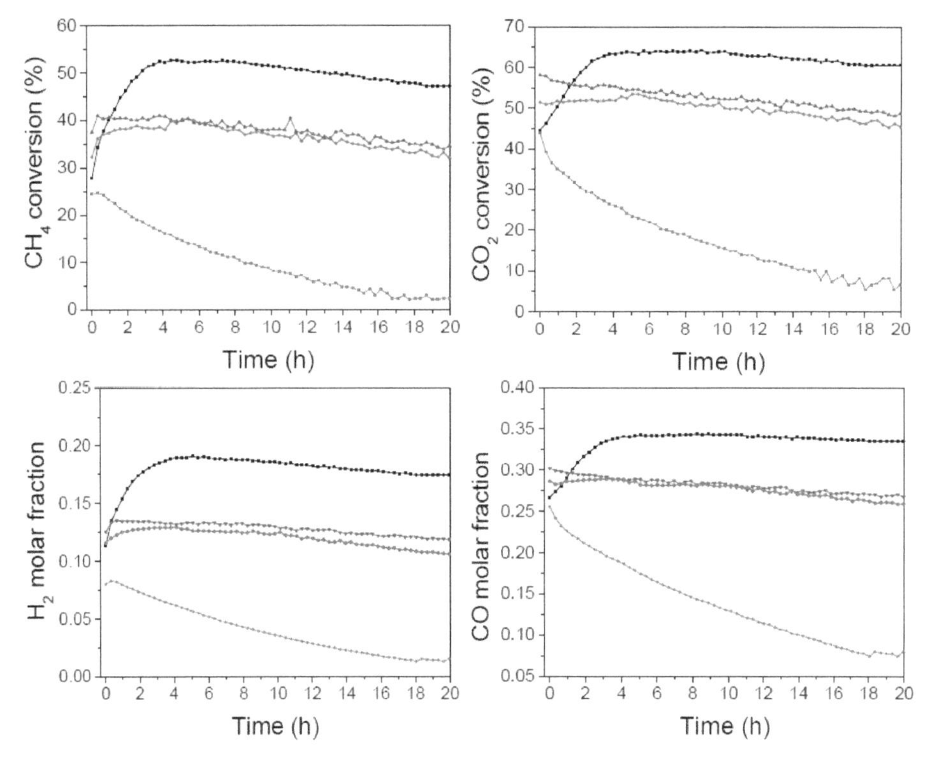

Figure 6. Reactant conversion, products molar fraction and H_2/CO ratio. $La_{1-x}Ba_xNiO_3$ (x = 0.0, 0.1, and 0.2) and NiO/La_2O_3. Feed: $CH_4/CO_2/Ar$ = 25/25/50 $mL \cdot min^{-1}$. T = 700 °C, GHSV = 200.000 $NL \cdot h^{-1} \cdot kg^{-1}$. (black line) $LaNiO_3$; (blue line) $La_{0,90}Ba_{0,1}NiO_3$; (red line) $La_{0,8}Ba_{0,2}NiO_3$; and (green line) NiO/La_2O_3.

The relative activity of the two systems is qualitatively in line with the different dispersion of Ni, as revealed by the XRD measurements. The average nickel crystallite size calculated from the diffraction line at 2θ = 44.4° by using the Scherrer equation corresponded in fact to 19 nm in $LaNiO_3$ catalyst and 40 nm in the NiO/La_2O_3 catalyst, as above mentioned.

It is beyond the scope of the present study to speculate on the reaction mechanism; however, a broad exam on DRM over NiLa-based catalysts supports the picture of CH_4 being activated over Ni sites (thus supporting an expected effect of Ni dispersion), while La-sites would be preferentially involved in CO_2 activation. According to Verykios et al. [26,36], the reaction pathway would involve the formation and interaction of C-Ni and $La_2CO_2O_3$ intermediates.

The Ba-containing formulations $La_{1-x}Ba_xNiO_3$ (x = 0.1 and 0.2) showed a somehow intermediate level of initial activity, shorter induction period than the Ba-free catalyst and appreciable stability; the conversions of CH_4 and CO_2 showed a moderate decline with time on stream, as the molar fractions of H_2 and CO. The decay of activity was attributed to the insertion of barium which leads to changes in the catalysts' structure as well as the presence of $BaCO_3$ that may cover part of the active sites and inhibit the interaction with the reactants to a certain degree.

4. Aging Tests

In order to further verify the catalysts' stability against more stressful treatments, sequential cycles were performed in which a DRM experiment at 700 °C was performed for 10 hours on stream, then the reactor was cooled down to ambient temperature and then re-heated at 700 °C and exposed again to the reacting mixture. Afterwards, an oxidizing treatment at 800 °C and a following reducing treatment at 700 °C were performed; finally, a third DRM experiment at 700 °C was performed and parameters were monitored along 10 hours on stream. The results are reported in Figures 7 and 8.

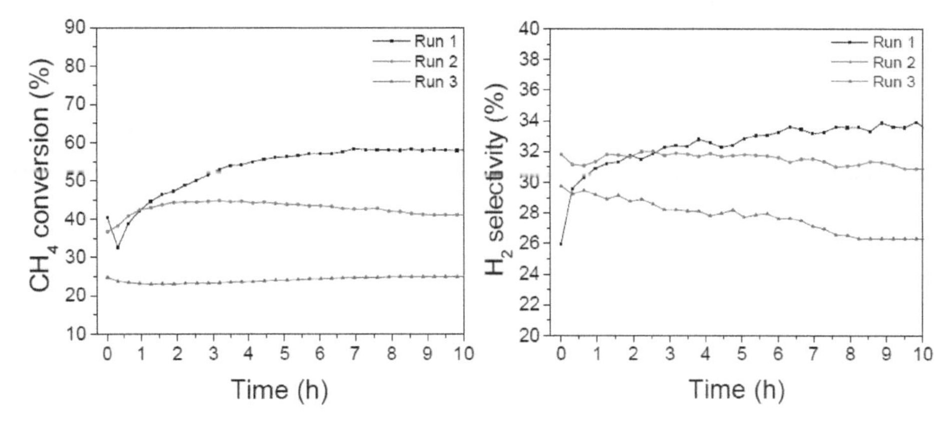

Figure 7. CH_4 conversion and H_2 selectivity as function of time of $LaNiO_3$ under successive catalytic runs.

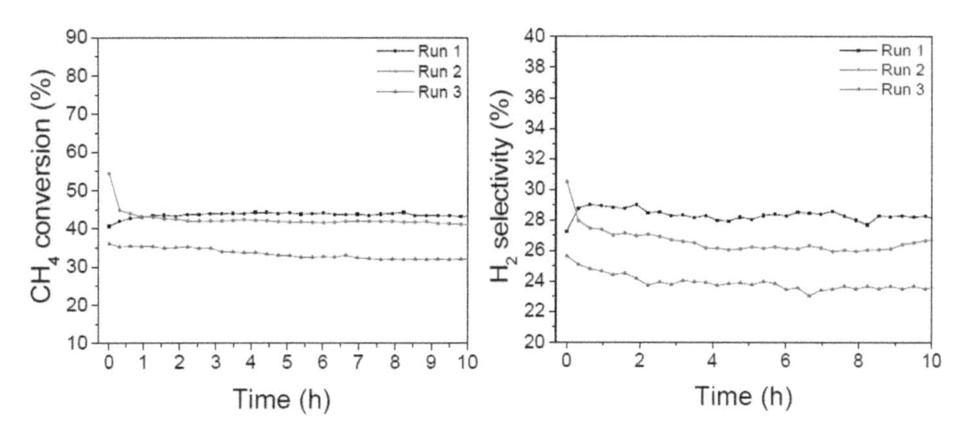

Figure 8. CH_4 conversion and H_2 selectivity as function of time of $La_{0.8}Ba_{0.2}NiO_3$ under successive catalytic runs.

The $LaNiO_3$ catalyst showed initially the highest activity (with a maximum of about 57% conversion after some hours on stream, consistently with the long-term experiments reported in Figure 6); in the second cycle the activity levelled off at about 40%, but after the oxidizing/reducing treatments the conversion dropped at about 25% with a final H_2 selectivity of 27%.

In the case of the $La_{0.8}Ba_{0.2}NiO_3$ catalyst, the initial conversion was lower but more stable (between 40–45% after 20 hours on stream) and after the oxidizing/reducing treatments it was affected to a lesser extent, since it declined to about 35% but kept almost constant along the following 10 hours on stream. The H_2 selectivity amounted to about 28% in the first 20 hours and dropped to 24% in the third DRM experiment.

These results suggest that the addition of barium promotes a higher resistance to poisoning and sintering processes that are considered side effects of multiple temperature treatments and to the presence of carbon-forming precursors.

5. Temperature Programmed Oxidation (TPO)

TPO experiments were performed on the spent catalysts after 20 h on stream. As can be seen in Figure 9, the aged $LaNiO_3$ catalyst showed an intense CO_2 formation, with a maximum close to 600 °C, which was assigned to oxidation of graphitic carbon.

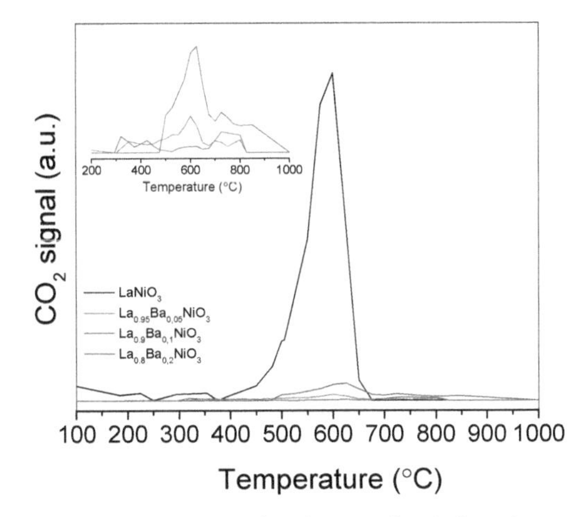

Figure 9. Temperature programmed oxidation under air flow of spent catalysts.

The amount of carbon deposited on catalysts during reaction, deduced from the CO_2 formation during TPO, is reported in Table 1.

Table 1. Carbon deposited on catalysts after 20 h under reaction stream.

Catalyst	$mg_{carbon}/g_{catalyst}$
$LaNiO_3$	1.64
$La_{0.95}Ba_{0.05}NiO_3$	0.08
$La_{0.9}Ba_{0.1}NiO_3$	0.16
$La_{0.8}Ba_{0.2}NiO_3$	0.03

$LaNiO_3$ showed the largest amount of carbon which is a consequence of its higher activity, at least in part. Barium-containing catalysts showed a less intense CO_2 peak and a substantial decrease in deposited carbon content (~55 lower than $LaNiO_3$), which could be a consequence of a lower activity. Moreover, although the $La_{0.9}Ba_{0.1}NiO_3$ and $La_{0.8}Ba_{0.2}NiO_3$ catalysts showed similar performances, the higher barium content promoted the lower amount of deposited carbon. It was proposed that catalysts obtained from perovskite reduction are regenerated via $La_2O_2CO_3$, formate species and carbon oxidation via adsorbed water [8,26]. Nevertheless, in order to explain the considerable decrease in carbon deposition for barium-containing catalysts, an extra carbon oxidation pathway is suggested.

According to the literature, $BaCO_3$ plays a key role in carbon catalytic oxidation. Barium carbonate and carbon species can interact to form a carbonate-carbon complex that decomposes rapidly to carbon monoxide [37,38]. The steps are presented by Equations (7) and (8):

$$BaO(s) + CO_2(g) \rightarrow BaCO_3(s) \quad (7)$$

$$BaCO_3(s) + C(s) \rightarrow BaO\ (s) + 2CO(g) \quad (8)$$

If this process was considered, it would be possible to explain the higher resistance of barium-containing catalysts to carbon deposition. It is worth mentioning that the CO_2 present in the feed may promote the $BaCO_3$ regeneration, since this phase is present in all temperature ranges to which the catalyst was exposed to the CH_4/CO_2 mixture (Section 2.3).

6. Materials and Methods

6.1. Precursor Synthesis

The perovskite precursors $La_{1-x}Ba_xNiO_3$ (x = 0.0, 0.05, 0.1, and 0.2) were prepared by citrate method [8]. A stoichiometric amount of $Ni(NO_3)_3 \cdot 6H_2O$ (Sigma-Aldrich, St. Louis, MO, USA) solution was added to a citric acid solution at 40 °C under constant stirring for 1 h. The molar ratio citric acid and metal of 1.5:1 was used. After that, the system was heated to 90 °C and the $La(NO_3)_3 \cdot 6H_2O$ (Sigma-Aldrich) and $Ba(NO_3)_3 \cdot 6H_2O$ (Sigma-Aldrich, St. Louis, MO, USA) solutions were added. Following the complexation, the solution was evaporated, and the resulting material pre-treated at 300 °C for 2 h to obtain a primary powder and calcined at 800 °C under air flow (50 mL min^{-1}) for 4 h. The reference precursor NiO/La_2O_3 (La_2O_3 Sigma-Aldrich, St. Louis, MO, USA) was prepared by incipient wetness impregnation. An aqueous solution of nickel nitrate, with 23 wt% Ni content, the same nickel content as the perovskite precursors, was added to a La_2O_3 suspension and stirred for 24 h. Then, the suspension was dried at 100 °C for 12 h, and the resulting solid calcined at 500 °C.

6.2. BET Specific Surface Area

Specific surface area measurements for the calcined samples were performed using the BET method. The analysis was performed at −196 °C using a NOVA–2000 volumetric adsorption system (Quanta Chrome Corporation, Boynton Beach, FL, USA). All the samples were pre-treated at 350 °C under vacuum for 2 h before measurement.

6.3. X-Ray Diffraction (XRD)

X-ray diffraction measurements of calcined precursors and of the reduced catalysts were performed on a Shimadzu XRD-6000 diffractometer using Cu Kα (λ = 1.5418 Å). The diffractograms were collected in 2θ range (20–60°) in steps of 0.25° min^{-1}. For the reduced catalysts diffraction pattern measurements, the samples were pre-reduced at 800 °C (10 °C min^{-1}) under 10.0 vol% $H_2/He_{(balance)}$ flow at a rate of 30 mL min^{-1}. The system was cooled down to room temperature under He flow and the diffractograms were collected. These characterization and reduction procedures were performed in an Anton–Paar reaction chamber HTK 1200 coupled to a Shimaden SR52 temperature controller (Shimaden, Nerima Ku, Japan).

6.4. In Situ X-ray Diffraction

In situ XRD experiments were performed to study the evolution of the phases under reducing and reaction atmospheres during temperature increase. The measurements were performed at Synchrotron Light Source–Campinas in a XPD beamline (Huber, Rimsting Germany). The diffractograms were collected in 2θ range (20–60°) in steps of 2° min^{-1} using λ = 1.5498 Å, which was suitable to identify the phases evolution as well as to obtain a good diffraction line resolution for phase identification. In the case of the reduction study, the precursors were exposed to 5.0 vol% $H_2/He_{(balance)}$ flow at a rate 30 mL min^{-1} from room temperature to 800 °C. For the catalyst study under reaction atmosphere, the precursors were reduced at 800 °C for 30 min and cooled down to room temperature under He flow. Then, the samples were exposed to a 25% CH_4/25% $CO_2/Ar_{(balance)}$ mixture from room temperature to 800 °C (10 °C min^{-1}) and the diffractograms were collected.

The average nickel crystallite size was calculated from the Ni diffraction line at 2θ = 44.4° at 100 °C by using the Scherrer equation.

6.5. Temperature-Programmed Reduction (TPR)

The TPR profiles were obtained by heating 100 mg of the precursors under 5.0 vol% $H_2/He_{(balance)}$ (30 mL min^{-1}) from room temperature to 800 °C at a rate of 10 °C min^{-1}. The experiments were performed in a fixed-bed quartz reactor and the hydrogen consumption was monitored using a multipurpose unit coupled with a gas analyzer Pfeiffer Vacuum Quadrupole Mass Spectrometer Balzers QMS 220 (Pfeifer Vacuum, Annecy, France). The mass fragment to monitor hydrogen signal was (m/z) = 2.

6.6. Reaction Conditions

Long-Term Reactions and Aging Tests

Dry reforming of methane was performed at atmospheric pressure in a stainless-steel tubular reactor. Initially, 30 mg of the precursors were diluted with 90 mg of quartz powder and reduced at 800 °C (10°C min^{-1}) for 1 h under ultra-pure hydrogen flow at 30 mL min^{-1}. After pre-reduction at 800 °C, the catalysts were cooled down to 700 °C and purged with N_2 flow. The reaction was carried out at 700 °C for 20 h under 100 mL min^{-1} 25% CH_4/25% $CO_2/Ar_{(balance)}$. The effluent gas was analyzed and quantified by using a Shimadzu 2014 chromatograph (Shimadzu, Kyoto, Japan) equipped with TCD detector and a Carboxen 1000 column (Sigma-Aldrich, St. Louis, MO, USA). The reactant conversions (X_j) and product molar fractions (n_i) were calculated according to Equations (9) and (10):

$$X_j\ (\%) = \frac{mol_{j_{(in)}} - mol_{j_{(out)}}}{mol_{j_{(in)}}} \times 100, \quad j = CH_4,\ CO_2 \tag{9}$$

$$n_i\ (\%) = \frac{n_i}{n_{total}}, \qquad i = CO,\ H_2 \tag{10}$$

In order to evaluate the catalyst resistance during successive runs, aging tests were performed in the same unit as used in long term experiments. After 20 h on stream at 700 °C, the catalysts were regenerated in the following steps: oxidation at 800 °C for 1 h under synthetic air flow at 30 mL min^{-1}, reduction at 700 °C for 1 h under 30 mL·min^{-1} of ultra-pure hydrogen and purge under N_2 flow at 700 °C. Subsequently, the catalysts were exposed to reaction conditions.

6.7. Temperature-Programmed Oxidation (TPO)

The amount of carbon deposited on the spent catalysts was analyzed by temperature-programmed oxidation. After the long-term tests, the catalysts were cooled down to room temperature under Ar flow, and then exposed to synthetic air flow (30 mL·min^{-1}) from room temperature to 1000 °C at a rate of 10 °C·min^{-1}. CO_2 formation was followed by chromatography using a Shimadzu GC-17A (Shimadzu, Kyoto, Japan) set-up equipped with FID and TCD detectors and a Carboxen 1006 column (Sigma-Aldrich, St. Louis, MO, USA).

7. Concluding Remarks

The bulk of the evidence collected in this study supports the hypothesis that the incorporation of Ba into NiLa perovskite-like structures lead to active catalysts for the dry reforming of methane with improved resistance to C-accumulation. The inclusion of Ba partly affects the dispersion and accessibility of Ni, since a moderate increase of Ni particle size was found by XRD analyses. This can explain the partial loss of activity at increasing Ba content; however, Ba-free formulations showed longer-term stability and superior robustness against thermal treatments.

Author Contributions: S.B. has conceived the project, designed the experiments, analyzed the data and written the paper. R.G. has designed and performed the experiments, analyzed the data and written the paper. A.B., R.F., R.J., and C.R. have designed the experiments, analyzed the data and written the paper. D.C. and M.S. have performed the experiments.

Funding: This research was funded by Laboratório Nacional de Luz Síncrotron, CNPQ, FAPESB and Petrobras.

Acknowledgments: The authors are grateful to CNPq, CAPES, FAPESB, FINEP, and Petrobras for the financial support.

Conflicts of Interest: The authors declare no conflict of interest.

References

1. *Eni, Sustainability Report-Eni for 2017;* Springer: Rome, Italy, 2017. [CrossRef]
2. *BP Statistical Review of World Energy 2018;* BP: London, UK, 2018. Available online: http://bp.com/energyoutlook (accessed on 29 March 2019).
3. Van Atten, C.; Saha, A.; Slawsky, L.; Russel, C.; Hellgren, L. *Benchmarking Air Emissions of the Largest Electric Power Producers in the United States;* Bradley, M.J. Associates: Concord, MA, USA; Washington, DC, USA, 2016; Available online: https://www.nrdc.org/resources/benchmarking-air-emissions-100-largest-electric-power-producers-united-states-june-2018 (accessed on 29 March 2019).
4. Arandia, A.; Remiro, A.; García, V.; Castaño, P.; Bilbao, J.; Gayubo, A. Oxidative steam reforming of raw bio-oil over supported and bulk ni catalysts for hydrogen production. *Catalysts* **2018**, *8*, 322. [CrossRef]
5. Hirano, T.; Xu, Y. Catalytic properties of a pure Ni coil catalyst for methane steam reforming. *Int. J. Hydrogen Energy* **2017**, *42*, 30621–30629. [CrossRef]
6. Bhavani, A.G.; Kim, W.Y.; Lee, J.W.; Lee, J.S. Influence of metal particle size on oxidative CO_2 reforming of methane over supported nickel catalysts: Effects of second-metal addition. *ChemCatChem* **2015**, *7*, 1445–1452. [CrossRef]
7. Cao, C.; Wang, Y.; Rozmiarek, R.T. Heterogeneous reactor model for steam reforming of methane in a microchannel reactor with microstructured catalysts. *Catal. Today* **2005**, *110*, 92–97. [CrossRef]
8. De Santana Santos, M.; Neto, R.C.R.; Noronha, F.B.; Bargiela, P.; da Rocha, M.G.C.; Resini, C.; Carbó-Argibay, E.; Fréty, R.; Brandão, S.T. Perovskite as catalyst precursors in the partial oxidation of methane: The effect of cobalt, nickel and pretreatment. *Catal. Today* **2018**, *299*, 229–241. [CrossRef]
9. Goscianska, J.; Pietrzak, R.; Matos, J. Catalytic performance of ordered mesoporous carbons modified with lanthanides in dry methane reforming. *Catal. Today* **2018**, *301*, 204–216. [CrossRef]
10. Akri, M.; Pronier, S.; Chafik, T.; Achak, O.; Granger, P.; Simon, P.; Trentesaux, M.; Batiot-Dupeyrat, C. Development of nickel supported La and Ce-natural illite clay for autothermal dry reforming of methane: Toward a better resistance to deactivation. *Appl. Catal. B Environ.* **2017**, *205*, 519–531. [CrossRef]
11. Rodrigues, L.M.T.S.; Silva, R.B.; Rocha, M.G.C.; Bargiela, P.; Noronha, F.B.; Brandão, S.T. Partial oxidation of methane on Ni and Pd catalysts: Influence of active phase and CeO_2 modification. *Catal. Today* **2012**, *197*, 137–143. [CrossRef]
12. Sudhakaran, M.S.P.; Hossain, M.; Gnanasekaran, G.; Mok, Y. Dry reforming of propane over γ-Al_2O_3 and nickel foam supported novel $SrNiO_3$ perovskite catalyst. *Catalysts* **2019**, *9*, 68. [CrossRef]
13. Barelli, L.; Bidini, G.; Cinti, G. Steam vs. dry reformer: Experimental study on a solid oxide fuel cell short stack. *Catalysts* **2018**, *8*, 599. [CrossRef]
14. Pakhare, D.; Spivey, J. A review of dry (CO_2) reforming of methane over noble metal catalysts. *Chem. Soc. Rev.* **2014**, *43*, 7813–7837. [CrossRef] [PubMed]
15. Usman, M.; Daud, W.M.A.W.; Abbas, H.F. Dry reforming of methane: Influence of process parameters—A review. *Renew. Sustain. Energy Rev.* **2015**, *45*, 710–744. [CrossRef]
16. Chein, R.Y.; Chen, Y.C.; Yu, C.T.; Chung, J.N. Thermodynamic analysis of dry reforming of CH_4 with CO_2 at high pressures. *J. Nat. Gas Sci. Eng.* **2015**, *26*, 617–629. [CrossRef]
17. Nikoo, M.K.; Amin, N.A.S. Thermodynamic analysis of carbon dioxide reforming of methane in view of solid carbon formation. *Fuel Process. Technol.* **2011**, *92*, 678–691. [CrossRef]
18. Wolfbeisser, A.; Sophiphun, O.; Bernardi, J.; Wittayakun, J.; Föttinger, K.; Rupprechter, G. Methane dry reforming over ceria-zirconia supported Ni catalysts. *Catal. Today.* **2016**, *277*, 234–245. [CrossRef]
19. Arora, S.; Prasad, R. An overview on dry reforming of methane: Strategies to reduce carbonaceous deactivation of catalysts. *RSC Adv.* **2016**, *6*, 108668–108688. [CrossRef]
20. Rabelo-Neto, R.C.; Sales, H.B.E.; Inocêncio, C.V.M.; Varga, E.; Oszko, A.; Erdohelyi, A.; Noronha, F.B.; Mattos, L.V. CO_2 reforming of methane over supported $LaNiO_3$ perovskite-type oxides. *Appl. Catal. B Environ.* **2018**, *221*, 349–361. [CrossRef]

21. Peña, M.A.; Fierro, J.L.G. Chemical structures and performance of perovskite oxides. *Chem. Rev.* **2001**, *101*, 1981–2017. [CrossRef] [PubMed]
22. Royer, S.; Duprez, D.; Can, F.; Courtois, X.; Batiot-Dupeyrat, C.; Laassiri, S.; Alamdari, H. Perovskites as substitutes of noble metals for heterogeneous catalysis: Dream or reality. *Chem. Rev.* **2014**, *114*, 10292–10368. [CrossRef]
23. Lin, Y.-C.; Hohn, K. Perovskite catalysts—A special issue on versatile oxide catalysts. *Catalysts* **2014**, *4*, 305–306. [CrossRef]
24. Dama, S.; Ghodke, S.R.; Bobade, R.; Gurav, H.R.; Chilukuri, S. Active and durable alkaline earth metal substituted perovskite catalysts for dry reforming of methane. *Appl. Catal. B Environ.* **2018**, *224*, 146–158. [CrossRef]
25. Gallego, S.; Batiot-dupeyrat, C.; Mondrago, F. Dual active-site mechanism for dry methane reforming over Ni/La_2O_3 Produced from $LaNiO_3$ Perovskite. *Ind. Eng. Chem. Res.* **2008**, *47*, 9272–9278. [CrossRef]
26. Zhang, Z.; Verykios, X.E.; Macdonald, S.M.; Affrossman, S. Comparative study of carbon dioxide reforming of methane to synthesis gas over Ni/La_2O_3 and conventional nickel-based catalysts. *J. Phys. Chem.* **1996**, *100*, 744–754. [CrossRef]
27. Alipour, Z.; Rezaei, M.; Meshkani, F. Effect of alkaline earth promoters (MgO, CaO, and BaO) on the activity and coke formation of Ni catalysts supported on nanocrystalline Al_2O_3 in dry reforming of methane. *J. Ind. Eng. Chem.* **2014**, *20*, 2858–2863. [CrossRef]
28. Lima, S.M.; Assaf, J.M.; Pena, M.A.; Fierro, J.L.G. Structural features of $La1\text{-}xCexNiO_3$ mixed oxides and performance for the dry reforming of methane. *Appl. Catal. A Gen.* **2006**, *311*, 94–104. [CrossRef]
29. Pecchi, G.; Reyes, P.; Zamora, R.; Campos, C.; Cadús, L.E.; Barbero, B.P. Effect of the preparation method on the catalytic activity of $La1\text{-}xCaxFeO_3$ perovskite-type oxides. *Catal. Today* **2008**, *133–135*, 420–427. [CrossRef]
30. Weinbruch, S.; Büttner, H.; Rosenhauer, M. The orthorhombic-hexagonal phase transformation in the system $BaCO_3\text{-}SrCO_3$ to pressures of 7000 bar. *Phys. Chem. Miner.* **1992**, *19*, 289–297. [CrossRef]
31. Broqvist, P.; Panas, I.; Grönbeck, H. Toward a realistic description of NOx storage in BaO: The aspect of $BaCO_3$. *J. Phys. Chem. B* **2005**, *109*, 9613–9621. [CrossRef] [PubMed]
32. Borges, R.P.; Ferreira, R.A.R.; Rabelo-Neto, R.C.; Noronha, F.B.; Hori, C.E. Hydrogen production by steam reforming of acetic acid using hydrotalcite type precursors. *Int. J. Hydrogen Energy* **2018**, *43*, 7881–7892. [CrossRef]
33. Qin, H.; Guo, C.; Wu, Y.; Zhang, J. Effect of La_2O_3 promoter on NiO/Al_2O_3 catalyst in CO methanation. *Korean J. Chem. Eng.* **2014**, *31*, 1168–1173. [CrossRef]
34. Zangouei, M.; Moghaddam, A.Z.; Arasteh, M. The influence of nickel loading on reducibility of NiO/Al_2O_3 catalysts synthesized by sol-gel method. *Chem. Eng. Res. Bull.* **2010**, *14*, 97–102. [CrossRef]
35. Gao, J.; Hou, Z.; Guo, J.; Zhu, Y.; Zheng, X. Catalytic conversion of methane and CO_2 to synthesis gas over a La_2O_3-modified SiO_2 supported Ni catalyst in fluidized-bed reactor. *Catal. Today* **2008**, *131*, 278–284. [CrossRef]
36. Zhang, Z.; Verykios, X.E. Carbon dioxide reforming of methane to synthesis gas over Ni/La_2O_3 catalysts. *Appl. Catal. A Gen.* **1996**, *138*, 109–133. [CrossRef]
37. Pérez-Florindo, A.; Cazorla-Amorós, D.; Linares-Solano, A. CO_2-carbon gasification catalyzed by alkaline-earths: Comparative study of the metal-carbon interaction and of the specific activity. *Carbon* **1993**, *31*, 493–500. [CrossRef]
38. Ersolmaz, C.; Falconer, J.L. Catalysed carbon gasification with $Ba^{13}CO_3$. *Fuel* **1986**, *65*, 400–406. [CrossRef]

Article

A Case Study for the Deactivation and Regeneration of a V_2O_5-WO_3/TiO_2 Catalyst in a Tail-End SCR Unit of a Municipal Waste Incineration Plant

Stefano Cimino [1,*], Claudio Ferone [2], Raffaele Cioffi [2], Giovanni Perillo [3] and Luciana Lisi [1,*]

1 Istituto Ricerche sulla Combustione, CNR, 80125 Napoli, Italy
2 Department of Engineering, University Parthenope of Napoli, 80143 Napoli, Italy; claudio.ferone@uniparthenope.it (C.F.); raffaele.cioffi@uniparthenope.it (R.C.)
3 Wessex Institute of Technology, Southampton SO40 7AA, UK; giovanni.perillo@uniparthenope.it
* Correspondence: stefano.cimino@cnr.it (S.C.); l.lisi@irc.cnr.it (L.L.)

Received: 5 May 2019; Accepted: 15 May 2019; Published: 20 May 2019

Abstract: In this work, we set out to investigate the deactivation of a commercial V_2O_5-WO_3/TiO_2 monolith catalyst that operated for a total of 18,000 h in a selective catalytic reduction unit treating the exhaust gases of a municipal waste incinerator in a tail end configuration. Extensive physical and chemical characterization analyses were performed comparing results for fresh and aged catalyst samples. The nature of poisoning species was determined with regards to their impact on the DeNOx catalytic activity which was experimentally evaluated through catalytic tests in the temperature range 90–500 °C at a gas hourly space velocity of 100,000 h^{-1} (NO = NH_3 = 400 ppmv, 6% O_2). Two simple regeneration strategies were also investigated: thermal treatment under static air at 400–450 °C and water washing at room temperature. The effectiveness of each treatment was determined on the basis of its ability to remove specific poisoning compounds and to restore the original performance of the virgin catalyst.

Keywords: DeNOx; MW incinerator; deactivation; ammonium sulfates; regeneration; washing

1. Introduction

Selective catalytic reduction (SCR) of nitrogen oxides with ammonia is one of the most widely used technologies for reducing NOx from both stationary and mobile sources [1,2]. This technology and, in particular, its application to the exhaust of fossil fuel-fired power plants, is quite mature. Nevertheless, more recent applications to industrial and municipal waste (MW) incinerators pose new issues related to the heterogeneity and variability of the fuel with the consequent release of different pollutants which have to be suitably treated before emission to the atmosphere [3,4]. In addition to SO_x, CO, and particulate matter, exhaust gases deriving from the combustion of solid wastes contain large amounts of alkali and alkaline-earth metals (potassium, sodium, calcium, magnesium), phosphorous, and chlorine [4]. All these pollutants must be controlled within the limit values defined by the more and more stringent environmental regulations.

Catalysts for the SCR of NOx in exhaust gases from MW incinerator are generally of the same type employed in traditional power plants, namely V_2O_5-WO_3/TiO_2. Poisoning/deactivation of such catalysts during operation with the exhaust gases from coal-fired plants has been widely investigated, mainly concerning compounds such as SO_2, mercury, arsenic, alkaline, and alkaline earth oxides [4]. Under conventional operating conditions, the expected catalyst lifetime has been well defined and possible regeneration strategies proposed, as catalysts represent the main cost of the SCR process [2,5]. More recently, research has been extended to poisoning of SCR catalysts for NOx abatement in flue gases from MW and biomass combustion [3–7]. The SCR unit in a MW incinerator plant is possibly located just after the boiler as high dust system where temperatures are high enough (300–400 °C) to assure optimal catalytic performance. In this configuration, however, SO_2, fly ashes, acid gases and particulate, not yet removed, impact the SCR catalyst. In addition, alkali and alkaline-earth metals, mostly present into the ashes which are more abundant in the MW emissions, can further deactivate the catalyst strongly reducing its lifetime [4–7].

In general, alkali and alkaline-earth metals deactivate the catalyst reducing the number of acid sites [8–11]. Nevertheless, one of the major causes of poisoning, also for MW incineration, remains the formation of ammonium sulfate and bi-sulfate on the catalyst surface mostly when the operation temperature is below 290 °C [5,11] due to the reaction occurring between ammonia and SO_3 produced by SO_2 oxidation. In particular, sulfates poisoning can reduce surface area and porosity by pore clogging or can affect the surface acidity [11–13]. Kröcher and Elesener [13] reported that potassium reacted with vanadium forming potassium vanadate whereas calcium sulfate just physically masks the active sites forming a barrier layer. Nicosia et al. [10] also investigated the deactivating effect of K and Ca on traditional SCR catalysts which affected both Brønsted and Lewis acidity of V_2O_5 inhibiting both the NH_3 adsorption and the decomposition of the nitrosamide intermediate. The extent of deactivation also depends on the counterion of the alkali or alkaline-earth metal cation [14]. It was reported [10] that calcium causes a strong reduction of activity if coupled with an organic anion whilst the deactivation is almost negligible when anions from inorganic acid are present on the surface. Among its inorganic salts, calcium deactivates the SCR catalyst in the following order: $SO_4^{2-} > PO_4^{3-} > BO^{3-}$ [12].

The tail-end configuration of the SCR unit is less common. It involves gas re-heating due to the complex upstream system of emissions control with additional costs related to the fuel to increase the gas temperature up to 180–250 °C [5]. Nevertheless, the tail-end arrangement enables the use of a high-activity catalyst with small diameter channels, since particulates and SO_2 have already been removed upstream; moreover, the expected deactivation rate is low due to the absence of fly ash and other poisons so that catalyst life is substantially extended [5].

In this paper, we report a case study on the deactivation of a commercial V_2O_5-WO_3/TiO_2 monolith catalyst operated for 18,000 h in a SCR (DeNOx) unit with tail-end configuration treating the exhaust gases of a MW incinerator plant located in Southern Italy [15]. The SCR unit, operating with ammonia injection at around 200 °C, is placed downstream a purification train consisting of dry and semi-dry absorbers with lime and active carbon injection to remove HCl, HF, SO_x; active carbon adsorber to remove heavy metals and organo-chlorinated compounds; bag filters to eliminate particulate matter [15]. The possible deactivating mechanisms and poisoning compounds were identified by several characterization techniques trying to establish their origin; the extent of deactivation was estimated by kinetic measurements of the SCR catalytic activity. Simple regeneration strategies such as thermal treatments in air and washing in water were investigated on the basis of the results of chemical and physical characterization of the aged catalyst.

2. Results

The commercial SCR catalyst consisted of a V_2O_5-WO_3/TiO_2 honeycomb monolith sample with square channels, a cell density of 7.3 cell/cm^2, and an overall density of 500 g/dm^3. Glass fibers with a characteristic diameter around 10 μm were distributed in the bulk of the material to provide mechanical resistance to the extruded monolith.

XRF analysis of virgin monolith catalyst indicated that the loading of the two active metals (V and W) was 3.51% and 6.36% when expressed as oxides respectively (Table 1), which is typical of SCR catalysts for stationary applications [1,16]. SEM-EDS analysis of fresh monolith sample (not shown) indicated that vanadium, tungsten, and titanium were uniformly distributed, whereas the silica content was associated with the reinforcing glass fibers. The catalyst also contained small amounts of iron and sulfur (Table 1), which are generally added to enhance the catalytic performance [17–19].

Table 1. Chemical composition of the fresh catalyst by XRF analysis.

	TiO_2	WO_3	V_2O_5	SiO_2	Al_2O_3	CaO	Fe_2O_3	SO_4^{--}
%wt	85.53	6.36	3.51	2.01	1.10	1.14	0.30	0.38

The fresh monolith had a surface area of ca. 54 m^2/g (Table 2) that falls in the typical range of values (50–60 m^2/g) generally reported for commercial mesoporous TiO_2 in the anatase phase. In fact, XRD analysis (not shown) confirmed the presence of a pure TiO_2 anatase phase. Moreover, the results of pore size distribution (PSD) analysis reported in Figure 1 revealed the presence of mesopores with characteristic diameters in the range 60–300 Å.

Table 2. BET surface area and volume of pores of fresh, aged and regenerated selective catalytic reduction (SCR) catalysts.

Sample	Pre-Treatment	BET s.a. m^2/g	V cc/g
Fresh catalyst		53.8	0.25
Aged 18 kh		43.4	0.26
Regen. R400	2 h @400 °C	44.8	0.26
Regen. R450	2 h @450 °C	44.8	0.26
Regen. WW	30 min in water	48.4	0.30

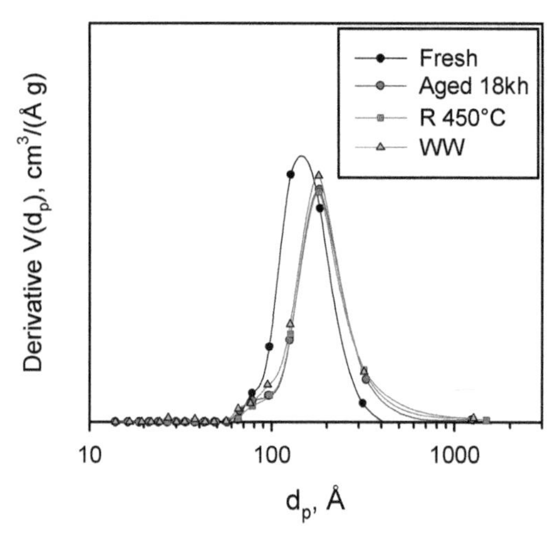

Figure 1. Pore size distribution for fresh, aged, and regenerated catalysts by heat treatment (R 450 °C) or washed in water (WW).

After 18 kh of operation, the BET specific surface area of the aged catalyst dropped by ca. 20% down to 43 m^2/g without any significant change in the total pore volume (Table 2), suggesting an increase of the average pore size. In fact, the PSD analysis indicates the partial loss of those mesopores in the range 80–170 Å (Figure 1), possibly due to occlusion and/or collapse.

The typical yellowish color of the virgin SCR catalyst turned into grey after 18 kh on stream, with some brownish shades in correspondence to the inlet edge the monolith (Figure 2a,b). However, the cross-section of the honeycomb channels was free due to the absence of massive ash deposition and to a rather limited occurrence of erosion phenomena (Figure 2a).

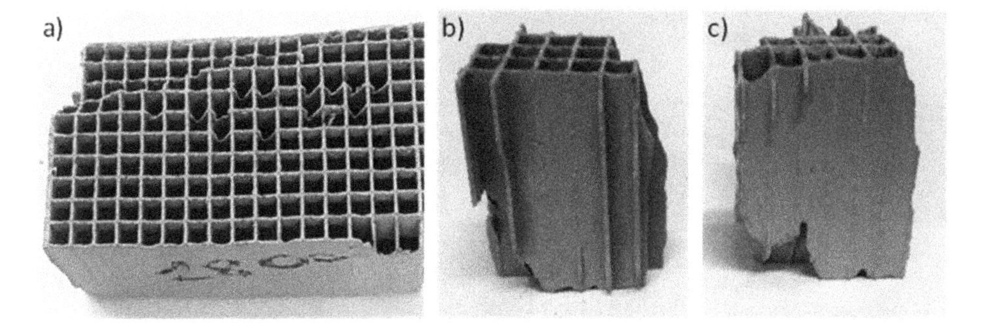

Figure 2. Appearance of the reaction-aged (18 kh) SCR honeycomb catalyst ((**a**,**b**): samples taken from the inlet section of the monolith) in comparison with a sample of the same catalyst after regeneration by washing in water (**c**).

SEM with EDS elemental analysis was performed at three different positions along the monolith channel close to the inlet section (1, 5, 10 mm from it) selecting two areas of 271 × 271 μm for each position; Table 3 reports the corresponding results for the elemental concentration of S and Ca.

Table 3. EDS elemental analysis of the surface concentration of sulfur and calcium at different locations along the channel of the reaction-aged SCR catalytic honeycomb (distance from inlet: a, a′ = 1 mm; b, b′ = 5 mm; c, c′ = 10 mm).

Location Along the Monolith		a	a′	b	b′	c	c′
Distance from inlet	[mm]	1		5		10	
Sulfur	[wt%]	4.3	2.6	2.2	2.3	1.3	1.1
Calcium	[wt%]	2.9	1.4	0.5	0.2	0.2	0.5
S/Ca	atomic	1.9	2.3	5.5	14.4	8.1	2.8

Calcium and sulfur were detected at relatively high concentrations close to the monolith inlet but their amount decreased along the channel length. This is better highlighted in Figure 3, where the elemental distribution is shown for zones a and b (Table 3). The overlapping of the most intense spots associated with sulfur and calcium confirms that indeed they were associated with the presence of $CaSO_4$ deposits. The additional formation of other sulfate species can be inferred by the higher atomic concentration of sulfur with respect to calcium detected in all investigated areas (Table 3).

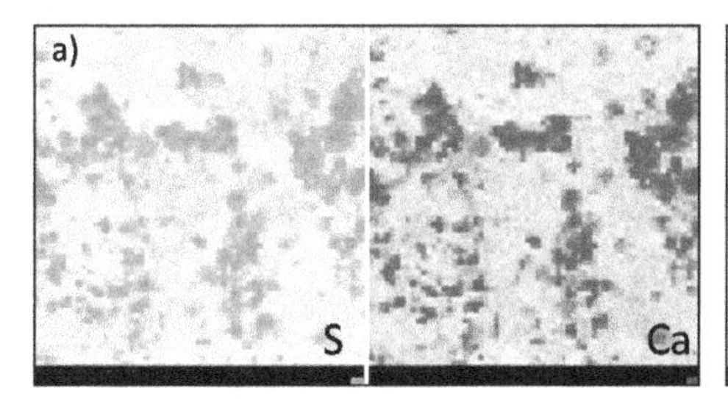

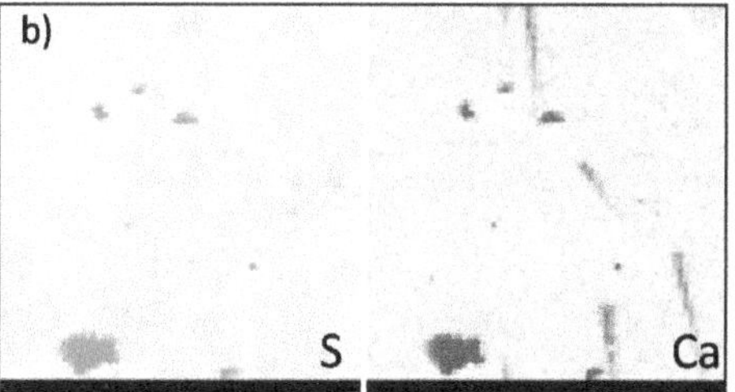

Figure 3. Sulfur (**green**) and calcium (**blue**) distribution maps at two locations ((**a**): 0.1 cm and (**b**): 0.5 cm from inlet) along a central channel of the 18 kh reaction-aged SCR honeycomb catalyst.

The larger content of calcium in the aged catalyst was most likely due to the dragging of fine particles from the upstream fabric filters that follow a desulfurization unit based on the injection of hydrated lime [15]. Notably, the presence of $CaSO_4$ surface deposits can derive directly from dragging of $CaSO_4$ powders or by the reaction of $Ca(OH)_2$, which is preferentially deposited where sulfur is present at a higher level.

The surface acid properties of the fresh monolith catalyst were evaluated by performing NH_3 temperature programmed desorption (TPD) runs after saturating the sample with ammonia at 100 °C. The corresponding TPD profile, reported in Figure 4, is quite broad and it is the result of the overlap of at least two peaks. Ammonia desorption starts from the adsorption temperature (100 °C) and ends at about 470 °C indicating a wide distribution of acid sites with different strengths, which can be assigned to the anatase TiO_2 support [13]. By integration of the TPD curve, the total amount of NH_3 was calculated to equal 0.161 mmol/g, which agrees well with values reported for other commercial SCR catalysts taking into account the different exposed surface areas [8].

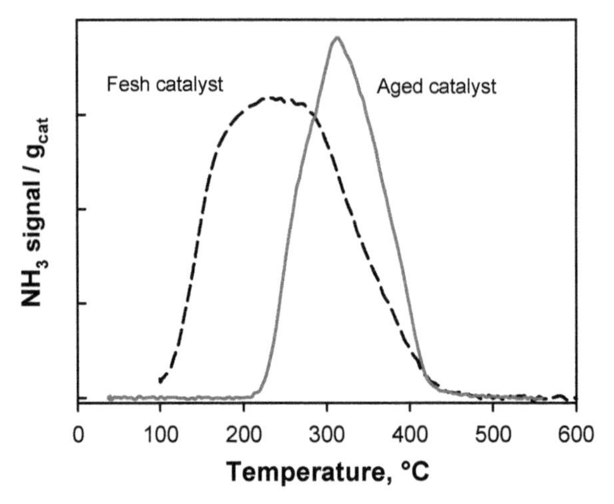

Figure 4. NH_3 temperature programmed desorption (TPD) of fresh and aged catalyst.

The NH_3 TPD of the aged sample carried out without pre-adsorbing ammonia at 100 °C (Figure 4), shows a more intense signal starting at about 210 °C, with a peak at 315 °C, and ending at the same temperature observed for the fresh catalyst. The ammonia adsorbed on the aged catalyst derived from its operation in the SCR reactor. Therefore, from the onset temperature of desorption, it can be argued that the last temperature the catalyst experienced in the SCR reactor was around 210 °C. The total amount of desorbed ammonia was lower (0.075 mmol/g) than the amount desorbed from the fresh sample due to the lack of NH_3 pre-adsorption step at 100 °C. Nevertheless, in the high-temperature range (above 300 °C), the NH_3 desorption peak for the aged catalyst exceeded that of the fresh catalyst. This result strongly suggests the decomposition of ammonium salts, like sulfate and bisulfate, probably formed by the reaction between NH_3 and SO_2/SO_3 or other superficial sulfates which were eventually deposited on the catalyst surface due to the low temperature of operation [7,11,18,20–22]. Notably, the content of gaseous SO_2 in the exhaust gas after the desulfurization unit is quite low (in the single digit ppmv range) [15].

The content of sulfates possibly adsorbed during the SCR operation was estimated by carrying out SO_2 TPD experiments under N_2 flow up to 900 °C on the aged catalyst in comparison with its fresh counterpart (Figure 5). Judging from its TPD trace it was confirmed that the virgin monolith contained some sulfur (see Table 1), likely in the form of sulfate species deliberately added during catalyst preparation in order to promote the SCR activity by increasing the number of acid sites for NH_3 adsorption [17]. By comparison with the TPD results of reference sulfate compounds, the SO_2 peaks between 400 and 600 °C were associated with the decomposition of sulfate species bonded to titania and of $VOSO_4$ species [20,23], while the decomposition of iron sulfates might contribute at higher temperatures [24].

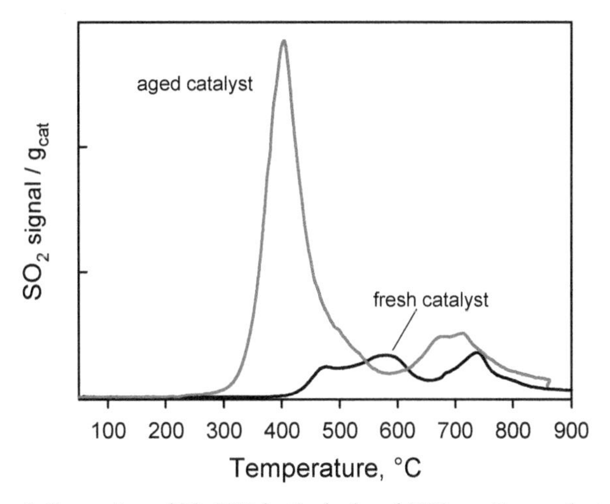

Figure 5. Comparison of SO_2 TPD for the fresh and 18 kh reaction-aged catalyst.

The onset temperature for SO_2 release from the virgin catalyst (400 °C) was a probable indication of the calcination temperature adopted during the preparation of the catalytic monoliths.

On the contrary, the reaction-aged catalyst showed an intense SO_2 desorption peak at low temperature starting from ca. 250 °C, which agrees with the accumulation of sulfates on the catalyst surface during operation. In fact, such a relatively low decomposition temperature is generally associated with the presence of ammonium (bi) sulfate [20], thus supporting the interpretation of the results from the NH_3 TPD test. However, when comparing the NH_3 and SO_2 desorption profiles from the aged catalyst it can be noticed that ammonia was released at a lower temperature with respect to SO_2 (Figures 4 and 5).

2.1. Regeneration Treatments

According to the characterization results of the reaction-aged catalyst, indicating the formation of ammonium sulfate and the deposition of $CaSO_4$ as the main possible causes of catalyst deactivation, two different types of regeneration treatments were investigated: (i) thermal treatment at 400 °C or 450 °C in static air; (ii) washing in (distilled) water at room temperature and subsequent drying. As mentioned before, the aged sample appeared grey in color in contrast to the green-yellowish fresh sample. Following both types of regeneration treatments, the original color was restored, as shown in Figure 2 for the case of water washed sample of the aged monolith.

Thermal regeneration of the aged catalyst at either 400 or 450 °C in air provided only limited recovery of the original surface area up to 45 m^2/g (Table 2). This is possibly caused by the volatilization/decomposition of some species formed and/or condensed on the catalyst during prolonged SCR operation. Nevertheless, the pore size distribution did not significantly change with respect to the aged catalyst (Figure 1), suggesting that the smallest mesopores were still partially blocked by relatively stable nano-particles or they were collapsed.

Washing in water was somehow more effective in restoring the surface area of the aged catalyst up to roughly 48 m^2/g (Table 2), though it also apparently increased the total pore volume, possibly due to the partial dissolution of some catalyst components. Notably, by washing the aged monolith in water it was possible to induce leaching of soluble species such as ammonium (bi) sulfate as well as the detachment of small debris and ashes largely insoluble in water, deposited on the outer surfaces of the monolith.

To acquire quantitative information on the sulfate salts formed on the spent catalyst, ion chromatography and ICP-MS experiments were performed on the supernatant solutions obtained by immersing samples of fresh, aged, and thermally regenerated catalysts in water.

In agreement with elemental analysis and SO_2-TPD experiments, it was found that some soluble sulfate species were present on the surface of the virgin catalyst, accounting for roughly 0.5% of its original weight. Moreover, no other anions were detected in significant concentration, thus excluding, in particular, the deposition of ammonium nitrates on the catalyst during its operation in the SCR unit at low temperature. As shown in Table 4, the quantity of soluble sulfates increases for the reaction-aged catalyst reaching 3% of the catalyst weight.

Table 4. Ion chromatography analysis showing the concentration of main anions in the supernatant water after washing samples of the fresh, reaction-aged, and thermally regenerated (R400) SCR catalyst. Results converted to weight% with respect to the catalyst sample.

	F	Cl	NO_3 % wt	PO_4	SO_4
Fresh Catalyst	0.00	0.00	0.00	0.01	0.53
Aged 18 kh	0.00	0.00	0.00	0.01	3.00
Regen. R400	0.00	0.00	0.00	0.00	1.92

The ICP-MS analysis of metal ions in the supernatant water revealed the presence of a significant lower concentration of dissolved calcium (about 0.005% wt.) with respect to the sulfates and the dissolution of a modest amount of vanadium (0.022% of the catalyst weight, corresponding to ca. 0.9% of the original vanadium loading).

Thermal treatment of the aged catalyst at 400 °C reduced the amount of water-soluble sulfates to ca. 1.9% wt. (Table 4), that was still higher than the original value measured for the virgin catalyst.

The effect of the regeneration treatments was further investigated by carrying out TG-MS experiments under N_2 flow either on fresh and aged catalysts as well as on reference sulfate salts. In Figure 6 the weight loss associated with the decomposition of surface salts is compared for the reaction-aged and regenerated catalysts. The aged catalyst showed a well detectable weight loss from roughly 300 °C whereas both regenerated catalysts experienced only a small weight change in the range of temperature investigated. In particular, the water washed catalyst (WW) showed a weak weight loss due to some residual water adsorbed on the surface, whereas the catalyst that was thermally regenerated at 400 °C (R400) displayed a small unexpected weight increase up to 400 °C and thereafter it approached the same final weight.

The inset graph in Figure 6 shows the decomposition of bulk $(NH_4)_2SO_4$ takes place between 260 °C and 460 °C with a characteristic two-step process resembling the trend of weight loss observed for the aged catalyst in the same temperature range. In order to further confirm this hypothesis, the TG analysis was also performed on a sample of the fresh catalyst to which 10 wt% of $(NH_4)_2SO_4$ was added by impregnation (with a water solution) followed by drying at 120 °C. As shown in Figure 6, the weight loss followed the same qualitative trend observed for the aged catalyst. Therefore, it can be argued that ammonium sulfate was one of the main species accumulated on the catalyst during 18 kh of SCR operation at low temperature. Its total amount on the catalyst accounted for a maximum of 1–1.5% wt. and it could be removed by either thermal regeneration treatments at temperatures above 400 °C or by washing with water.

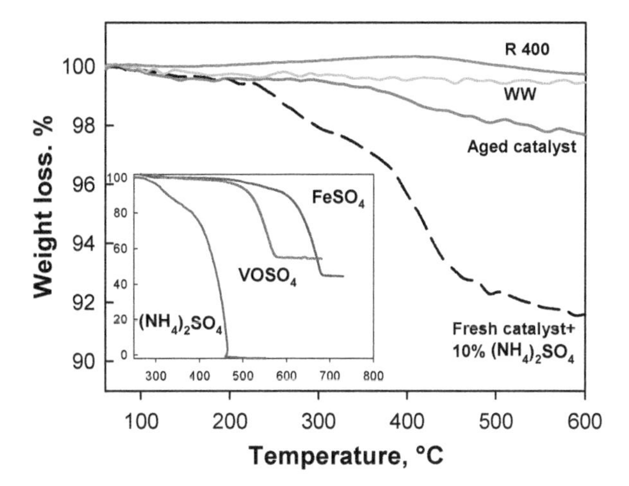

Figure 6. TG analysis (in N_2) of a fresh catalyst sample impregnated with 10 wt% $(NH_4)_2SO_4$, a reaction-aged catalyst sample, and the same catalyst after regeneration by either thermal treatment (R400) or water treatment (WW). The inset shows the TG profile recorded for the thermal decomposition of reference bulk sulfates.

Accordingly, the MS profile of SO_2 (m/z = 64) that evolved from the reaction-aged catalyst (Figure 7) revealed a main emission peak at about 410 °C followed by a long tail extending up to 850 °C with poorly resolved, low-intensity peaks at ca. 550 and 750 °C. By comparison with the SO_2 profile obtained from the fresh catalyst sample impregnated with 10% $(NH_4)_2SO_4$ it can be confirmed that decomposition of ammonium (bi) sulfate was responsible for the first emission peak around 400 °C. In agreement with the results of elemental analysis showing (Tables 1 and 4) some different sulfate species were already present on the fresh catalyst, it can be inferred that those sulfates that bonded to titania, $VOSO_4$, and iron sulfates, contributed to SO_2 emission in the temperature range from 400 °C to 800 °C. On the other hand, $CaSO_4$ deposits, whose presence on the surface of the aged catalyst was detected by SEM-EDS analysis, would require higher temperatures (in excess of 900 °C) to start decomposing under flowing N_2.

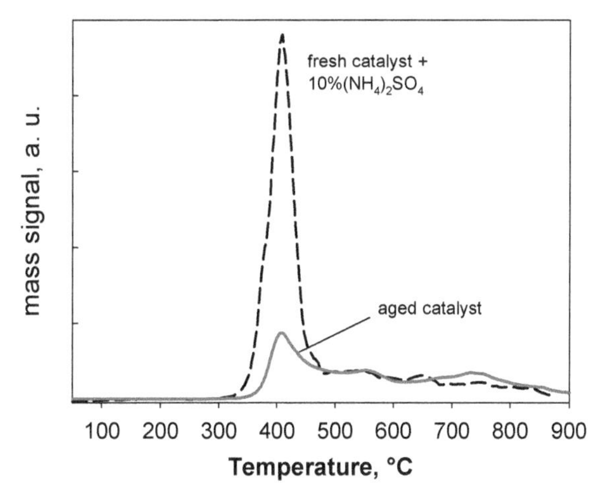

Figure 7. MS profiles of SO_2 (m/z = 64) recorded during the TG analysis (Figure 6) of the fresh catalyst sample impregnated with 10 wt% $(NH_4)_2SO_4$ and the 18 kh reaction-aged catalyst.

The absence SO_2 signals at high temperature for WW regenerated catalyst indicates that $CaSO_4$ deposits were detached and/or partly solubilized in water: indeed, a bulk solubility of 2.4 g/L at room temperature [25] is large enough to assure dissolution of small $CaSO_4$ surface particles.

This conclusion explains the reason why WW regeneration is more effective than thermal regeneration. In fact, the thermal treatment could only promote ammonium sulfate decomposition, which largely occurs at T < 400 °C, as also reported by Gan et al. [26], whereas WW solubilized also the other more stable sulfates and easily removed ash debris poorly attached to the monolith walls. This result is roughly in agreement with data reported in Table 4 showing that only 1/3 of the total sulfates were removed through thermal regeneration, whereas with the washing treatment it was possible to remove the remaining fraction.

In Figure 8, the FTIR spectra of the fresh, aged and regenerated catalysts are reported. The fresh catalyst showed, in addition to the small band at 980 cm^{-1} assignable to both hydrated vanadyl and wolframyl groups [16], also a large broad band consisting of overlapping signals with maxima at about 1135 and 1050 cm^{-1} attributed to bi-dentate sulfate on TiO_2 [16,27]. Sulfation of the TiO_2 support in the fresh sample was already deduced by elemental analysis and TPD-SO_2 results and it is deliberately performed to enhance the intrinsic SCR activity by increasing the number of Brønsted acid sites [17]. Aging under reaction for 18 kh caused the appearance of an additional sharp band at about 1400 cm^{-1}, a shoulder at about 1210 cm^{-1}, and a slight increase in the intensity of the band at 1050 cm^{-1}. These two last bands are related to further TiO_2 sulfation [16,26,27]. Moreover, the 1400 cm^{-1} band, detected in sulfated V_2O_5/TiO_2 catalysts, is attributed to the asymmetric bending vibrations of NH_4^+ [17,26] although, the superimposition of a band at 1383 cm^{-1}, assigned by Li et al. [11] to $VOSO_4$, cannot be excluded. In order to confirm that ammonium sulfate or bisulfate were present in the aged catalyst, the reference spectrum for the fresh catalyst impregnated with 10% wt. $(NH_4)_2SO_4$ was also recorded and it is reported in Figure 8. In agreement with TG-MS results, similar but more intense signals were found for this reference sample compared to the aged catalyst, thus confirming the deposition of ammonium sulfate and/or bisulfate during SCR operation.

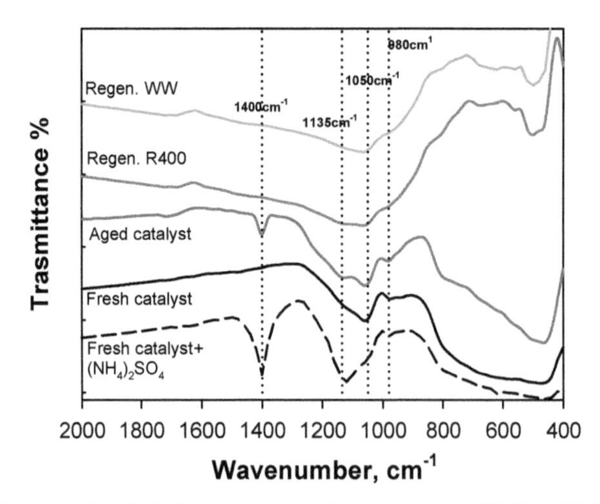

Figure 8. FTIR spectra of the fresh, 18 kh reaction-aged and regenerated (R400 and WW) SCR catalysts as compared to the reference sample of fresh catalyst impregnated with 10 wt% $(NH_4)_2SO_4$.

The absence of the band at 1400 cm^{-1} in the FTIR spectra of regenerated catalysts and the strong reduction of those bands in the range 1000–1230 cm^{-1} indicate that it was possible to (completely) remove those ammonium sulfate deposits by either a thermal treatment at 400 °C or by washing in water. In fact, ammonium sulfates are highly soluble in water and can be thermally decomposed at temperatures around 400–450 °C [27].

Characterization results showed that sulfates were accumulated on the aged catalyst, in spite of the tail end arrangement of the SCR unit guarantees rather low concentrations of SO_x in the inlet feed to the reactor. One possible source is related to the entrainment of small $CaSO_4$ particles escaping the upstream filters. Moreover, it has been reported that sulfation occurs at 200 °C via the weak adsorption of SO_2 molecules on the catalyst (favored at low temperature) as SO_3^{2-} species that are further oxidized by vanadia to form bridged bidentate sulfates bound to titania sites [20]. Thereafter, ammonium (bi) sulfate species deposit on the catalyst through the reaction between adjacent adsorbed NH_3 and sulfate species. In fact, ammonium sulfate and bisulfate are stable at the low operating temperatures (around 200 °C) typical of this SCR unit, whereas the ammonia consumption rate by the SCR reaction is not fast enough to inhibit the formation and accumulation of those compounds, which cover the active sites and plug the pores [5,17,18].

On the other hand, calcium was also found to accumulate on the catalyst, and it can reduce the activity by decreasing of both Lewis and Brønsted sites [9] and/or masking the catalyst surface [5]. However, in the present case, the reaction-aged honeycomb catalyst remained relatively clean as it was exposed to a dust free flue gas. Moreover, Odenbrand [12] noticed, for V_2O_5-WO_3/TiO_2 SCR catalyst utilized in a diesel power plant, that deactivation by $CaSO_4$ in the real system was lower than that simulated by impregnation of the catalyst with $CaSO_4$. They explained this difference supposing that a surface layer of $CaSO_4$ deposits on the catalyst during the use on the engine whereas the impregnation from solution promotes the introduction of calcium sulfate into the pores of the catalyst.

Notably, a relatively high calcium content was detected by EDS analysis on the outer surface of 18 kh aged honeycomb catalyst (particularly close to its inlet section), but the concentration of Ca dissolved in the supernatant solution obtained after washing this sample in water was rather low (well below the solubility of $CaSO_4$). This suggests the presence of poorly soluble species such as $CaCO_3$, that is also possibly formed in the upstream desulfurization unit, or it can be formed in situ together with ammonium sulfates by the reaction of $CaSO_4$ with NH_3 in the presence of CO_2 and water.

2.2. Catalytic Testing

Results of catalytic tests on fresh, aged, and regenerated catalysts are shown in Figure 9a–d. The NO conversion profile as a function of temperature showed a typical broad maximum in the range between 300–400 °C (Figure 9a). At higher temperatures, the direct ammonia oxidation reaction started to proceed at significant rates consuming the reactant for the SCR reaction of NO while also leading to the formation of different nitrogen oxides (including NO) apart from N_2, thus apparently decreasing NO conversion. Accordingly, ammonia conversion increased monotonically along with the reaction temperature approaching 100% for T > 300 °C without any further decrease (Figure 9b). In line with many literature reports, the fresh V_2O_5-WO_3/TiO_2 catalyst was very selective towards the formation of N_2 up to 300 °C (Figure 9c), whereas at higher temperatures the selectivity dropped due to the undesired formation of increasing quantities of N_2O in addition to N_2.

A clear reduction of catalytic activity was observed for the aged catalyst used for a total of 18kh in the industrial reactor. Both the NH_3 and NO conversion curves relevant to the aged catalyst sample were generally shifted towards higher temperatures in their ascending brands (Figure 9a,b). A 100% selectivity to N_2 was still measured up to 300 °C (Figure 9c), but, for higher temperatures, the aged catalyst showed a more pronounced tendency to form N_2O.

All regeneration treatments were able to restore most of the original catalytic performance, with a small but still measurable advantage for the washed catalyst.

The kinetic constant for the NO consumption rate and the corresponding apparent activation energy were estimated from integral reactor conversion data under the common assumptions of an ideal isothermal plug flow behavior and a first-order dependency on NO concentration [16,28].

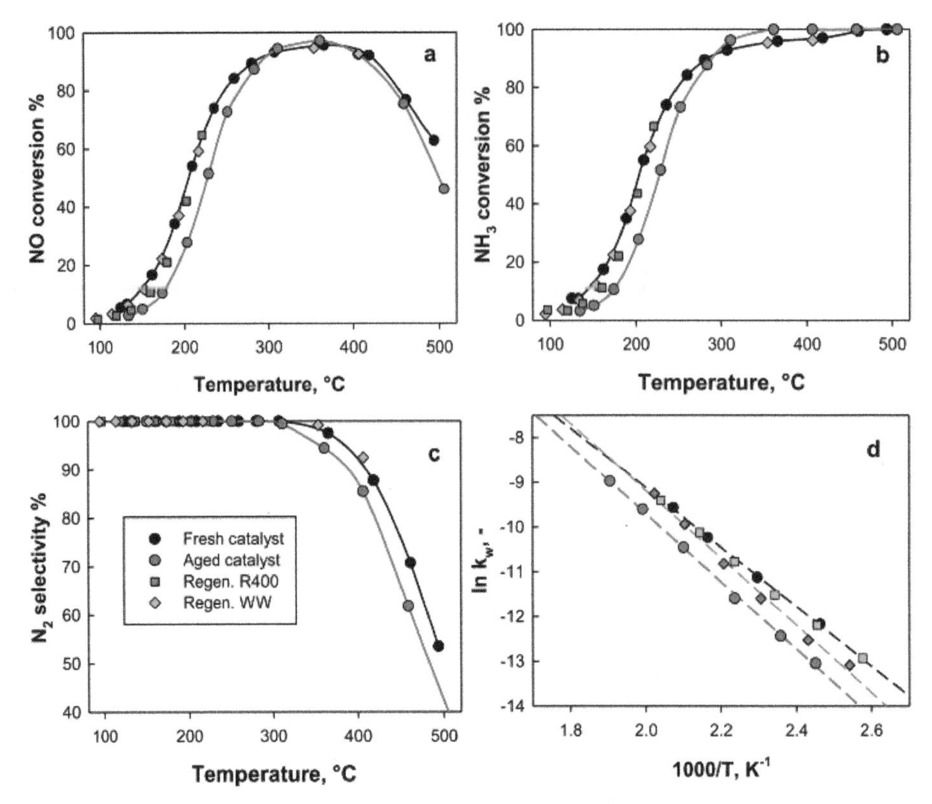

Figure 9. NO (**a**) and NH_3 (**b**) conversion and N_2 selectivity (**c**) as a function of the reaction temperature during the NH_3-SCR tests on fresh, aged, and regenerated (R400 and WW) catalyst; (**d**)) Arrhenius plots showing the corresponding values of the NO reaction rate constant as a function of inverse temperature.

Figure 9d reports Arrenhius plots for the specific (mass-based) NO consumption rates: data sets obtained for each catalyst in the low to mid conversion range showed a linear trend, thus suggesting that they were representative of a predominantly kinetic regime. Apparent activation energy values (E_a) were estimated from the slope of the corresponding Arrhenius plots and are reported in Table 5.

Table 5. Kinetic parameters E_a and $k_w{}^0$, values of the kinetic constant and relative activity at 200 °C for fresh, aged, and regenerated catalysts.

	Ea kJ/mol	ln $k_w{}^0$ $m^3/(g\ s)$	k_w (200 °C) $m^3/(g\ s)$	Relative Activity %
Fresh catalyst	54.3	3.89	5.02E-05	100
Aged 18 kh	63.0	5.44	2.58E-05	51
Regen. R400	62.6	5.88	4.38E-05	87
Regen. WW	54.8	4.00	4.84E-05	96

The fresh catalyst displayed an apparent activation energy equal to 54 kJ/mol, in general agreement with results for analogous systems in the literature [11,17]. Upon aging under SCR conditions, E_a value increased up to 63 kJ/mol, whereas the kinetic constant evaluated at 200 °C decreased by almost 50% from 5.02×10^{-5} to 2.58×10^{-5} $m^3/(g\ s)$ (Table 5). This figure should be regarded as the maximum degree of deactivation experienced by the reaction-aged catalyst after 18 kh of operation since it was calculated for a sample taken from the uppermost inlet section of a honeycomb catalyst module placed in the first layer of the SCR reactor. It is confirmed that catalyst deactivation proceeded at a relatively low rate due to the tail-end installation of the SCR unit. Nevertheless, an increase of only 10–15 °C in the operating temperature of the SCR reactor was required to compensate for the activity loss due catalyst deactivation has a significant adverse economic impact due to the increased operating costs to re-heat the flue gases exiting the $DeSO_x$ unit and filters [5].

Notably, with the washing treatment in water, it was possible to restore the original value of the apparent activation energy; moreover, the kinetic constant calculated for the regenerated WW catalyst was equal (within experimental error) to that of the fresh system (Table 5). On the other hand, the thermal regeneration at 400 °C was slightly less effective in recovering the catalytic activity at 200 °C, with a calculated kinetic constant equal to 87% of the initial value. However, the thermal treatment was not able to restore the apparent activation energy that remained at 62.6 kJ/mol.

At least two main causes of deactivation were identified on the basis of the characterization of fresh and spent catalyst: ammonium sulfate and/or bisulfate and calcium.

Ammonium bisulfate causes the deactivation of the catalyst at low temperature but it can activate it at a higher temperature (350–450 °C) through sulfation of TiO_2 [6]. In agreement with this assertion, the aged catalyst performed worse than its fresh counterpart up to 300 °C; however, during those SCR tests at higher temperatures, ammonium (bi) sulfates started to be desorbed and decomposed (also by reaction with NO [20]), so that the overall catalytic performance was almost restored.

In addition, alkaline earth metals are reported to deactivate SCR catalysts, although to a lesser extent compared to alkaline metals [13]. In particular, Kröcher and Elesener [13] reported that calcium, as $CaSO_4$, has a poor deactivation effect but if calcium is not or only partly sulfated it has a stronger deactivation potential. They found that calcium does not form vanadates, like potassium, but can deposit on vanadia sites or also on titania or tungsta sites, reducing the ammonia surface coverage acting as a buffer for active sites.

Both types of regeneration treatments mostly removed ammonium sulfates (see spectra in Figure 8). Washing in water caused limited leaching of soluble vanadia species from the catalyst and removed most of its sulfur content. Those effects did not significantly limit the recovery of activity with respect to the virgin catalyst. On the contrary, the residual presence of calcium compounds (sulfates and carbonates) probably limited the effectiveness of the thermal regeneration treatment. In particular, a comparison between the apparent activation energy values for the fresh, aged, and regenerated catalysts suggests that ammonium (bi)sulfate reduced the number of available active catalytic sites for the SCR reaction, whereas Ca compounds induced an additional but limited poisoning effect at low temperature.

3. Materials and Methods

3.1. SCR Catalyst

The industrial V_2O_5-WO_3/TiO_2 SCR catalyst studied in this work was obtained in its virgin form and after a total of 18,000 h of continuous operation for the purification of the exhaust gases of a municipal solid waste to energy plant located in Italy, which adopts a tail-end configuration for the DeNOx unit operating at low temperature [15] after the $DeSO_x$ and Electrostatic precipitation units.

Catalytic elements consisted of extruded honeycomb monoliths (15 cm × 15 cm × 105 cm) with parallel channels with square section. Samples were cut out (length = 3 cm) from the front section (gas inlet) of the original monoliths and, after being eventually ground in an agate mortar, they were subjected to the characterization analyses and catalytic tests.

3.2. Regeneration Procedures

Similar samples (ca. 1 cm × 2 cm × 3 cm) taken from the inlet section of the reaction-aged monolith catalyst were regenerated according to two different procedures: (i) thermal treatment in static air at

400 °C or 450 °C for 2 h; (ii) washing in distilled water (7.5 g of catalyst per cm^3 of water) at room temperature for 30 min (first 15 min in ultra-sound bath) followed by drying in air at 120 °C.

3.3. *Physical and Chemical Characterization*

The chemical analysis of the catalysts was performed by X-ray fluorescence (XRF) using a Bruker M4 Tornado instrument (Billerica, MA, USA).

Scanning Electron Microscopy (SEM) was performed with a Phenomworld instrument (Eindhoven, The Netherland), equipped with energy dispersion microprobe (EDS) for chemical analysis.

The textural properties were determined by N_2 adsorption at 77 K from p/p_0 10^{-5} using a Quantachrome Autosorb 1-C analyzer (Quantachrome Instruments, Boynton Beach, FL, USA) using the BET method for the calculation of the specific surface area, and the Barrett–Joyner–Halenda (BJH) method applied to the desorption branch for pore size distribution (PSD).

Powder X-ray diffraction (XRD) analysis was performed with a Rigaku Miniflex 600 instrument (Tokyo, Japan) operated with the following acquisition parameters: radiation $Cu_{k\alpha}\lambda$ = 1.54060 nm; range 2θ 3–90°; step 0.02°; time per step 0.1 s.

FT-IR analysis was performed on disks of KBr mixed with 2% wt. of the powdered catalyst sample using a Perkin Elmer Spectrum GX spectrometer (Waltham, MA, USA) equipped with a liquid-N_2 cooled MCT detector, with a spectral resolution of 4 cm^{-1}, averaging each spectrum over 64 scans. All spectra for fresh, aged and regenerated catalysts were ratioed against the spectrum of pure TiO_2 anatase.

Thermo-gravimetric analysis coupled with mass-spectometry (TG-MS) was performed in a Setaram Labsys Evo TGA-DTA-DSC 1600 (Setaram Instrumentation, Caluire, France) flow microbalance connected via a heated capillary to a Pfeiffer Thermostar G mass spectrometer for the analysis of the evolved gases.

Temperature programmed desorption (TPD) of NH_3 or SO_2 was carried out in the same experimental rig used for catalytic activity tests (see below). For each test 125 mg of powdered catalyst was used. In the NH_3-TPD of the fresh catalyst, the sample was initially saturated at 100 °C with a flow of NH_3/He mixture (530 ppmv); thereafter, NH_3 was desorbed by heating at 10 °C/min up to 650 °C under He flow (20 Std dm^3/h). For the NH_3-TPD analysis of the reaction-aged catalyst, ammonia pre-saturation was not performed.

In the case of SO_2-TPD, samples of the virgin or aged catalyst were heated at 10 °C/min up to 1000 °C under N_2 flow while monitoring the SO_2 released due to the decomposition of sulfates present on the catalyst with an UV continuous analyzer (ABB Advanced Optima Limas 11 UV, Zürich, Switzerland).

The concentration of ionic species released from fresh, aged, and regenerated catalyst samples during washing in the aqueous phase was evaluated by means of an 883 Basic IC Plus ionic chromatograph (Metrohm) and an Agilent 7500 ICP-MS. To this aim ca. 50 mg of each catalyst sample were contacted at room temperature with 10 mL of distilled water.

3.4. *Selective Catalytic Reduction of NO*

The catalytic tests were carried out in a lab-scale fixed bed tubular quartz reactor (10 mm internal diameter) placed in a tubular furnace and operated in the temperature range 90–550 °C as described in [14,29], using 0.125 mg of powdered catalyst (125–200 μm particle diameter).

High-purity gas streams (NO in He, NH_3 in He, O_2, He) from cylinders were regulated by four independent mass flow controllers (BROOKS MFC SLA5850S) and mixed at nearly atmospheric pressure before entering the reactor at a total flow rate of 25 Std dm^3/h, corresponding to a gas hourly space velocity of 100,000 h^{-1}. The feed composition was NO = NH_3= 400 ppmv (NO_2 impurity < 4 ppmv), 6% O_2 balanced with He.

Gas analysis was performed by two continuous analyzers with independent detectors respectively for (i) NH_3 (Tunable Diode Laser Spectroscopy, GAS 3000R Geit-Europe); (ii) NO, N_2O (ND-IR) and

NO_2 (UV) (Emerson X-Stream XEGP). A Sycapent™ (P_2O_5) trap was used to remove water and NH_3 from the gas stream before entering the NOx analyzer. N_2 concentration in the products was estimated by N-balance (excluding the eventual formation of other N-bearing species apart from those measured). NH_3 and NO conversions and N_2 selectivity were calculated as defined in [14].

4. Conclusions

The deactivation of a commercial V_2O_5-WO_3/TiO_2 honeycomb catalyst operated 18,000 h in a tail-end selective catalytic reduction De-NOx unit treating the exhaust gases from a MW incinerator in Southern Italy was investigated as a case study. The SCR catalyst generally operated at around 200 °C with a relatively clean inlet feed after a desulphurization unit with CaO/CaOH and fabric filters. Nevertheless, catalyst deactivation was observed to cause as much as a 50% reduction in the rate of reaction evaluated at 200 °C (worst scenario).

Results of the characterization of fresh and spent catalyst suggested at least two main causes of deactivation: The formation of ammonium sulfate and/or bisulfate and the deposition of calcium sulfate. The former species slowly formed through the reaction between adjacent adsorbed NH_3 and sulfate species; the latter was related to the entrainment of small $CaSO_4$ particles escaping the filters located upstream the SCR unit.

The effect of ammonium (bi) sulfate was the reduction of the number of available active catalytic sites for the SCR reaction, whereas Ca compounds induced an additional but limited poisoning effect at low temperature, inducing a small increase in the apparent activation energy of the SCR reaction of NO.

Furthermore, laboratory catalyst regeneration studies were carried out comparing the effectiveness to restore DeNOx-activity of thermal treatment in air at 400/450 °C or washing in water at room temperature. Both regeneration treatments were able to remove ammonium (bi) sulfate from the aged catalyst, which was identified as the main cause of deactivation. A residual presence of calcium compounds (sulfates/carbonates) was deduced only in the case of thermally regenerated samples. As a consequence, by washing the aged catalyst in water it was possible to almost completely recover its original SCR activity at low temperature, in spite of a negligible loss of active vanadium species due to solubilization. On the other hand, the thermally regenerated catalyst recovered 87% of the original rate of SCR at 200 °C, although the corresponding value of the apparent activation energy remained close to that measured for the aged catalyst, i.e., 10–15% higher than for the virgin catalyst.

Author Contributions: Data curation, S.C., C.F. and L.L.; investigation, S.C., C.F., and L.L.; supervision, R.C., G.P., S.C. and L.L.; writing – original draft, S.C. and L.L.; writing – review and editing, S.C.

Funding: This research received no external funding

Acknowledgments: Ing. S. Malvezzi (A2A) is kindly acknowledged for providing catalyst samples and valuable support throughout the project.

Conflicts of Interest: The authors declare no conflict of interest.

References

1. Zhang, J.; Li, X.; Chen, P.; Zhu, B. Research Status and Prospect on Vanadium-Based Catalysts for NH_3-SCR Denitration. *Materials* **2018**, *11*, 1632. [CrossRef] [PubMed]
2. Mladenović, M.; Paprika, M.; Marinković, A. Denitrification techniques for biomass combustion. *Renew. Sustain. Energy Rev.* **2018**, *82*, 3350–3364. [CrossRef]
3. Kuo, J.-H.; Lin, C.-L.; Chen, J.-C.; Tseng, H.-H.; Wey, M.-Y. Emission of carbon dioxide in municipal solid waste incineration in Taiwan: A comparison with thermal power plants. *Int. J. Greenh. Gas Control* **2011**, *5*, 889–898. [CrossRef]
4. Yan, Q.; Yang, R.; Zhang, Y.; Umar, A.; Huang, Z.; Wang, Q. A Comprehensive Review on Selective Catalytic Reduction Catalysts for NOx Emission Abatement from Municipal Solid Waste Incinerators. *Environ. Prog. Sustain. Energy* **2016**, *35*, 1061–1069. [CrossRef]

5. Argyle, M.; Bartholomew, C. Heterogeneous Catalyst Deactivation and Regeneration: A Review. *Catalysts* **2015**, *5*, 145–269. [CrossRef]
6. Brandin JG, M.; Odenbrand, C.U.I. Deactivation and Characterization of SCR Catalysts Used in Municipal Waste Incineration Applications. *Catal. Lett.* **2018**, *148*, 312–327. [CrossRef]
7. Brandin JG, M.; Odenbrand, C.U.I. Poisoning of SCR Catalysts used in Municipal Waste Incineration Applications. *Top. Catal.* **2017**, *60*, 1306–1316. [CrossRef]
8. Lisi, L.; Lasorella, G.; Malloggi, S.; Russo, G. Single and combined deactivating effect of alkali metals and HCl on commercial SCR catalysts. *Appl. Catal. B Environ.* **2004**, *50*, 251–258. [CrossRef]
9. Nicosia, D.; Czekaj, I.; Kröcher, O. Chemical deactivation of V_2O_5/WO_3–TiO_2 SCR catalysts by additives and impurities from fuels, lubrication oils and urea solution Part II. *Characterization study of the effect of alkali and alkaline earth metals Appl. Catal. B Environ.* **2008**, *77*, 228–236.
10. Nicosia, D.; Elsener, M.; Kröcher, O.; Jansohn, P. Basic investigation of the chemical deactivation of V_2O_5/WO_3-TiO_2 SCR catalysts by potassium, calcium, and phosphate. *Top. Catal.* **2007**, *42–43*, 333–336. [CrossRef]
11. Li, C.; Shen, M.; Wang, J.; Wang, J.; Zhai, Y. New Insights into the Role of WO_3 in Improved Activity and Ammonium Bisulfate Resistance for NO Reduction with NH_3 over V−W/Ce/Ti Catalyst. *Ind. Eng. Chem. Res.* **2018**, *57*, 8424–8435. [CrossRef]
12. Odenbrand, C.U.I. $CaSO_4$ deactivated V_2O_5-WO_3/TiO_2 SCR catalyst for a diesel power plant. Characterization and simulation of the kinetics of the SCR reactions. *Appl. Catal. B Environ.* **2018**, *234*, 365–377. [CrossRef]
13. Kröcher, O.; Elesener, M. Chemical deactivation of V_2O_5/WO_3–TiO_2 SCR catalysts by additives and impurities from fuels, lubrication oils, and urea solution I. Catalytic studies. *Appl. Catal. B Environ.* **2008**, *75*, 215–227. [CrossRef]
14. Cimino, S.; Totarella, G.; Tortorelli, M.; Lisi, L. Combined poisoning effect of K^+ and its counter-ion (Cl^- or NO_3^-) on MnO_x /TiO_2 catalyst during the low temperature NH_3-SCR of NO. *Chem. Eng. J.* **2017**, *330*, 92–101. [CrossRef]
15. a2a Impianti di Termovalorizzazione. Available online: https://www.a2a.eu/it/gruppo/i-nostri-impianti/termovalorizzatori/acerra (accessed on 16 May 2019).
16. Forzatti, P. Present status and perspectives in de-NO*x* SCR catalysis. *Appl. Catal. A Gen.* **2001**, *222*, 221–236. [CrossRef]
17. Guo, X.; Bartholomew, C.; Hecker, W.; Baxter, L.L. Effects of sulfate species on V_2O_5/TiO_2 SCR catalysts in coal and biomass-fired systems. *Appl. Catal. B Environ.* **2009**, *92*, 30–40. [CrossRef]
18. Guo, K.; Fan, G.; Gu, D.; Yu, S.; Ma, K.; Liu, A.; Tan, W.; Wang, J.; Du, X.; Zou, W.; et al. Pore Size Expansion Accelerates Ammonium Bisulfate Decomposition for Improved Sulfur Resistance in Low-Temperature NH_3-SCR. *ACS Appl. Mater. Interf.* **2019**, *11*, 4900–4907. [CrossRef] [PubMed]
19. Wu, G.; Li, J.; Fang, Z.; Lan, L.; Wang, R.; Gong, M.; Chen, Y. $FeVO_4$ nanorods supported TiO_2 as a superior catalyst for NH3–SCR reaction in a broad temperature range. *Catal. Commun.* **2015**, *64*, 75–79. [CrossRef]
20. Li, C.; Shen, M.; Yu, T.; Wang, J.; Wang, J.; Zhai, Y. The mechanism of ammonium bisulfate formation and decomposition over V/WTi catalysts for NH_3-selective catalytic reduction at various temperatures. *Phys. Chem. Chem. Phys.* **2017**, *19*, 15194–15206. [CrossRef]
21. Zhu, Y.; Hou, Q.; Shreka, M.; Yuan, L.; Zhou, S.; Feng, Y.; Xia, C. Ammonium-Salt Formation and Catalyst Deactivation in the SCR System for a Marine Diesel Engine. *Catalysts* **2018**, *9*, 21. [CrossRef]
22. Zhang, Y.S.; Li, C.; Wang, C.; Yu, J.; Xu, G.; Zhang, Z.G.; Yang, Y. Pilot-Scale Test of a V_2O_5-WO_3/TiO_2-Coated Type of Honeycomb DeNOx Catalyst and Its Deactivation Mechanism. *Ind. Eng. Chem. Res.* **2019**, *58*, 828–835. [CrossRef]
23. Saur, O. The structure and stability of sulfated alumina and titania. *J. Catal.* **1986**, *99*, 104–110. [CrossRef]
24. Kanari, N.; Menad, N.-E.; Ostrosi, E.; Shallari, S.; Diot, F.; Allain, E.; Yvon, J. Thermal Behavior of Hydrated Iron Sulfate in Various Atmospheres. *Metals* **2018**, *8*, 1084. [CrossRef]
25. Partridge, E.P.; White, A.H. The Solubility of Calcium Sulfate From 0 to 200°. *J. Am. Chem. Soc.* **1929**, *51*, 360–370. [CrossRef]
26. Gan, L.; Guo, F.; Yu, J.; Xu, G. Improved Low-Temperature Activity of V_2O_5-WO_3/TiO_2 for Denitration Using Different Vanadium Precursors. *Catalysts* **2016**, *6*, 25. [CrossRef]

27. Ye, D.; Qu, R.; Song, H.; Zheng, C.; Gao, X.; Luo, Z. Investigation of the promotion effect of WO3 on the decomposition and reactivity of NH_4HSO_4 with NO on V_2O_5–WO_3/TiO_2 SCR catalysts. *RSC Adv.* **2016**, *6*, 55584–55592. [CrossRef]
28. Dumesic, J.A.; Topsoe, N.-Y.; Topsoe, H.; Slabiak, T. Kinetics of Selective Catalytic Reduction of Nitric Oxide by Ammonia over Vanadia/Titania. *J. Catal.* **1996**, *163*, 409–417. [CrossRef]
29. Gargiulo, N.; Caputo, D.; Totarella, G.; Lisi, L.; Cimino, S. Me-ZSM-5 monolith foams for the NH_3 -SCR of NO. *Catal. Today* **2018**, *304*, 112–118. [CrossRef]

Article

Decomposition of Al_2O_3-Supported $PdSO_4$ and $Al_2(SO_4)_3$ in the Regeneration of Methane Combustion Catalyst: A Model Catalyst Study

Niko M. Kinnunen [1,*], Ville H. Nissinen [1], Janne T. Hirvi [1], Kauko Kallinen [2], Teuvo Maunula [2], Matthew Keenan [3] and Mika Suvanto [1]

1 Department of Chemistry, University of Eastern Finland, P.O. Box 111, FI-80101 Joensuu, Finland; vnissine@uef.fi (V.H.N.); janne.hirvi@uef.fi (J.T.H.); mika.suvanto@uef.fi (M.S.)
2 Dinex Finland Oy, Global Catalyst Competence Centre, P.O. Box 20, FI-41331 Vihtavuori, Finland; kki@dinex.fi (K.K.); tma@dinex.fi (T.M.)
3 Shoreham Technical Centre, Ricardo UK Ltd., Shoreham-by-Sea, West Sussex BN43 5FG, UK; matthew.keenan@ricardo.com
* Correspondence: niko.kinnunen@uef.fi; Tel.: +358-5034-240-84

Received: 22 March 2019; Accepted: 3 May 2019; Published: 8 May 2019

Abstract: Exhaust gas aftertreatment systems play a key role in controlling transportation greenhouse gas emissions. Modern aftertreatment systems, often based on Pd metal supported on aluminum oxide, provide high catalytic activity but are vulnerable to sulfur poisoning due to formation of inactive sulfate species. This paper focuses on regeneration of Pd-based catalyst via the decomposition of alumina-supported aluminum and palladium sulfates existing both individually and in combination. Decomposition experiments were carried out under hydrogen (10% H_2/Ar), helium (He), low oxygen (0.1% O_2/He), and excess oxygen (10% O_2/He). The structure and composition of the model catalysts were examined before and after the decomposition reactions via powder X-ray diffraction and elemental sulfur analysis. The study revealed that individual alumina-supported aluminum sulfate decomposed at a higher temperature compared to individual alumina-supported palladium sulfate. The simultaneous presence of aluminum and palladium sulfates on the alumina support decreased their decomposition temperatures and led to a higher amount of metallic palladium than in the corresponding case of individual supported palladium sulfate. From a fundamental point of view, the lowest decomposition temperature was achieved in the presence of hydrogen gas, which is the optimal decomposition atmosphere among the studied conditions. In summary, aluminum sulfate has a two-fold role in the regeneration of a catalyst—it decreases the Pd sulfate decomposition temperature and hinders re-oxidation of less-active metallic palladium to active palladium oxide.

Keywords: sulfur deactivation; catalyst deactivation; aluminum sulfate; palladium sulfate; regeneration

1. Introduction

Exhaust gas aftertreatment systems (ATSs) have been used since the 1950s to remove at least a portion of the exhaust gases emitted from internal combustion engines of vehicles [1]. Nowadays, tight emission limits and durability requirements guide vehicle developers to implement regeneration procedures into the vehicle operation.

Exhaust gas ATS catalysts consist of a support material, active metal(s), and promoter(s). Alumina as a support material for exhaust gas ATS catalysts has been known to be the best alternative owing to its good thermal stability and high surface area [2,3]. Meanwhile, palladium is currently used as an active metal or as a promoter in almost all catalysts in vehicles exhaust gas ATSs, except in selective catalytic reduction (SCR) systems, which are commonly based on Fe- and Cu-promoted zeolites [4]. Diesel oxidation catalysts (DOC) are palladium-promoted and platinum-rich in composition. A three-way catalyst (TWC) is palladium-rich and promoted with a small amount of rhodium, whereas methane oxidation catalysts (MOC) are palladium-rich and promoted with a small amount of platinum [1,5]. The reason for the popularity of palladium relies on its good thermal stability compared to platinum [6], but sometimes it is also mandatory due to activity such as in methane combustion catalysis [3,7].

One deactivation mechanism of exhaust gas ATS catalysts is sulfur poisoning. Even a low amount of sulfur deactivates the catalysts and, thus, increases vehicle-emitted exhaust gas emissions. Sulfur originates from fuels and lubricant oils. Standards in Europe have been regulated in such a way that sulfur level in fuels can be 10 ppm at the highest [8]. The sulfur content in natural gas and its biological counterpart, bio-gas, is even less than 1 ppm. However, in combustion, sulfur compounds form SO_2, which further oxidizes to SO_x (x = 3 or 4) over a catalyst. The SO_x accumulates on the catalyst, forming stable and less active sulfates, such as aluminum sulfate [9,10] and palladium sulfate [3,11–14], during long-time operation.

Alumina sulfate formation in various conditions and catalyst systems have been previously studied owing to its role in many catalytic reactions and processes [15,16]. However, there has been no systematic research about the stability of Al_2O_3-supported aluminum sulfate or its role in the decomposition of palladium sulfate. In addition, the literature concerning Al_2O_3-supported palladium sulfate or sulfur-poisoned methane combustion catalyst is limited to regeneration studies under different gas atmospheres and varying temperatures [13,17–20]. No systematic and quantitative knowledge about the stability of the supported sulfates and catalyst structure after decomposition is available.

The study focuses on the stability and structure of individual and combined Al_2O_3-supported aluminum and palladium sulfates. The stabilities of the samples will be measured under reductive conditions (H_2), in the absence of oxygen (He), low-oxygen atmosphere, and high-oxygen (lean) conditions. Sulfur content and structure of the catalysts are compared before and after decomposition reactions.

2. Results and Discussion

Regeneration of $PdSO_4/Al_2O_3$ methane combustion model catalyst was studied under both dry and wet synthetic exhaust gas. Figure 1 shows CH_4 conversion and SO_2 concentration during the regeneration procedure with and without water vapor in the exhaust gas stream. In both cases, a decrease in CH_4 conversion can be detected during the regeneration procedure, due to lack of O_2 in the reaction gas feed. However, SO_2 release and, thus, decomposition of sulfate can be detected only if water vapor is present in the exhaust gas stream. This is due to steam reforming and water–gas shift reactions [21,22], which provides H_2 for the sulfate to decompose. Without water vapor in the exhaust gas stream, CH_4 is the only reducing agent, and it is not as effective as H_2. Decay in methane conversion after regeneration in the presence of water vapor is due to the loss of oxygen from the active PdO phase [23].

Figure 2 explains in greater detail the regeneration behavior of the methane combustion catalyst. It shows a detailed step-wise illustration of the reactions that may occur during the steady-state regeneration of the methane combustion catalyst, and it thus justifies the method that can be used in the fundamental study below. Point 1 to 2: Steady-state methane conversion under lean-burn conditions over an Al_2O_3-supported $PdSO_4$ (PS) catalyst. Point 2 to 3: Oxygen from the exhaust gas is compensated with N_2, initiating steam reforming and possibly water–gas shift reactions, which provides H_2 gas for low-temperature sulfate decomposition [21,22]. The activity of the catalyst against steam reforming and possibly water–gas shift reactions decays faster under steady-state rich conditions, which can be seen in Figure 2 as a decrease in CH_4 conversion. SO_2 also reaches its maximum at Point 3. A decrease in non-sulfur-poisoned methane conversion under rich conditions was observed earlier by other researchers [18,21,24] and it was concluded to be reversible if O_2 was introduced periodically after rich conditions were attained. Temperature-programmed reduction (TPR) is a feasible technique to provide more detailed information about the decomposition of sulfates and the catalyst structure at this point. Due to the low rate of the steam reforming reaction, which can be observed as low CH_4 conversion, H_2 regeneration almost stops and the SO_2 concentration begins to decrease (Point 3 to 4). Temperature-programmed desorption (TPD) and oxidation (TPO) with low oxygen are relevant techniques to enhance the fundamental understanding of the catalyst decomposition and structure under these conditions. Lean operation conditions are restored at Point (4), which re-oxidizes and recovers the active form of the catalyst at Point 5, as discussed in our previous work [25].

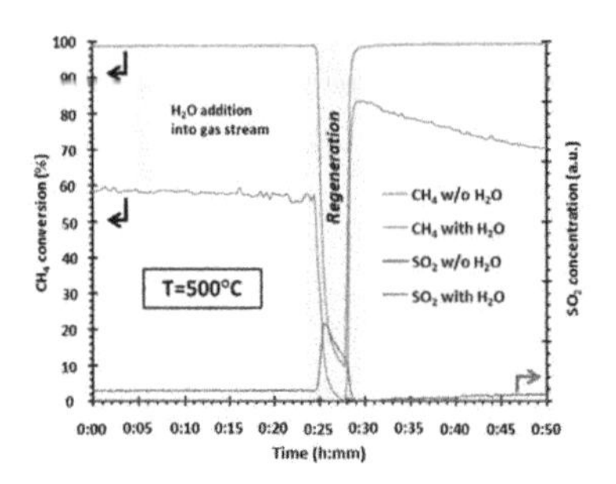

Figure 1. Effect of water vapor in the exhaust gas on regeneration of a model $PdSO_4/Al_2O_3$ methane combustion catalyst.

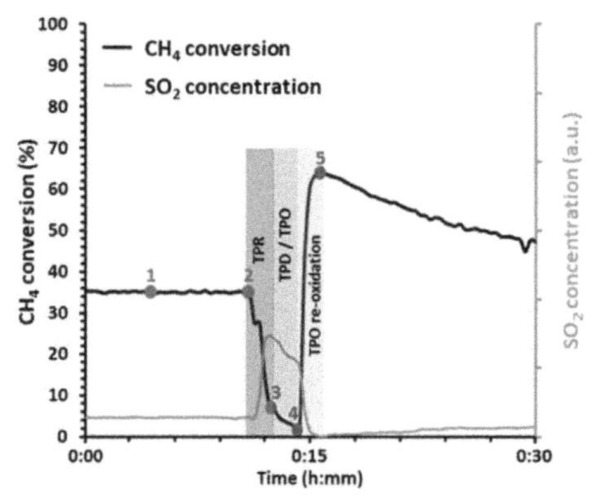

Figure 2. Detailed step-wise description of the lean-burn CH_4 combustion catalyst regeneration procedure. (TPR = Temperature Programmed Reduction, TPD = Temperature Programmed Desorption, and TPO = Temperature Programmed Oxidation).

2.1. *Thermal Decomposition of Al_2O_3-Supported $PdSO_4$ and $Al_2(SO_4)_3$ under Hydrogen Gas*

Temperature-programmed reduction (TPR), desorption (TPD), and oxidation (TPO) were used to clarify the decomposition of sulfur compounds. The TPR experiments were carried out under blended gas of 10% H_2 and Ar. The thermal decompositions of Al_2O_3-supported $PdSO_4$ (PS) and $Al_2(SO_4)_3$ (AS) model catalysts individually and in combination under a hydrogen gas mixture are represented in Figure 3. Al_2O_3-supported $PdSO_4$ decomposes at a lower temperature compared to individual supported $Al_2(SO_4)_3$. Hydrogen consumption started around 250 °C, but no sulfur release was observed with a quadrupole mass spectrometer detector before the temperature range of 400–500 °C. The result is well in line with previous observations [26]. The decomposition of sulfate combinations began, however, at about a temperature of 100 °C lower than that for the individual one, and the amount of $Al_2(SO_4)_3$ in the catalyst affected this directly—the higher the $Al_2(SO_4)_3$ content, the lower the decomposition temperature. This may indicate that the presence of $Al_2(SO_4)_3$ de-stabilizes $PdSO_4$. One explanation could be the exothermic decomposition reaction of sulfates that promotes itself when the reaction proceeds. Moreover, the high overall sulfate content of the catalyst may decrease the support material effect and thus make the sulfates more like a bulk sulfate, which decomposes at a lower temperature than do support-material-stabilized sulfates. The elemental analysis results of the samples, tabulated in Table 1, showed that the sulfur content of each TPR-treated catalyst was 0.30 wt.% if $PdSO_4$ was included in the catalyst. The residual sulfur content corresponds to the stoichiometric sulfur amount in the Pd_4S structure. Powder X-ray diffraction (PXRD) measurements were recorded to confirm the Pd_4S structure (Figure 4). Based on the PXRD results, if $Al_2(SO_4)_3$ is present in the catalyst, more crystalline Pd_4S is formed during TPR treatment. The result gives an indication that Pd_4S may be one factor that is affecting catalyst regeneration.

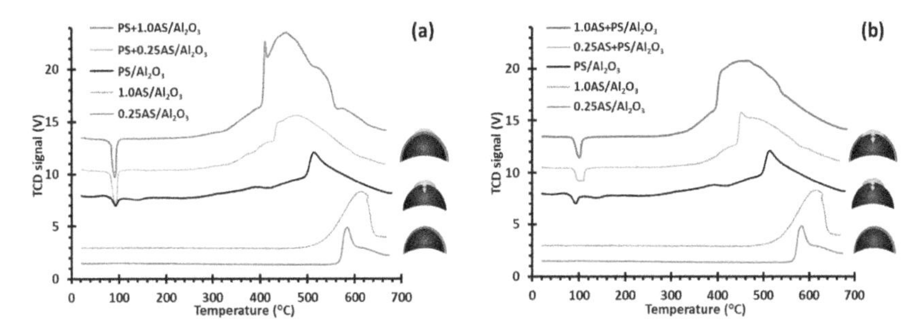

Figure 3. Thermal decomposition under hydrogen gas of (**a**) alumina-supported combined sulfates $PdSO_4$ (PS) + 0.25 $Al_2(SO_4)_3$ (AS)/Al_2O_3 and PS + 1.0 AS/Al_2O_3, and (**b**) alumina-supported combined sulfates 0.25 AS + PS/Al_2O_3 and 1.0 AS + PS/Al_2O_3 together with individual alumina-supported sulfates PS/Al_2O_3, 0.25 AS/Al_2O_3 and 1.0 AS/Al_2O_3. Experiments were carried out under 10% H_2 in an argon gas atmosphere.

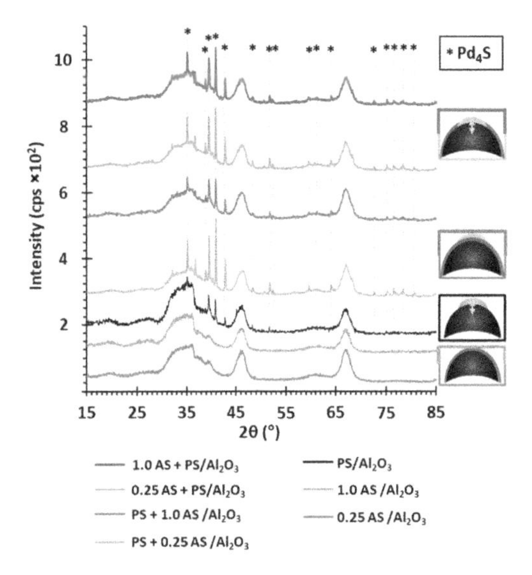

Figure 4. Powder X-ray diffraction results of alumina-supported sulfates together with characteristic peaks of Pd_4S (*) after temperature-programmed reduction treatment under hydrogen gas.

Table 1. Designation of the catalysts, sulfur contents, and PdO and Pd peak areas and crystallite sizes.

Catalyst	Treatment	Sulfur Content (wt.-%) [1]	Relative Sulfur Content (%) [2]	PdO Peak Area	PdO Crystallite Size (nm)	Pd Peak Area	Pd Crystallite Size (nm)
PS/Al_2O_3 [3]	**Prepared**	**0.88**	**100**	-	-	-	-
	10% O_2/He	0.41	47	148.1	12.1	71.5	12.3
	0.1% O_2/He	0.11	13	121.3	5.7	118.8	14.4
	He TPD	0.06	7	-	-	143.0	20.7
	TPR	0.30	34	-	-	-	-
0.25 AS/Al_2O_3 [4]	**Prepared**	**0.29**	**100**	-	-	-	-
	10% O_2/He	0.23	79	-	-	-	-
	0.1% O_2/He	0.11	38	-	-	-	-
	He TPD	0.08	28	-	-	-	-
	TPR	0.05	16	-	-	-	-
1.0 AS/Al_2O_3 [4]	**Prepared**	**0.97**	**100**	-	-	-	-
	10% O_2/He	0.48	49	-	-	-	-
	0.1% O_2/He	0.30	31	-	-	-	-
	He TPD	0.23	24	-	-	-	-
	TPR	0.07	7	-	-	-	-
PS + 0.25 AS/Al_2O_3 [3,4]	**Prepared**	**1.39**	**100**	-	-	-	-
	10% O_2/He	0.24	17	40.7	34.8	75.2	96.0
	0.1% O_2/He	0.05	4	18.3	20.1	114.7	110.4
	He TPD	0.03	2	-	-	121.6	122.8
	TPR	0.30	22	-	-	-	-
PS + 1.0 AS/Al_2O_3 [3,4]	**Prepared**	**2.10**	**100**	-	-	-	-
	10% O_2/He	0.35	17	56.7	20.5	25.8	74.7
	0.1% O_2/He	0.09	4	24.5	17.7	64.7	66.5
	He TPD	0.07	3	-	-	63.2	48.0
	TPR	0.30	14	-	-	-	-
0.25 AS + PS/Al_2O_3 [3,4]	**Prepared**	**1.03**	**100**	-	-	-	-
	10% O_2/He	0.30	29	45.1	33.1	56.1	94.0
	0.1% O_2/He	0.10	10	16.0	23.4	85.6	101.2
	He TPD	0.07	6	-	-	90.9	111.5
	TPR	0.30	29	-	-	-	-
1.0 AS + PS/Al_2O_3 [3,4]	**Prepared**	**1.71**	**100**	-	-	-	-
	10% O_2/He	0.40	23	53.2	23.0	49.5	73.5
	0.1% O_2/He	0.12	7	7.7	38.8	71.5	85.7
	He TPD	0.07	4	-	-	67.6	109.8
	TPR	0.30	18	-	-	-	-

[1] Sulfur content of the sample before treatment (prepared) or after treatment (10% O_2/He, 0.1% O_2/He, He TPD, TPR). [2] Relative sulfur contents were calculated (sulfur content of a treated sample/sulfur content of a prepared sample) *100%. [3] Palladium loading of the catalyst is 4% in a metallic state. [4] 0.25 and 1.0 refer to the percentage sulfur content of $Al_2(SO_4)_3/Al_2O_3$.

2.2. Thermal Decomposition of Al_2O_3-Supported $PdSO_4$ and $Al_2(SO_4)_3$

The thermal decomposition results for the PS/Al_2O_3 model catalyst in the absence of oxygen (He TPD), under low oxygen concentration (0.1% O_2/He), and in excess oxygen (10% O_2/He) are shown in Figure 5a. Two separate decomposition steps were observed in each case, as concluded in our previous study [27]. However, the first step of the $PdSO_4$ decomposition reaction was observed to decay as a function of the oxygen concentration. The analysis relies on the mass signals of O_2 and SO_2 recorded during the thermal decomposition. It has also been suggested that in the absence of oxygen, sulfate species could decompose via a one-step mechanism [26]. The decomposition of Al_2O_3-supported $PdSO_4$ was initiated at 600 °C in He gas, whereas the corresponding temperature under 10% O_2/He was 800 °C. The oxygen concentration also affected the quantitative sulfur content—the lower oxygen concentration corresponded to better sulfur removal (Table 1) due to the longer time period at temperatures required for the decomposition. FTIR spectra of the PS/Al_2O_3 model catalyst recorded after different treatments (Figure 5b) supported the result as the relative intensity of the asymmetric stretching vibration of sulfate decays as the oxygen concentration of the gas mixture decreases.

The formation of $Al_2(SO_4)_3$ over an Al_2O_3-supported methane oxidation catalyst has been reported [9,10], and it is thus crucial to know its decomposition temperature relative to that of $PdSO_4$

and, further, its role in the decomposition of $PdSO_4$. The decomposition experiments for the bulk $Al_2(SO_4)_3$ and AS/Al_2O_3 model catalysts were carried out as in the case of the PS/Al_2O_3 model catalyst under three different gas atmospheres. The unsupported bulk $Al_2(SO_4)_3$ decomposed completely between 600 °C and 800 °C [28]. The thermal decomposition of Al_2O_3-supported $Al_2(SO_4)_3$ under different gas atmospheres is shown in Figure 6a,b. The decomposition of Al_2O_3-supported $Al_2(SO_4)_3$ took place at a higher temperature than did that of bulk $Al_2(SO_4)_3$ or Al_2O_3-supported $PdSO_4$. Overall, the trends in the decomposition temperatures of the bulk and supported $Al_2(SO_4)_3$ showed strong support interaction. The decrease in the oxygen concentration of the feed gas resulted in higher removal of sulfur during the decomposition of supported $Al_2(SO_4)_3$ based on FTIR data (Figure 6c,d) and quantitative sulfur analysis (Table 1). However, the removal of sulfur was evidently less than that in the case of the PS/Al_2O_3 model catalyst. It can be concluded that individual and Al_2O_3-supported $Al_2(SO_1)_3$ is more stable than $PdSO_4$, and the decomposition order of the supported sulfates is the same as that observed already in TPR, despite the gas atmosphere, even though the decomposition temperature under a hydrogen gas mixture is lower. Moreover, it is worth noting that the decomposition route of AS/Al_2O_3 was different from that of PS/Al_2O_3, as no O_2 peak was observed before simultaneous release of O_2 and SO_2.

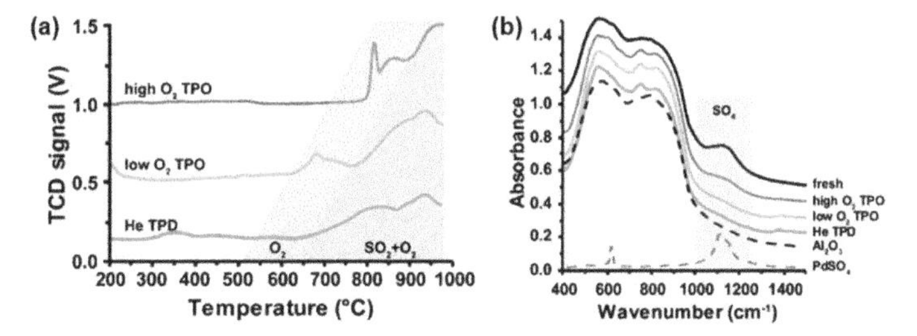

Figure 5. (**a**) Thermal decomposition curves of the PS/Al_2O_3 model catalyst and (**b**) FTIR spectra of the PS/Al_2O_3 model catalyst before and after decomposition, accompanied by reference spectra of bulk $PdSO_4$ and Al_2O_3.

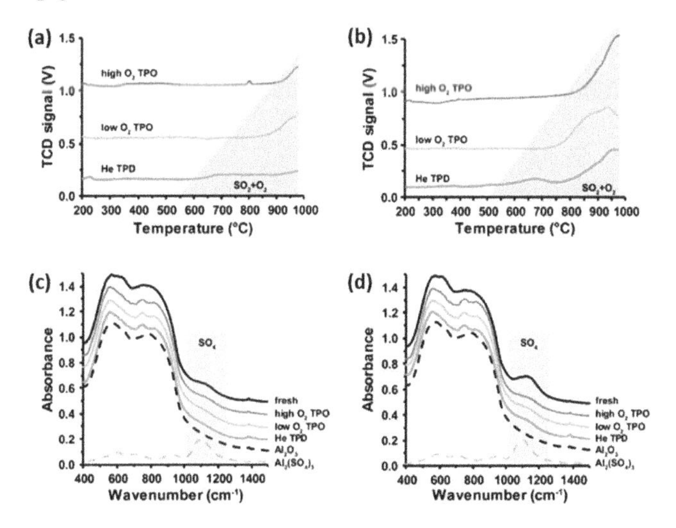

Figure 6. Thermal decomposition curves of (**a**) 0.25 AS/Al_2O_3 and (**b**) 1.0 AS/Al_2O_3 model catalysts together with FTIR spectra of (**c**) 0.25 AS/Al_2O_3 and (**d**) 1.0 AS/Al_2O_3 model catalysts before and after decomposition, accompanied by reference spectra of bulk $Al_2(SO_4)_3$ and Al_2O_3.

2.3. Thermal Decomposition of Al_2O_3-Supported $PdSO_4$ and $Al_2(SO_4)_3$ Combinations

The co-existence of $PdSO_4$ and $Al_2(SO_4)_3$ species on sulfur-poisoned lean-burn methane oxidation catalyst is conceivable [2,9,10,12]. Thus, the effect of $Al_2(SO_4)_3$ on the decomposition of $PdSO_4$ was examined under three dry gas atmospheres (Figures 7 and 8). The stability of sulfate species decreased as a function of the decreasing oxygen concentration in the feed gas in the same way as described above for the PS/Al_2O_3 and AS/Al_2O_3 model catalysts. The decomposition began at a temperature 100 °C lower than that for PS/Al_2O_3, showing that $Al_2(SO_4)_3$ de-stabilized the $PdSO_4$ phase, promoting its decomposition. The relative sulfur release of the PS + X AS/Al_2O_3 model catalysts (X = 0.25 or 1.0) was greater than that together for the PS/Al_2O_3 and X AS/Al_2O_3 (X = 0.25 or 1.0) model catalysts (Table 1). This also indicated the favorable effect of $PdSO_4$ on the decomposition of $Al_2(SO_4)_3$.

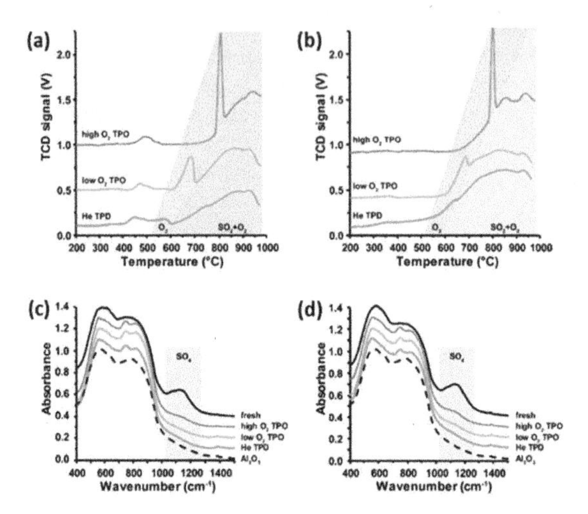

Figure 7. Thermal decomposition curves of (**a**) PS + 0.25 AS/Al_2O_3 and (**b**) PS + 1.0 AS/Al_2O_3 model catalysts and (**c**) FTIR spectra of PS + 0.25 AS/Al_2O_3 and (**d**) PS + 1.0 AS/Al_2O_3 model catalysts before and after decomposition, accompanied by a reference spectrum of Al_2O_3.

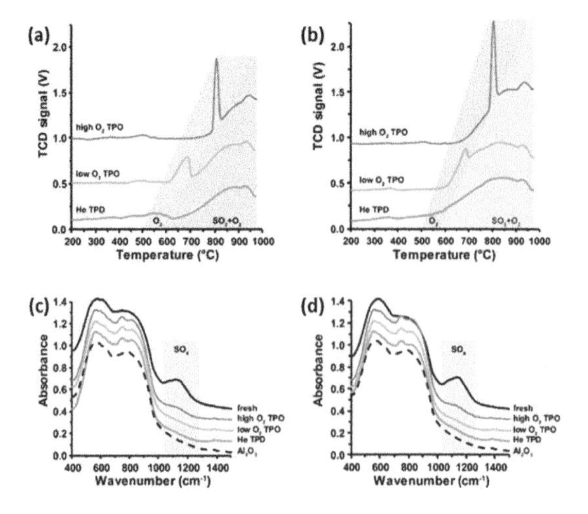

Figure 8. Thermal decomposition curves of (**a**) 0.25 AS + PS/Al_2O_3 and (**b**) 1.0 AS + PS/Al_2O_3 model catalysts and (**c**) FTIR spectra of 0.25 AS + PS/Al_2O_3 and (**d**) 1.0 AS + PS/Al_2O_3 model catalysts before and after decomposition, accompanied by a reference spectrum of Al_2O_3.

2.4. State of Palladium after Thermal Decomposition of Al_2O_3-Supported $PdSO_4$ and $Al_2(SO_4)_3$

The powder X-ray results presented in Figure 9 reveal that a decrease in oxygen concentration in the feed gas resulted in a higher content of metallic palladium in the regenerated PS/Al_2O_3 model catalyst. The qualitative PdO and Pd peak areas and crystallite sizes are tabulated in Table 1. The decomposition of the PS/Al_2O_3 model catalyst under helium gas resulted in the formation of only metallic Pd. Earlier, Hoyos et al. [13] concluded that decomposition under an inert gas atmosphere led to PdO, although the presence of an active phase was not directly observed. Qualitatively powder X-ray diffraction results showed that if $Al_2(SO_4)_3$ was present in the catalyst, a high content of metallic Pd was observed. This indicates that $Al_2(SO_4)_3$ stabilized the metallic Pd phase and/or lead to sintering of metallic Pd particles, possibly prevented re-oxidation of metallic Pd to the active PdO form. Overall, the formation of metallic Pd is undesirable, because it is known to be inactive in low-temperature methane oxidation under lean-burn conditions and it is also known to be vulnerable to sintering [29–31]. Overall, a high oxygen concentration in the feed gas resulted in a prominent amount of PdO in the model catalyst after the thermal decomposition of the sulfate species, whereas a low oxygen concentration in regeneration led to a high metallic Pd content.

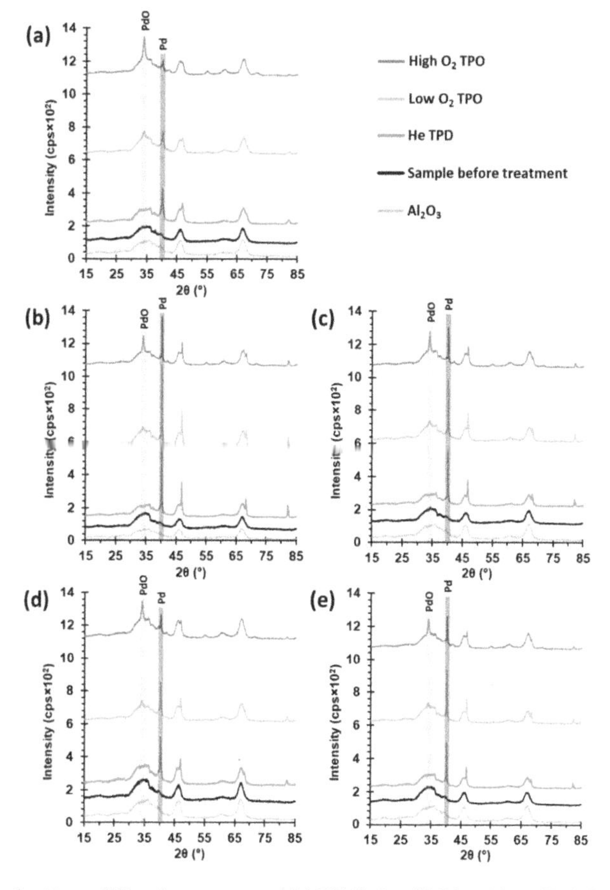

Figure 9. Powder X-ray diffraction patterns of (**a**) PS/Al_2O_3, (**b**) PS + 0.25 AS/Al_2O_3, (**c**) 0.25 AS + PS/Al_2O_3, (**d**) PS + 1.0 AS/Al_2O_3, and (**e**) 1.0 AS + PS/Al_2O_3 model catalysts after decomposition under varied oxygen concentrations. The characteristic main peaks of active PdO and inactive metallic Pd are highlighted in the figure.

3. Materials and Methods

3.1. Catalysts

Modern commercially available methane combustion catalyst after sulfur poisoning treatment contains 0.97 wt. % of sulfur [27]. A modern sulfur-poisoned methane combustion catalyst was used as a reference in activity experiments for model catalysts to justify their similar performance and behavior after sulfur poisoning treatment. To model and mimic the catalyst composition, a series of catalysts were prepared by using $PdSO_4$ (Sigma Aldrich, Saint Louis, MO, USA, CAS: 13566-03-5), $Al_2(SO_4)_3 \times 18H_2O$ (Merck, Darmstad, Germany, CAS: 7784-31-8), and Al_2O_3 (Sasol, Hamburg, Germany) as starting materials. Bulk $PdSO_4$ and $Al_2(SO_4)_3$ compounds were used in model catalyst preparation to control quantitatively the sulfur amount and sulfate structure. If the sulfating were to be done in the gas phase with SO_2 gas, the formed sulfate species and amounts could be hard to control. The preparation of model catalyst powders was carried out at room temperature, and detailed preparation procedures are described below for each of the model catalysts. The amounts of added $PdSO_4$ correspond to 1 wt.% of sulfur and 4 wt.% of palladium loading, whereas X indicates the amount of sulfur in $Al_2(SO_4)_3$-containing catalysts, and it is 0.25 or 1 wt.% of sulfur. The catalyst was provided by the Dinex Ecocat Oy. Sulfates are abbreviated as follows to clarify the names of the catalysts: PS refers to $PdSO_4$ and AS refers to $Al_2(SO_4)_3$.

PS/Al_2O_3: $PdSO_4$ precursor was mixed into water at room temperature. Al_2O_3 powder was added into the mixture of $PdSO_4$ and water to disperse $PdSO_4$ over Al_2O_3. The mixture was stirred at room temperature with ultrasonic treatment. The catalyst powder of the $PdSO_4/Al_2O_3$ model catalyst was obtained by evaporating the water at room temperature and finishing the drying in an oven at 90 °C.

X AS/Al_2O_3 (X = 0.25 or 1.0): $Al_2(SO_4)_3 \times 18H_2O$ precursor was dissolved into water at room temperature. Al_2O_3 powder was added into the $Al_2(SO_4)_3 \times 18H_2O$ solution to disperse $Al_2(SO_4)_3$ over Al_2O_3. The mixture was stirred for 2 h at room temperature. Model catalyst powders with different $Al_2(SO_4)_3$ contents were obtained by evaporating the water at room temperature and finishing the drying in an oven at 90 °C for 1 h.

PS + X AS/Al_2O_3 (X = 0.25 or 1.0): $PdSO_4$ and X $Al_2(SO_4)_3/Al_2O_3$ (X = 0.25 or 1.0) powder precursors were mixed into water. The mixture was stirred for 2 h at room temperature with ultrasonic treatment. The catalyst powders of $PdSO_4$ + X $Al_2(SO_4)_3/Al_2O_3$ (X = 0.25 or 1) were obtained by evaporating water at room temperature and finishing the drying in an oven at 90 °C for 1 h.

X AS + PS/Al_2O_3 (X = 0.25 or 1.0): $Al_2(SO_4)_3$ and $PdSO_4/Al_2O_3$ powder precursors were mixed into a water solution. The mixture was stirred for 2 h at room temperature with ultrasonic treatment. The catalyst powders with different $Al_2(SO_4)_3$ loadings were obtained by evaporating water at room temperature and finishing the drying in an oven at 90 °C for 1 h. The catalyst composition was the same as above, but the addition order was different, which leads to different natures of $PdSO_4$ on the Al_2O_3 support material as illustrated in Scheme 1.

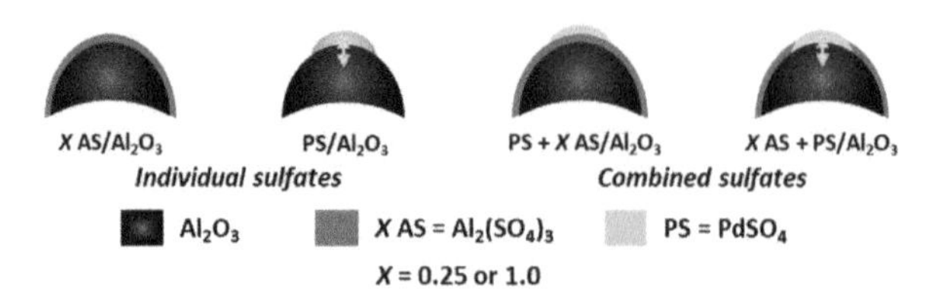

Scheme 1. Illustration of model catalyst composition.

3.2. Characterization Techniques

Regenerations of the model methane combustion catalyst were obtained under dry and wet synthetic exhaust gas. Synthetic exhaust gas consisted of 2000 ppm of CH_4, 2000 ppm of CO, 500 ppm of NO, 6% of CO_2, 10% of H_2O if wet, and 10% of O_2 balanced with N_2. Water vapor in dry gas regeneration was compensated with additional N_2 to maintain a total gas flow rate of 1.18 L min^{-1}. The amount of catalyst in the measurements was 0.2 g. A Gasmet™ DX-4000 Multigas FTIR spectrometer (Gasmet technologies, Helsinki, Finland) was used for gas analysis during regeneration.

Temperature-programmed reduction (TPR) experiments were done under a blend gas of 10% H_2 in Ar with a Quantachrome Autosorb iQ device (Quantachrome Corporation, Boynton Beach, FL, USA). A sample (100 mg) was heated under the measurement gas and the TCD signal was recorded from room temperature up to 700 °C with a heating rate of 10 °C min^{-1}. A cold trap was used in the measurements. A quadrupole mass spectrometer (MS) was utilized to examine the gaseous products on a qualitative level during the measurements.

An Elementar varioMICRO cube (Elementar, Langenselbold, Germany) device was used in sulfur analyses. Sulfanilamide was used as a reference and for the calibration. The catalyst sample amount varied from 10 mg to 30 mg in the measurements.

The stability of the sulfur compounds was studied via a temperature-programmed oxidation (TPO) hysteresis technique. A sample of 100 mg was heated from room temperature to 1000 °C with a heating rate of 10 °C min^{-1} under continuous flow of helium, 0.1% O_2/He, or 10% O_2/He gases. The gas flow rate was 20 mL min^{-1}. The sample was cooled down to 250 °C under the same gas atmosphere as that used during heating. No pretreatment was done prior to the measurement and a cold trap was used in the measurement. A thermal conductive detector (TCD) and quadrupole mass spectrometer (MS) were used to analyze the gaseous products.

Temperature-programmed desorption measurements with helium (He TPD) gas were conducted with similar parameters as the TPO measurements, but instead of oxygen blend gas, only helium was used.

The IR spectra of the solid catalyst samples were recorded using a Bruker Vertex 70 spectrometer (Bruker, Karlsruhe, Germany) using the pressed pellet technique. Catalyst samples were finely ground and diluted in KBr. The number of scans was 32 and the resolution was 2 cm^{-1}.

A Bruker-AXD D8 Advance diffractometer (Bruker, Karlsruhe, Germany) was used in the powder X-ray diffraction measurements of catalyst samples. Cu Kα radiation was used in the measurements. The diffraction pattern at a range of 2θ from 15° to 85° was recorded with a scanning speed of 0.11° min^{-1} and a step size of 0.04°. Bragg–Brentano geometry was utilized in the experiments. TOPAS software (Bruker, Karlsruhe, Germany) was applied to estimate crystallite sizes and peak areas of metallic Pd and PdO [32].

4. Conclusions

The Al_2O_3-supported $PdSO_4/Al_2(SO_4)_3$ model catalysts were studied under various gas atmospheres in order to study their decomposition routes as well as their texture before and after each treatment. The lowest sulfate decomposition temperature was achieved in treatment under a hydrogen gas atmosphere in all the cases, being the optimal conditions among the studied cases. Aluminum sulfate decomposed almost completely under hydrogen gas, whereas in the case of palladium sulfate, the decomposition was always incomplete, leading to Pd_4S formation. The TPD always resulted in decomposition of $PdSO_4$ to metallic palladium. Combining both aluminum and palladium sulfates into the same catalyst led to a decrease in the decomposition temperature and a high portion of metallic palladium under all the gas atmospheres.

Author Contributions: Conceptualization, N.M.K.; methodology, N.M.K.; formal analysis, N.M.K. and V.H.N.; data analysis, N.M.K., V.H.N., J.T.H.; investigation, N.M.K. and V.H.N.; writing—original draft preparation, N.M.K. and V.H.N.; writing—review and editing, N.M.K., V.H.N., J.H.T., K.K., T.M., M.K. and M.S.

Funding: This research was funded by European Commission (Horizon 2020), Grant Agreement no. 653391 Heavy-Duty Gas engines integrated into Vehicles, HDGAS-project.

Acknowledgments: The research leading to these results received funding from the European Union's Horizon 2020 research and innovation programme under Grant Agreement no. 653391 (HDGAS-project). Laboratory technicians Taina Nivajärvi, Urpo Ratinen, and Martti Lappalainen are acknowledged for their expertise and guidance in supporting experiments and their help with the activity reactor.

Conflicts of Interest: The authors declare no conflict of interest.

References

1. Keenan, M. Exhaust Emissions Control: 60 Years of Innovation and Development. *SAE Tech. Pap.* **2017**. [CrossRef]
2. Gélin, P.; Primet, M. Complete oxidation of methane at low temperature over noble metal based catalysts: A review. *Appl. Catal. B Environ.* **2002**, *39*, 1–37. [CrossRef]
3. Gélin, P.; Urfels, L.; Primet, M.; Tena, E. Complete oxidation of methane at low temperature over Pt and Pd catalysts for the abatement of lean-burn natural gas fuelled vehicles emissions: Influence of water and sulphur containing compounds. *Catal. Today* **2003**, *83*, 45–57. [CrossRef]
4. Villamaina, R.; Nova, I.; Tronconi, E.; Maunula, T.; Keenan, M. The Effect of CH_4 on NH_3-SCR Over Metal-Promoted Zeolite Catalysts for Lean-Burn Natural Gas Vehicles. *Top. Catal.* **2018**, *61*, 1974–1982. [CrossRef]
5. Maunula, T.; Kallinen, K.; Savimäki, A.; Wolff, T. Durability Evaluations and Rapid Ageing Methods in Commercial Emission Catalyst Development for Diesel, Natural Gas and Gasoline Applications. *Top. Catal.* **2016**, *59*, 1049–1053. [CrossRef]
6. Kinnunen, N.M.; Hirvi, J.T.; Suvanto, M.; Pakkanen, T.A. Methane combustion activity of Pd–PdO_x–Pt/Al_2O_3 catalyst: The role of platinum promoter. *J. Mol. Catal. A Chem.* **2012**, *356*, 20–28. [CrossRef]
7. Roth, D.; Gélin, P.; Primet, M.; Tena, E. Catalytic behaviour of Cl-free and Cl-containing Pd/Al_2O_3 catalysts in the total oxidation of methane at low temperature. *Appl. Catal. A Gen.* **2000**, *203*, 37–45. [CrossRef]
8. European standard for diesel fuel, EN 590:2009; European Commission Standard; 2009. Available online: https://ec.europa.eu/growth/single-market/european-standards_en (accessed on 7 May 2019).
9. Jones, J.M.; Dupont, V.A.; Brydson, R.; Fullerton, D.J.; Nasri, N.S.; Ross, A.B.; Westwood, A.V.K. Sulphur poisoning and regeneration of precious metal catalysed methane combustion. *Catal. Today* **2003**, *81*, 589–601. [CrossRef]
10. Honkanen, M.; Kärkkäinen, M.; Kolli, T.; Heikkinen, O.; Viitanen, V.; Zeng, L.; Jiang, H.; Kallinen, K.; Huuhtanen, M.; Keiski, R.L.; et al. Accelerated deactivation studies of the natural-gas oxidation catalyst—Verifying the role of sulfur and elevated temperature in catalyst aging. *Appl. Catal. B Environ.* **2016**, *182*, 439–448. [CrossRef]
11. Honkanen, M.; Wang, J.; Kärkkäinen, M.; Huuhtanen, M.; Jiang, H.; Kallinen, K.; Keiski, R.L.; Akola, J.; Vippola, M. Regeneration of sulfur-poisoned Pd-based catalyst for natural gas oxidation. *J. Catal.* **2018**, *358*, 253–265. [CrossRef]
12. Mowery, D.L.; McCormick, R.L. Deactivation of alumina supported and unsupported PdO methane oxidation catalyst: The effect of water on sulfate poisoning. *Appl. Catal. B Environ.* **2001**, *34*, 287–297. [CrossRef]
13. Hoyos, L.J.; Praliaud, H.; Primet, M. Catalytic combustion of methane over palladium supported on alumina and silica in presence of hydrogen sulfide. *Appl. Catal. A Gen.* **1993**, *98*, 125–138. [CrossRef]
14. Venezia, A.M.; Di Carlo, G.; Pantaleo, G.; Liotta, L.F.; Melaet, G.; Kruse, N. Oxidation of CH_4 over Pd supported on TiO_2-doped SiO_2: Effect of Ti(IV) loading and influence of SO_2. *Appl. Catal. B Environ.* **2009**, *88*, 430–437. [CrossRef]
15. Laperdrix, E.; Justin, I.; Costentin, G.; Saur, O.; Lavalley, J.; Aboulayt, A.; Ray, J.; Nédez, C. Comparative study of CS_2 hydrolysis catalyzed by alumina and titania. *Appl. Catal. B Environ.* **1998**, *17*, 167–173. [CrossRef]
16. Laperdrix, E.; Sahibed-dine, A.; Costentin, G.; Saur, O.; Bensitel, M.; Nédez, C.; Mohamed Saad, A.B.; Lavalley, J.C. Reduction of sulfate species by H_2S on different metal oxides and promoted aluminas. *Appl. Catal. B Environ.* **2000**, *26*, 71–80. [CrossRef]
17. Arosio, F.; Colussi, S.; Groppi, G.; Trovarelli, A. Regeneration of S-poisoned Pd/Al_2O_3 catalysts for the combustion of methane. *Catal. Today* **2006**, *117*, 569–576. [CrossRef]
18. Castellazzi, P.; Groppi, G.; Forzatti, P.; Finocchio, E.; Busca, G. Activation process of Pd/Al_2O_3 catalysts for CH_4 combustion by reduction/oxidation cycles in CH_4-containing atmosphere. *J. Catal.* **2010**, *275*, 218–227. [CrossRef]

19. Ordóñez, S.; Hurtado, P.; Diez, F.V. Methane catalytic combustion over Pd/Al_2O_3 in presence of sulphur dioxide: Development of a regeneration procedure. *Catal. Lett.* **2005**, *100*, 27–34. [CrossRef]
20. Yu, T.-C.; Shaw, H. The effect of sulfur poisoning on methane oxidation over palladium supported on γ-alumina catalysts. *Appl. Catal. B Environ.* **1998**, *18*, 105–114. [CrossRef]
21. Bounechada, D.; Groppi, G.; Forzatti, P.; Kallinen, K.; Kinnunen, T. Effect of periodic lean/rich switch on methane conversion over a Ce–Zr promoted $Pd\text{-}Rh/Al_2O_3$ catalyst in the exhausts of natural gas vehicles. *Appl. Catal. B Environ.* **2012**, *119*, 91–99. [CrossRef]
22. Salaün, M.; Kouakou, A.; Da Costa, S.; Da Costa, P. Synthetic gas bench study of a natural gas vehicle commercial catalyst in monolithic form: On the effect of gas composition. *Appl. Catal. B Environ.* **2009**, *88*, 386–397. [CrossRef]
23. Gremminger, A.T.; de Carvalho, H.W.P.; Popescu, R.; Grunwaldt, J.-D.; Deutschmann, O. Influence of gas composition on activity and durability of bimetallic $Pd\text{-}Pt/Al_2O_3$ catalysts for total oxidation of methane. *Catal. Today* **2015**, *258*, 470–480. [CrossRef]
24. Arosio, F.; Colussi, S.; Trovarelli, A.; Groppi, G. Effect of alternate CH_4-reducing/lean combustion treatments on the reactivity of fresh and S-poisoned $Pd/CeO_2/Al_2O_3$ catalysts. *Appl. Catal. B Environ.* **2008**, *80*, 335–342. [CrossRef]
25. Kinnunen, N.M.; Kallinen, K.; Maunula, T.; Keenan, M.; Suvanto, M. Fundamentals of sulfate species in methane combustion catalyst operation and regeneration—A simulated exhaust gas study. *Catalysts* **2019**, *9*, 417. [CrossRef]
26. Nissinen, V.H.; Kinnunen, N.M.; Suvanto, M. Regeneration of a sulfur-poisoned methane combustion catalyst: Structural evidence of Pd_4S formation. *Appl. Catal. B Environ.* **2018**, *237*, 110–115. [CrossRef]
27. Kinnunen, N.M.; Hirvi, J.T.; Kallinen, K.; Maunula, T.; Keenan, M.; Suvanto, M. Case study of a modern lean-burn methane combustion catalyst for automotive applications: What are the deactivation and regeneration mechanisms? *Appl. Catal. B Environ.* **2017**, *207*, 114–119. [CrossRef]
28. Neyestanaki, A.K.; Klingstedt, F.; Salmi, T.; Murzin, D.Y. Deactivation of postcombustion catalysts, a review. *Fuel* **2004**, *83*, 395–408. [CrossRef]
29. Ciuparu, D.; Lyubovsky, M.R.; Altman, E.; Pfefferle, L.D.; Datye, A. Catalytic Combustion of Methane over Palladium-Based Catalysts. *Catal. Rev.* **2002**, *44*, 593–649. [CrossRef]
30. Euzen, P.; Le Gal, J.H.; Rebours, B.; Martin, G. Deactivation of palladium catalyst in catalytic combustion of methane. *Catal. Today* **1999**, *47*, 19–27. [CrossRef]
31. Grunwaldt, J.D.; van Vegten, N.; Baiker, A. Insight into the structure of supported palladium catalysts during the total oxidation of methane. *Chem. Commun.* **2007**, 4635–4637. [CrossRef]
32. Mighell, A.D. *TOPAS V2.0: General Profile Analysis Software for Powder Diffraction Data*; Bruker AXS: Karlsruhe, Germany, 2000.

Article

Fundamentals of Sulfate Species in Methane Combustion Catalyst Operation and Regeneration—A Simulated Exhaust Gas Study

Niko M. Kinnunen [1,*], Kauko Kallinen [2], Teuvo Maunula [2], Matthew Keenan [3] and Mika Suvanto [1]

1. Department of Chemistry, University of Eastern Finland, P.O. Box 111, FI-80101 Joensuu, Finland; mika.suvanto@uef.fi
2. Dinex Finland Oy, Global Catalyst Competence Centre, P.O. Box 20, FI-41331 Vihtavuori, Finland; kki@dinex.fi (K.K.); tma@dinex.fi (T.M.)
3. Ricardo UK Ltd., Shoreham Technical Centre, Shoreham-by-Sea, West Sussex BN43 5FG, UK; matthew.keenan@ricardo.com

* Correspondence: niko.kinnunen@uef.fi; Tel.: +358-5034-240-84

Received: 22 March 2019; Accepted: 26 April 2019; Published: 3 May 2019

Abstract: Emission regulations and legislation inside the European Union (EU) have a target to reduce tailpipe emissions in the transportation sector. Exhaust gas aftertreatment systems play a key role in low emission vehicles, particularly when natural gas or bio-methane is used as the fuel. The main question for methane operating vehicles is the durability of the palladium-rich aftertreatment system. To improve the durability of the catalysts, a regeneration method involving an efficient removal of sulfur species needs to be developed and implemented on the vehicle. This paper tackles the topic and its issues from a fundamental point of view. This study showed that $Al_2(SO_4)_3$ over Al_2O_3 support material inhibits re-oxidation of Pd to PdO, and thus hinders the formation of the low-temperature active phase, PdO_x. The presence of $Al_2(SO_4)_3$ increases light-off temperature, which may be due to a blocking of active sites. Overall, this study showed that research should also focus on support material development, not only active phase inspection. An active catalyst can always be developed, but the catalyst should have the ability to be regenerated.

Keywords: catalytic methane combustion; exhaust gas; catalyst durability; Liquefied natural gas; biogas; vehicle emission control

1. Introduction

Current and future emission regulations and legislation of fossil fuels inside the European Union aspire to decrease the tailpipe emissions of transportation. The exhaust gas aftertreatment system plays a key role in low emission vehicles, and one of the main issues is its durability. Exhaust gas aftertreatment systems of heavy-duty applications in Europe already have a durability requirement, that is, 700,000 km, or seven years maximum [1].

Natural gas and bio-methane will be the next generation alternative fuels in the transportation sector, generating low overall emissions. The fuels can be stored in liquid form, thus increasing their energy capacity, which generates interest in the sector. In addition, the gases emit less CO_2 per energy equivalent compared to regular diesel fuel, which decreases their carbon footprint. However, a small amount of un-burnt methane, the main constituent of natural and bio-gas, always slips from an engine to its exhaust gas. Due to the higher global warming potential of methane, compared to CO_2, its emissions must be converted with a catalyst.

The catalyst, at a reasonably low operation temperature, in an aftertreatment system of natural gas or bio-methane application, is palladium rich, and supported on Al_2O_3, if high methane conversion activity is needed [2–6]. However, palladium-rich lean-burn methane combustion catalysts are known to be sensitive to sulfur poisoning. Sulfur originates from lubricant oil and natural gas, and it oxidizes further during the burning process to SO_2, and on a catalyst, it further reacts with oxygen from SO_2 to SO_3. It accumulates in the presence of water vapor over a catalyst as $PdSO_4$ and $Al_2(SO_4)_3$ [7–9]. Attempts have been made to solve the disadvantage of sulfur poisoning on methane conversion activity by modifying washcoat materials [2,10–12], and by varying noble metal content and their combinations [13–15]. Fundamental studies of poisoned and aged methane oxidation catalysts have been conducted to understand the formation of the inactive form of the catalyst. In general, the formation of $PdSO_4$ has been concluded to be the reason for the deactivation of the catalyst [8,16–18]. Aluminum oxide has been known to be the best support for the palladium-rich methane combustion catalyst, because it hinders any poisoning of the catalyst by forming $Al_2(SO_4)_3$ [2,16,19]. The latest results show that in fact, $PdSO_4$ itself is not a poison, but it sensitizes the catalyst for water inhibition [20].

Deactivation of the exhaust gas aftertreatment system is still a challenge, and thus solutions are needed to meet the future durability requirements [21–23]. A possible solution to increase the life-time of the exhaust gas aftertreatment system could be a regeneration of the catalyst. However, the number of studies conducted in the field of regeneration so far is low, which has also been noted in a recent review article [24]. Based on the published research, sulfur removal in the presence of excess oxygen requires high temperature, at least 650 °C, and it has been concluded to be always incomplete [3,25]. The regeneration of the sulfur-poisoned catalyst by heating under vacuum [17] or treating with hydrogen gas, occurs at a remarkably lower temperature [26]. A small improvement in methane conversion activity has been achieved already after regeneration under hydrogen at 350 °C [25]. However, a better response to methane conversion activity has been achieved when the poisoned catalyst has been treated at 600 °C under the same gas atmosphere [27]. An alternative reductive method to regenerate a sulfur-poisoned catalyst, besides thermal or hydrogen treatments, has been presented by Arosio et al. [28]. They successfully used reductive methane pulses to partially regenerate the catalyst at 550 °C; such temperature could also be achieved in a real engine, but complete regeneration was achieved at 600 °C, which requires additional thermal energy, and may cause a fuel penalty. A reason for partial regeneration of the sulfur-poisoned catalyst has been proposed in a recent study [29], where $PdSO_4$ has been observed to decompose under a reductive atmosphere to Pd_4S. Hence, small quantities of sulfur will always remain in the regenerated catalyst. They concluded also that alternately reductive (rich) and oxidative (lean) pulses result in better sulfur removal in a regenerated catalyst compared to rich-only conditions.

This study focuses on the decomposition of sulfur-poisoned methane combustion catalyst with model catalysts. This paper answers the following research questions: How does the presence of $Al_2(SO_4)_3$ affect decomposition of $PdSO_4$? What is the state of palladium after the regeneration?

2. Results and Discussion

2.1. Methane Conversion Activity and Regeneration of Model Catalysts under Simulated Exhaust Gas

The activities of the model catalysts were evaluated in powder form with simulated exhaust gas in methane combustion. The effect of $Al_2(SO_4)_3$ (AS) on the performance of $PdSO_4$ (PS) in complete methane oxidation is presented in Figure 1, together with the sulfur-poisoned modern methane combustion catalyst as a reference [20]. The activity of the commercial reference catalyst is between the activities of the PS/Al_2O_3 and 0.25 AS+PS/Al_2O_3 catalysts. The similarity of CH_4 conversion curves of the commercial reference catalyst and model catalysts justifies the use of the model catalyst in further fundamental studies, when regeneration and texture of the catalysts will be examined. The shape analysis of the curves revealed that the methane oxidation reaction was temperature-controlled for the PS/Al_2O_3 model catalyst.

An addition of $Al_2(SO_4)_3$ into the catalyst led to a loss of active sites [30]. Hence, the temperature that is required to initiate methane conversion increased if $Al_2(SO_4)_3$ was added into the model catalyst.

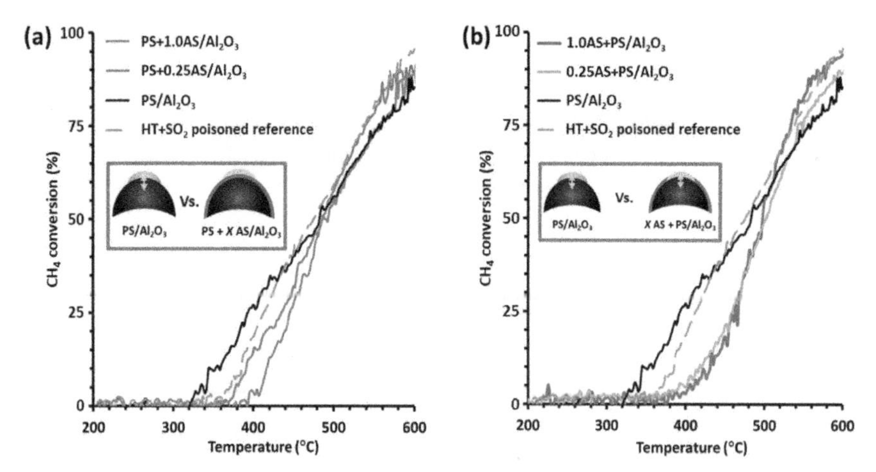

Figure 1. Methane conversion curves of (**a**) PS/Al_2O_3, PS + X AS/Al_2O_3, and (**b**) X AS + PS/Al_2O_3 (X = 0.25 or 1.0) model catalysts together with sulfur-poisoned commercial reference [20] before regeneration under simulated exhaust gas (indicated with an orange dashed line in the figures). See Scheme 1 for details about the catalysts.

It was decided to perform regeneration at 500 °C due to the threshold temperature observed by several researchers [25,29]. Simulated exhaust gas contains reducing agents such as CH_4, CO and NO, which can be expected to decrease the decomposition temperature of the sulfur species. The best regeneration was observed in the case of the PS/Al_2O_3 model catalyst, for which the methane conversion at 500 °C doubled to 60%, due to the regeneration (Figure 2). The co-existence of $PdSO_4$ and $Al_2(SO_4)_3$ was observed to have a disadvantageous effect on regeneration of the sulfated methane oxidation catalyst. After regeneration the catalysts PS + 1.0 AS/Al_2O_3 and 1.0 AS + PS/Al_2O_3 show higher methane conversion activities than the catalysts PS + 0.25 AS/Al_2O_3 and 0.25 AS + PS/Al_2O_3. A reason relies on the higher sulfur content of the samples, resulting in the catalysts bulk-kind-of sulfates, which decomposes in regeneration at lower temperature [31]. In fact, the PS + 0.25 AS/Al_2O_3 model catalyst even showed a slight decrease in activity even though sulfur species were decomposed at least partially during the regeneration. The decrease in activity after regeneration may indicate that $PdSO_4$ does not decompose during the process, or the active phase does not form after the regeneration. However, SO_2 release can be detected in all the cases, and thus decomposition of inactive $PdSO_4$ to metallic Pd and Pd_4S can be noted to have occurred. Thus, a potential reason for the decrease in activity could be the lack of active phase re-formation (PdO_X) of the low-temperature methane combustion catalyst. Overall, the simulated exhaust gas regeneration results allow us to deduce the following hypothesis: "*After regeneration, $Al_2(SO_4)_3$ inhibits Pd re-oxidation to PdO, which leads to an activity decrease in low temperature methane oxidation, when lean operation conditions are returned.*"

2.2. Palladium State after Regeneration under Simulated Exhaust Gas

As shown in Figure 2, the presence of $Al_2(SO_4)_3$ in the model catalysts decreased methane conversion activity after regeneration under realistic operation conditions. Inspection of the crystalline palladium state after the regeneration treatment, was carried out by powder X-ray diffraction in order to support the hypothesis. The powder X-ray diffraction method may be used in this case, because regeneration affects the surface of less active $PdSO_4$. Thus, the regenerated active Pd/PdO may form on top of less active $PdSO_4$.

The powder X-ray diffractograms of the catalyst in Figure 3 show that the peak of crystalline metallic Pd is pronounced if the catalyst contained $Al_2(SO_4)_3$. Closer inspection of the peak data, shown in Table 1, reveals that the presence of $Al_2(SO_4)_3$ results in a high amount of metallic Pd.

The results support the conclusions that $PdSO_4$ decomposed during the regeneration process, and the lower activity after regeneration could be due to less active metallic Pd phase formation. Thus, we rely on the fact that $Al_2(SO_4)_3$ may hinder the re-oxidization of metallic Pd after regeneration, thus inhibiting the formation of active PdO_x.

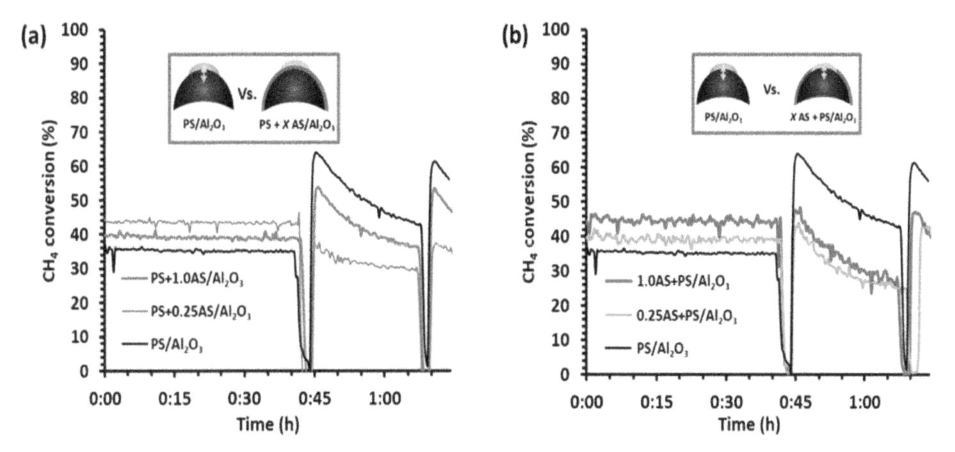

Figure 2. Methane conversion during steady-state operation and regeneration at 500 °C under simulated exhaust gas for (**a**) PS/Al_2O_3, PS + X AS/Al_2O_3 and (**b**) X AS + PS/Al_2O_3 (X = 0.25 or 1.0) model catalysts. See Scheme 1 for details about the catalysts.

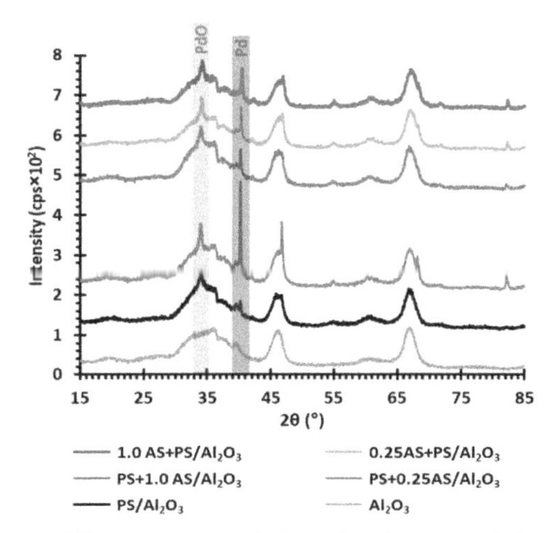

Figure 3. Powder X-ray diffraction patterns of PS/Al_2O_3, PS + X AS/Al_2O_3 (X = 0.25 or 1.0) and X AS + PS/Al_2O_3 (X = 0.25 or 1.0) model catalysts after regeneration at 500 °C under simulated exhaust gas. See Scheme 1 for details about the catalysts.

Table 1. Pd and PdO peak areas and crystallite sizes of model catalysts after regeneration under simulated exhaust gas.

Catalyst [1]	PdO(101) Peak Area [2]	PdO(101) Crystallite Size (nm) [2]	Pd(111) Peak Area [2]	Pd(111) Crystallite Size (nm) [2]	PdO(101):Pd(111) [3]
PS/Al_2O_3	55.3	4.7	17.3	13.1	3.20
PS + 0.25 AS/Al_2O_3	41.0	15.6	52.1	111.0	0.79
PS + 1.0 AS/Al_2O_3	51.5	11.2	20.0	50.9	2.58
0.25 AS + PS/Al_2O_3	27.0	23.6	24.7	60.8	1.09
1.0 AS + PS/Al_2O_3	31.1	13.9	20.6	56.8	1.51

[1] See Scheme 1 for details about the catalysts. [2] Peak areas and crystallite size were measured for the catalysts after regeneration under simulated exhaust gas. [3] PdO(101):Pd(111) ratio was calculated based on corresponding peak areas of X-ray diffraction patterns. Alumina signals were used as internal references.

The observation can be further supported and confirmed with TPO re-oxidation measurements. Re-oxidation of metallic Pd to active PdO can be observed as a downward peak in Figure 4 in a temperature range of between 470 °C and 700 °C. The presence of Pd_4S should bear in mind as detected in the latest experiments if the catalyst is heated under hydrogen gas [29,31]. Because steam reforming and water gas shift reactions form hydrogen during regeneration in simulated exhaust gas, Pd_4S structure is possible to form during the regeneration period. The most feasible re-oxidation product of Pd_4S may be $PdSO_4$, PdO and metallic Pd. Due to the stoichiometry of Pd_4S structure, in respect of Pd atoms, re-oxidation may form one $PdSO_4$, and three Pd units may form PdO or remain in the metallic Pd state. Quantitative oxygen uptake was determined by integrating the peaks and the values represented in Table 2. Individual $PdSO_4$ supported on Al_2O_3 (PS/Al_2O_3) had the highest oxygen uptake, being 19.2 μmol g_{cat}^{-1}, corresponding to an O:Pd mole ratio of 0.43. This means that regenerated metallic Pd oxidizes only partially into an active PdO form. Oxygen uptakes of model catalysts including $Al_2(SO_4)_3$ were lower than that of PS/Al_2O_3, between 9.0 and 15.7 μmol g_{cat}^{-1}, corresponding to O:Pd mole ratios of 0.22–0.36, revealing the fact that $Al_2(SO_4)_3$ hinders oxidation of metallic Pd back to an active PdO form. The addition of $Al_2(SO_4)_3$ increases the re-oxidation temperature of metallic Pd, and thus the formation of $Al_2(SO_4)_3$ might be undesirable in the CH_4 catalyst, even though conclusions in literature have been contradictory [2,6]. The results confirm the above hypothesis that $Al_2(SO_4)_3$ hinders the re-oxidation of metallic Pd to active PdO. Inhibition of the re-oxidation was the strongest for the PS + 0.25 AS/Al_2O_3 model catalyst, which explained the observed decrease in activity after the regeneration.

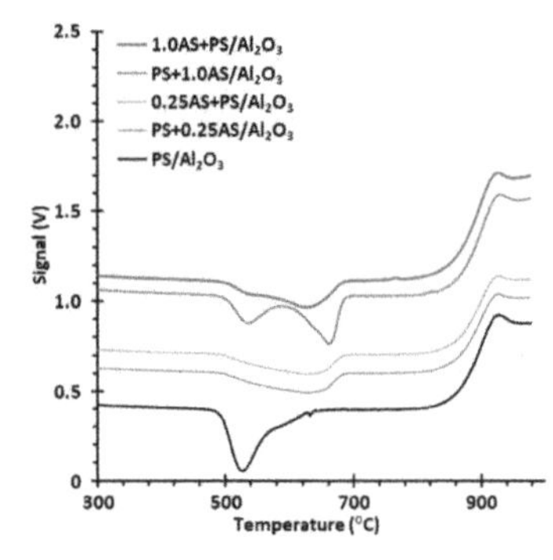

Figure 4. Re-oxidation of metallic Pd to an active PdO form after thermal decomposition. Re-oxidation was studied by decreasing the temperature after thermal decomposition of model catalysts under a gas blend of 10% O_2 in He. See Scheme 1 for details about the catalysts.

Table 2. Quantitative O_2 uptake of the catalysts during cooling down.

Catalyst [1]	Oxygen Uptake (μmol g_{cat}^{-1})	O:Pd Mole Ratio	Regenerated CH_4 Conversion (%) [2]
PS/Al_2O_3	19.2	0.43	63.7
PS + 0.25 AS/Al_2O_3	11.1	0.25	37.6
PS + 1.0 AS/Al_2O_3	15.7	0.36	53.4
0.25 AS + PS/Al_2O_3	9.0	0.22	42.7
1.0 AS + PS/Al_2O_3	11.4	0.22	47.0

[1] See Scheme 1 for details about the catalysts. [2] CH_4 conversion is the maximum that has been achieved after the regeneration procedure (Figure 2).

To summarize the results, the relation between regenerated CH_4 conversion is illustrated in Figure 5, together with three indicators: O:Pd mole ratio (Table 2), oxygen uptake of the catalyst (Table 2) and PdO(101):Pd(111) ratio (Table 1). The trends of all three indicators show that addition of $Al_2(SO_4)_3$ into the $PdSO_4$-containing catalyst hinders the formation of the active phase, such as PdO after regeneration, and thus activity in CH_4 conversion decreases, compared to the case of the individual $PdSO_4$ model catalyst. Thus we conclude that the formation of $Al_2(SO_4)_3$ is not beneficial for the low temperature methane combustion catalyst, because it causes a decrease in the methane conversion activity of the catalyst after regeneration. Unforeseen correlation of methane combustion activity of the model catalysts, together with the PdO(101):Pd(111) ratio could be explained, at least partially, by the formation of a regenerated, active Pd/PdO structure on top of less active $PdSO_4$. Regeneration studies were done at 500 °C, which may be low to re-oxidize regenerated metallic Pd to PdO_x. However, higher regeneration temperature may enhance re-oxidation, but it exposes the catalyst to sintering, and thus may decrease its methane conversion activity.

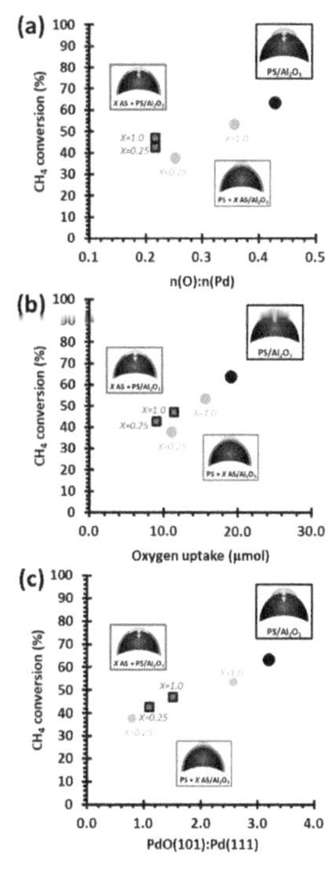

Figure 5. Regenerated CH_4 conversion under simulated exhaust gas as a function of (**a**) O:Pd mole ratio, (**b**) oxygen uptake of the catalyst and (**c**) PdO(101):Pd(111) ratio. See Scheme 1 for details about the catalysts.

3. Materials and Methods

3.1. Catalysts

The modern commercially available methane combustion catalyst contains 0.97 wt.% of sulfur after sulfur poisoning treatment [20]. A modern sulfur-poisoned methane combustion catalyst was used as a reference in activity experiments for model catalysts to justify their similar performance and behavior after sulfur poisoning treatment. The catalyst was provided by the Dinex Finland Oy. To model and mimic the catalyst composition, a series of catalysts were prepared (Table 3 and Scheme 1) by using $PdSO_4$ (Sigma Aldrich, Saint Louis, MO, USA, CAS: 13566-03-5), $Al_2(SO_4)_3 \times 18H_2O$ (Merck, Darmstad, Germany, CAS: 7784-31-8), and Al_2O_3 (Sasol) as starting materials. Bulk $PdSO_4$ and $Al_2(SO_4)_3$ compounds were used in model catalyst preparation to quantitatively control the amount of sulfur and structure of sulfate. If the sulfating were done in the gas phase with SO_2 gas, the formed sulfates species and amounts could be hard to control. Impregnation of $PdSO_4$ and $Al_2(SO_4)_3$ was carried out in cold water by mixing at least for 2 h. After impregnation, the solid was dried at room temperature, and to finalize the catalyst it was heated at 90 °C under air. Detailed preparation procedures are described in our previous work [31]. Amounts of added $PdSO_4$ correspond to 1 wt.% of sulfur and 4 wt.% of palladium loading, whereas X indicates the amount of sulfur in $Al_2(SO_4)_3$-containing catalysts, and it is 0.25 or 1.0 wt.% of sulfur. Sulfates are abbreviated as follows to clarify the names of the catalysts: PS refers to $PdSO_4$, and AS refers to $Al_2(SO_4)_3$.

Table 3. Model catalyst compositions.

Catalyst	Total Sulfur Content of the Catalyst (wt.%)
PS/Al_2O_3 [1]	0.88
PS + 0.25 AS/Al_2O_3 [1,2]	1.39
PS + 1.0 AS/Al_2O_3 [1,2]	2.10
0.25 AS + PS/Al_2O_3 [1,2]	1.03
1.0 AS + PS/Al_2O_3 [1,2]	1.71

[1] Amount of PS ($PdSO_4$) corresponds to 4 wt.% active metal loading of Pd and 1.0 wt.% loading of sulfur. [2] Amount of AS ($Al_2(SO_4)_3$) corresponds to 0.25 or 1.0 wt.% loadings of sulfur.

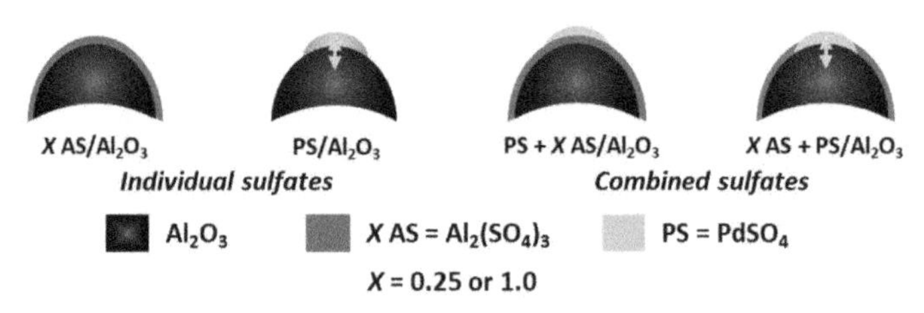

Scheme 1. Illustration of model catalyst ideas and composition [31].

3.2. Characterization Techniques

Methane conversion activities were measured for five $PdSO_4$-containing model catalysts. Regeneration experiments were carried out under steady-state conditions with a laboratory reactor at 500 °C in the presence of simulated exhaust gas. Gasmet™ DX-4000 Multigas FTIR (Gasmet technologies, Helsinki, Finland) was used as a detector in both light-off and regeneration experiments. An amount of 0.2 g model catalyst powder was used in the experiments. The exhaust gas composition used in the experiments was 2000 ppm of CO, 2000 ppm of CH_4, 500 ppm of C_3H_8, 500 ppm of NO, 10 ppm of SO_2, 6% of CO_2, 10% of O_2 and a balancing amount of N_2. The total gas flow rate of 1180 cm^3 min^{-1} corresponded to a space velocity of 354,000 cm^3 $gcat^{-1}$ h^{-1} through the model catalyst powder. Regenerations were carried out by replacing oxygen from the exhaust gas with N_2 in order to maintain a constant gas flow rate. Otherwise, the composition of the gas mixture remained the same.

Powder X-ray diffractograms of catalyst samples were recorded with a Bruker-AXD D8 Advance diffractometer (Bruker, Karlsruhe, Germany) using Cu Kα radiation. The diffraction pattern at a range of 2θ from 15° to 85° was recorded with a scanning speed of 0.11° min^{-1}, and a step size of 0.04°. Bragg–Brentano geometry was utilized in the experiments. TOPAS software was utilized in estimating palladium and palladium oxide crystallite sizes of the model catalysts and peak areas [32].

Re-oxidation model catalysts were studied with a Quantachrome Autosorb iQ device using a temperature programmed oxidation (TPO) hysteresis technique. A sample of 100 mg was heated from room temperature to 1000 °C with a heating rate of 10 °C min^{-1} under continuous flow of 10% O_2/He gas. The gas flow rate was 20 mL min^{-1}. To obtain re-oxidation, the sample was cooled down to 250 °C under the same gas atmosphere. No pretreatment was done prior to the measurement and cold trap was used in the measurement.

Elemental analyses were carried out with an Elementar varioMICRO cube device (Elementar, Langenselbold, Germany). Sulfanilamide was used both for calibrating the device, and also as a reference compound for sulfur in the measurements. The mass of the samples varied between 10 mg and 30 mg.

4. Conclusions

The role of $Al_2(SO_4)_3$ on the regeneration of the low-temperature methane combustion catalyst was studied in the presence of simulated exhaust gas. This study showed that the presence of $Al_2(SO_4)_3$ over Al_2O_3 support material inhibits the re-oxidation of metallic Pd back to its active form of PdO_x. Hence, the low temperature activity of the regenerated catalyst decreases, and does not necessarily increase after regeneration. The outcome was supported with powder X-ray measurements and finally confirmed with the TPO re-oxidation method. These aspects should be taken into account when developing a regeneration method for an aftertreatment system of natural gas or bio-methane fueled engines. From the catalyst development point of view, this study shows that we should also focus on support materials, not only on the active phase, because a good catalyst can always be developed, but the catalyst should have the ability to be regenerated.

Author Contributions: Conceptualization, N.M.K.; methodology, N.M.K.; formal analysis, N.M.K.; data analysis, N.M.K.; investigation, N.M.K.; writing—original draft preparation, N.M.K and M.K.; writing—review and editing, N.M.K, K.K., T.M., M.K. and M.S.

Funding: This research was funded by European Commission (Horizon 2020), Grant Agreement no. 653391, Heavy-Duty Gas engines integrated into Vehicles (HDGAS project).

Acknowledgments: The research leading to these results has received funding from the European Union's Horizon 2020 research and innovation programme under Grant Agreement no. 653391 (HDGAS-project). Laboratory technicians Taina Nivajärvi, Urpo Ratinen, and Martti Lappalainen are acknowledged for their expertise and guidance in supporting experiments and help with the activity test reactor.

Conflicts of Interest: The authors declare no conflict of interest.

References

1. Regulation (EC) No 595/2009 of the European Council. Available online: https://eur-lex.europa.eu/LexUriServ/LexUriServ.do?uri=OJ:L:2009:188:0001:0013:EN:PDF (accessed on 30 April 2019).
2. Gélin, P.; Primet, M. Complete oxidation of methane at low temperature over noble metal based catalysts: A review. *Appl. Catal. B Environ.* **2002**, *39*, 1–37. [CrossRef]
3. Gélin, P.; Urfels, L.; Primet, M.; Tena, E. Complete oxidation of methane at low temperature over Pt and Pd catalysts for the abatement of lean-burn natural gas fuelled vehicles emissions: Influence of water and sulphur containing compounds. *Catal. Today* **2003**, *83*, 45–57. [CrossRef]
4. Antony, A.; Asthagiri, A.; Weaver, J.F. Pathways and kinetics of methane and ethane C–H bond cleavage on PdO(101). *J. Chem. Phys.* **2013**, *139*, 104702. [CrossRef]
5. Ciuparu, D.; Lyubovsky, M.R.; Altman, E.; Pfefferle, L.D.; Datye, A. Catalytic Combustion of Methane over Palladium-Based Catalysts. *Catal. Rev.* **2002**, *44*, 593–649. [CrossRef]

6. Lampert, J.K.; Kazi, M.S.; Farrauto, R.J. Palladium catalyst performance for methane emissions abatement from lean burn natural gas vehicles. *Appl. Catal. B Environ.* **1997**, *14*, 211–223. [CrossRef]
7. Mowery, D.L.; McCormick, R.L. Deactivation of alumina supported and unsupported PdO methane oxidation catalyst: The effect of water on sulfate poisoning. *Appl. Catal. B Environ.* **2001**, *34*, 287–297. [CrossRef]
8. Chenakin, S.P.; Melaet, G.; Szukiewicz, R.; Kruse, N. XPS study of the surface chemical state of a Pd/(SiO_2 + TiO_2) catalyst after methane oxidation and SO_2 treatment. *J. Catal.* **2014**, *312*, 1–11. [CrossRef]
9. Honkanen, M.; Wang, J.; Kärkkäinen, M.; Huuhtanen, M.; Jiang, H.; Kallinen, K.; Keiski, R.L.; Akola, J.; Vippola, M. Regeneration of sulfur-poisoned Pd-based catalyst for natural gas oxidation. *J. Catal.* **2018**, *358*, 253–265. [CrossRef]
10. Escandón, L.S.; Niño, D.; Díaz, E.; Ordóñez, S.; Díez, F.V. Effect of hydrothermal ageing on the performance of Ce-promoted PdO/ZrO_2 for methane combustion. *Catal. Commun.* **2008**, *9*, 2291–2296. [CrossRef]
11. Kinnunen, N.; Kinnunen, T.; Kallinen, K. *Improved Sulfur Resistance of Noble Metal Catalyst for Lean-Burn Natural Gas Applications*; SAE Technical Paper 2013-24-0155; SAE International: Warrendale, PA, USA, 2013.
12. Venezia, A.M.; Di Carlo, G.; Pantaleo, G.; Liotta, L.F.; Melaet, G.; Kruse, N. Oxidation of CH_4 over Pd supported on TiO_2-doped SiO_2: Effect of Ti(IV) loading and influence of SO_2. *Appl. Catal. B Environ.* **2009**, *88*, 430–437. [CrossRef]
13. Corro, G.; Cano, C.; Fierro, J.L.G. A study of Pt–Pd/γ-Al_2O_3 catalysts for methane oxidation resistant to deactivation by sulfur poisoning. *J. Mol. Catal. A Chem.* **2010**, *315*, 35–42. [CrossRef]
14. Wilburn, M.S.; Epling, W.S. Sulfur deactivation and regeneration of mono- and bimetallic Pd-Pt methane oxidation catalysts. *Appl. Catal. B Environ.* **2017**, *206*, 589–598. [CrossRef]
15. Yashnik, S.A.; Chesalov, Y.A.; Ishchenko, A.V.; Kaichev, V.V.; Ismagilov, Z.R. Effect of Pt addition on sulfur dioxide and water vapor tolerance of Pd-Mn-hexaaluminate catalysts for high-temperature oxidation of methane. *Appl. Catal. B Environ.* **2017**, *204*, 89–106. [CrossRef]
16. Honkanen, M.; Kärkkäinen, M.; Kolli, T.; Heikkinen, O.; Viitanen, V.; Zeng, L.; Jiang, H.; Kallinen, K.; Huuhtanen, M.; Keiski, R.L.; et al. Accelerated deactivation studies of the natural-gas oxidation catalyst—Verifying the role of sulfur and elevated temperature in catalyst aging. *Appl. Catal. B Environ.* **2016**, *182*, 439–448. [CrossRef]
17. Hoyos, L.J.; Praliaud, H.; Primet, M. Catalytic combustion of methane over palladium supported on alumina and silica in presence of hydrogen sulfide. *Appl. Catal. A Gen.* **1993**, *98*, 125–138. [CrossRef]
18. Ordóñez, S.; Hurtado, P.; Sastre, H.; Diez, F.V. Methane catalytic combustion over Pd/Al_2O_3 in presence of sulphur dioxide: Development of a deactivation model. *Appl. Catal. A Gen.* **2004**, *259*, 41–48. [CrossRef]
19. Jones, J.M.; Dupont, V.A.; Brydson, R.; Fullerton, D.J.; Nasri, N.S.; Ross, A.B.; Westwood, A.V.K. Sulphur poisoning and regeneration of precious metal catalysed methane combustion. *Catal. Today* **2003**, *81*, 589–601. [CrossRef]
20. Kinnunen, N.M.; Hirvi, J.T.; Kallinen, K.; Maunula, T.; Keenan, M.; Suvanto, M. Case study of a modern lean-burn methane combustion catalyst for automotive applications: What are the deactivation and regeneration mechanisms? *Appl. Catal. B Environ.* **2017**, *207*, 114–119. [CrossRef]
21. Keenan, M.; Pickett, R.; Tronconi, E.; Nova, I.; Kinnunen, N.; Suvanto, M.; Maunula, T.; Kallinen, K.; Baert, R. The Catalytic Challenges of Implementing a Euro VI Heavy Duty Emissions Control System for a Dedicated Lean Operating Natural Gas Engine. *Top. Catal.* **2018**, *62*, 273–281. [CrossRef]
22. Maunula, T.; Kallinen, K.; Kinnunen, N.; Keenan, M.; Wolff, T. Methane Abatement and Catalyst Durability in Heterogeneous Lean-Rich and Dual-Fuel Conditions. *Top. Catal.* **2019**, *62*, 315–323. [CrossRef]
23. Maunula, T.; Kallinen, K.; Savimäki, A.; Wolff, T. Durability Evaluations and Rapid Ageing Methods in Commercial Emission Catalyst Development for Diesel, Natural Gas and Gasoline Applications. *Top. Catal.* **2016**, *59*, 1049–1053. [CrossRef]
24. Chen, J.; Arandiyan, H.; Gao, X.; Li, J. Recent Advances in Catalysts for Methane Combustion. *Catal. Surv. Asia* **2015**, *19*, 140–171. [CrossRef]
25. Arosio, F.; Colussi, S.; Groppi, G.; Trovarelli, A. Regeneration of S-poisoned Pd/Al_2O_3 catalysts for the combustion of methane. *Catal. Today* **2006**, *117*, 569–576. [CrossRef]
26. Ordóñez, S.; Hurtado, P.; Diez, F.V. Methane catalytic combustion over Pd/Al_2O_3 in presence of sulphur dioxide: Development of a regeneration procedure. *Catal. Lett.* **2005**, *100*, 27–34. [CrossRef]
27. Yu, T.-C.; Shaw, H. The effect of sulfur poisoning on methane oxidation over palladium supported on γ-alumina catalysts. *Appl. Catal. B Environ.* **1998**, *18*, 105–114. [CrossRef]

28. Arosio, F.; Colussi, S.; Trovarelli, A.; Groppi, G. Effect of alternate CH_4-reducing/lean combustion treatments on the reactivity of fresh and S-poisoned $Pd/CeO_2/Al_2O_3$ catalysts. *Appl. Catal. B Environ.* **2008**, *80*, 335–342. [CrossRef]
29. Nissinen, V.H.; Nissinen, N.; Suvanto, M. Regeneration of a sulfur-poisoned methane combustion catalyst: Structural evidence of Pd_4S formation. *Appl. Catal. B Environ.* **2018**, *237*, 110–115. [CrossRef]
30. Heck, R.M.; Farrauto, R.J.; Gulati, S.T. *Catalytic Air Pollution Control: Commercial Technology*, 3rd ed.; Wiley: Hoboken, NJ, USA, 2009.
31. Kinnunen, N.M.; Nissinen, V.H.; Hirvi, J.T.; Kallinen, K.; Maunula, T.; Keenan, M.; Suvanto, M. Decomposition of Al_2O_3 supported $PdSO_4$ and $Al_2(SO_4)_3$ in regeneration of model methane combustion catalyst: A fundamental study. *Catalysts* **2019**. submitted.
32. Bruker AXS TOPAS V2.0: General Profile Analysis Software for Powder Diffraction Data. 2000.

Article

Byproduct Analysis of SO_2 Poisoning on NH_3-SCR over MnFe/TiO_2 Catalysts at Medium to Low Temperatures

Tsungyu Lee and Hsunling Bai *

Institute of Environmental Engineering, National Chiao Tung University, Hsinchu 300, Taiwan; ashley.ev98g@nctu.edu.tw

* Correspondence: hlbai@nctu.edu.tw; Tel.: +886-3-5731868; Fax: +886-3-5725958

Received: 18 February 2019; Accepted: 11 March 2019; Published: 15 March 2019

Abstract: The byproducts of ammonia-selective catalytic reduction (NH_3-SCR) process over MnFe/TiO_2 catalysts under the conditions of both with and without SO_2 poisoning were analyzed. In addition to the NH_3-SCR reaction, the NH_3 oxidation and the NO oxidation reactions were also evaluated at temperatures of 100–300 °C to clarify the reactions occurred during the SCR process. The results indicated that major byproducts for the NH_3 oxidation and NO oxidation tests were N_2O and NO_2, respectively, and their concentrations increased as the reaction temperature increased. For the NH_3-SCR test without the presence of SO_2, it revealed that N_2O was majorly from the NH_3-SCR reaction instead of from NH_3 oxidation reaction. The byproducts of N_2O and NO_2 for the NH_3-SCR reaction also increased after increasing the reaction temperature, which caused the decreasing of N_2-selectivity and NO consumption. For the NH_3-SCR test with SO_2 at 150 °C, there were two decay stages during SO_2 poisoning. The first decay was due to a certain amount of NH_3 preferably reacted with SO_2 instead of with NO or O_2. Then the catalysts were accumulated with metal sulfates and ammonium salts, which caused the second decay of NO conversion. The effluent N_2O increased as poisoning time increased, which was majorly from oxidation of unreacted NH_3. On the other hand, for the NH_3-SCR test with SO_2 at 300 °C, the NO conversion was not decreased after increasing the poisoning time, but the N_2O byproduct concentration was high. However, the SO_2 led to the formation of metal sulfates, which might inhibit NO oxidation reactions and cause the concentration of N_2O gradually decreased as well as the N_2-selectivity increased.

Keywords: Selective Catalytic Reduction (SCR); SO_2 poisoning; Low-temperature catalyst; nitrogen oxides; nitrous oxide

1. Introduction

Nitrogen oxides (NO_x, NO and NO_2) produced from stationary sources are major air pollutants that lead to environmental concerns such as photochemical smog and acid rain [1]. The most effective technology for the removal of NO_x emission from coal-fired power plants is ammonia-selective catalytic reduction (NH_3-SCR; SCR hereafter) [2]. The traditional SCR catalysts are active within the temperature window of 300–400 °C [3,4]. Even though some of the traditional catalyst compositions such as V_2O_5–WO_3/ TiO_2 or Fe-zeolite-based catalysts can lower down their working temperature window to be as low as 250 or even 200 °C [5–10], there is still a strong demand in developing SCR catalysts to be active at less than 200 °C and placing them downstream of the electrostatic precipitator and desulfurizer [11–13].

Literature data showed that Mn-based catalysts have good activity for low-temperature SCR [14–17]. Moreover, in iron containing SCR catalysts, the introduction of Mn could obviously enhance the low-temperature activity, probably due to the fact that synergistic effect between iron and manganese species [18]. It was reported that the MnFe/TiO_2 could improve the activity, stability and SO_2 durability of the SCR catalysts using NH_3 as the reducing agent [19–23].

In the past, there have been extensive studies usingFourier-transform infrared spectroscopy (FTIR) for understanding the mechanism of SCR reaction on the surface of MnFe catalysts [24–28]. In addition, several different types of reaction mechanisms have been proposed including the typical Eley–Rideal mechanism and Langmuir–Hinshelwood mechanism. For the Eley–Rideal mechanism, it is assumed that the gaseous NO directly reacts with an activated ammonia surface complex [29]. On the other hand, the Langmuir–Hinshelwood reaction mechanism involves that a surface NO complex reacts with an activated ammonia [3,30]. Moreover, Yang et al. [31] used in situ diffuse reflectance infrared Fourier transform spectroscopy (DRIFTS-FTIR) to reveal the mechanism of low-temperature SCR reaction over the MnFe spinel. The results indicated that the contribution of Eley–Rideal mechanism to NO conversion increased after increasing the reaction temperature.

Although many studies have been done on the SCR mechanism, to the authors' best knowledge, there has been no work on clarifying reactions occurred in the medium to low-temperature SCR system with and without the presence of SO_2. Therefore, this study employed MnFe/TiO_2 catalyst to study the oxidation of NH_3, the oxidation of NO and the NH_3-SCR reaction with and without the presence of SO_2 at 100–300 °C. The reactants and byproducts of gaseous NO, NH_3, N_2O, and NO_2 in the effluent streams as well as solid byproducts of ammonium salts and metal sulfates on the catalysts were analyzed. The results can offer useful information to understand the reaction pathway at different operation conditions for the application of medium to low-temperature SCR catalysts.

2. Results and Discussion

2.1. Oxidation Reactions of NH_3 and NO without SO_2

To understand the products and byproducts of oxidation reactions, the NH_3 and NO oxidation reactions over the MnFe/TiO_2 catalyst were studied within the temperature range of 100–300 °C, which was the most active temperature region for MnFe/TiO_2 catalyst in SCR reaction [32]. In addition, the N-balance of NH_3 and NO oxidation tests were calculated by Equations (1) and (2), respectively.

$$\text{N-balance in NH}_3\text{ oxidation} = \frac{[NH_3]_{out} + [N_2O]_{out} + [NO]_{out} + [NO_2]_{out}}{[NH_3]_{in}} \times 100\% \quad (1)$$

$$\text{N-balance in NO oxidation} = \frac{[NO]_{out} + [NO_2]_{out}}{[NO]_{in}} \times 100\% \quad (2)$$

The results of N-balance and outlet gas concentrations in the NH_3 oxidation reaction without SO_2 over MnFe/TiO_2 catalyst at different reaction temperatures are shown in Figure 1. It can be seen that the major oxidation product of NH_3 was N_2O (i.e., Equation (3) shown later in the Materials and Method section) instead of NO (Equation (4)). In addition, N_2O increased from 31 ppm to 219 ppm after increasing the reaction temperature from 100 °C to 300 °C. On the other hand, the effluent NO concentration was only 36 ppm at most, which occurred at reaction temperature of 300 °C; but no NO_2 was found in the effluent at all tested temperatures.

The results of N-balance and outlet gas concentrations in the NO oxidation reaction over MnFe/TiO_2 catalyst at different reaction temperatures are shown in Figure 2, it can be seen that NO could be oxidized to NO_2, and the concentration of NO_2 was significantly increased from 10 to 309 ppmv for reaction temperatures from 100 to 300 °C.

The above results indicated that NH_3 and NO oxidations (Equations (3)–(5)) increased after increasing the reaction temperature, and the major products of NH_3 oxidation and NO oxidation were N_2O and NO_2, respectively. In addition, it can be observed that both the N-balance results shown in Figures 1 and 2 were very high at 97~100% in the 100–300 °C range. This indicates that we can detect almost all the reaction species.

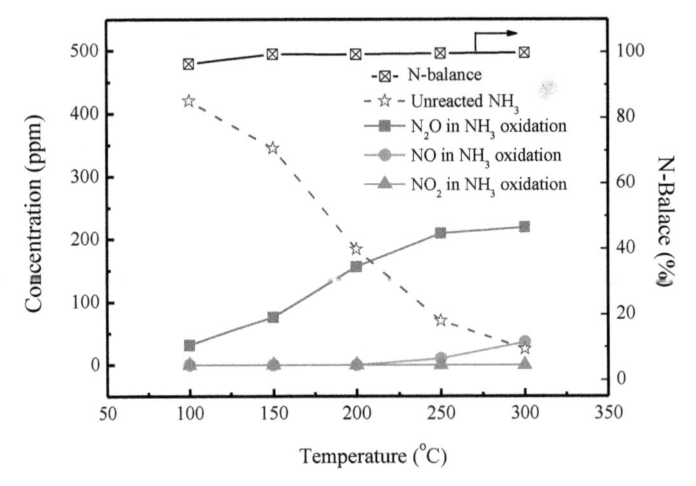

Figure 1. N-balance and outlet gas concentrations in the NH_3 oxidation reaction without SO_2 over $MnFe/TiO_2$ catalyst at different reaction temperatures. Reaction conditions: $[NH_3]$ = 500 ppm, $[O_2]$ = 10%, balanced with N_2, GHSV = 50,000 h^{-1}.

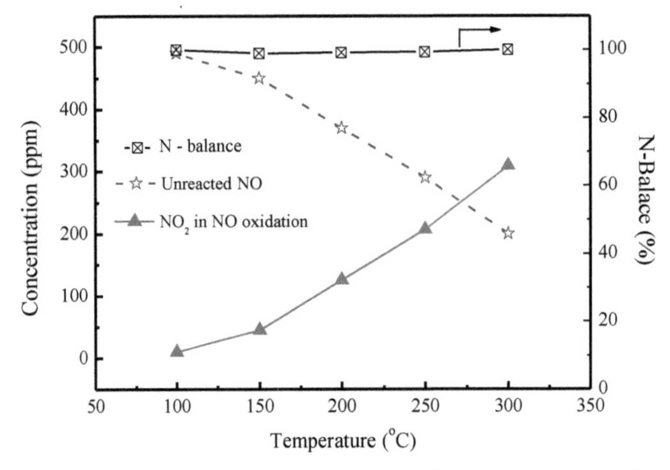

Figure 2. N-balance and outlet gas concentrations in the NO oxidation reaction without SO_2 over $MnFe/TiO_2$ catalyst at different reaction temperatures. Reaction conditions: [NO] = 500 ppm, $[O_2]$ = 10%, balanced with N_2, GHSV = 50,000 h^{-1}.

2.2. SCR Reactions without SO_2

Figure 3 shows the NO consumption (in ppmv) and outlet gas concentrations of N_2O, NO_2 and NH_3 over the $MnFe/TiO_2$ catalyst for reaction temperatures of 100–300 °C. It can be seen that the $MnFe/TiO_2$ catalyst maintained high NO consumption in the SCR reaction. When raising reaction temperature from 100 to 200 °C, the NO consumption in the SCR reaction increased slightly from 450 ppm to 495 ppm, and the outlet concentration of NH_3 was slightly decreased from 63 to 0 ppm.

At the reaction temperature of 200 °C, the concentrations of N_2O and NO_2 were 175 and 0 ppm, respectively, with 98% of NO consumption in the SCR reactions (500 ppm NO reacted with 500 ppm NH_3). On the other hand, the N_2O from NH_3 oxidation (500 ppm NH_3 reacted with O_2) at the reaction temperature of 200 °C was 156 ppm (as indicated by Figure 1). Therefore, it can be seen that in the SCR system, lower N_2O percentage ($\frac{175\text{ppm } N_2O}{500 \text{ ppm } NO + 500 \text{ ppm } NH_3}$ = 17.5%) was being produced as compared to that during only NH_3 oxidation ($\frac{156 \text{ ppm } N_2O}{500 \text{ ppm } NH_3}$ = 31.2%).

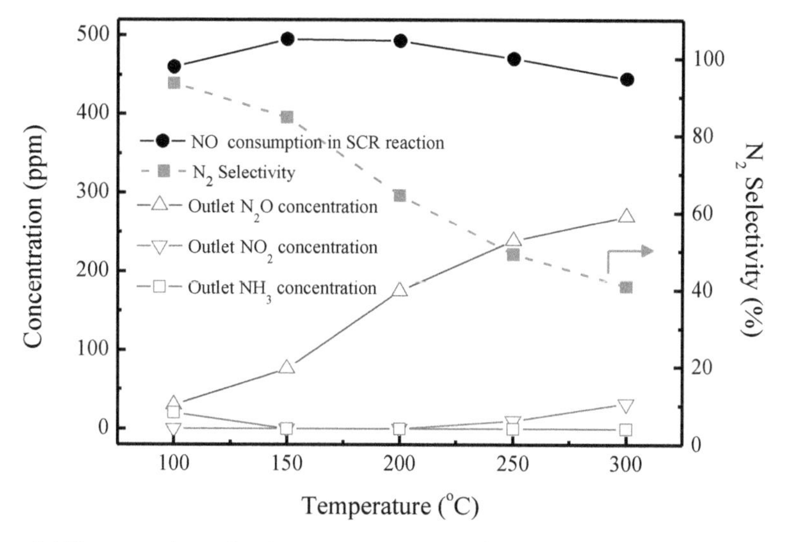

Figure 3. NO consumption and outlet gas concentrations in the NH_3-SCR reaction without SO_2 over MnFe/TiO_2 catalyst at different reaction temperatures. Reaction conditions: Reaction temperature = 100~300 °C, [NO] = 500 ppm, [NH_3] = 500 ppm, [O_2] = 10%, balanced with N_2, and GHSV = 50,000 h^{-1}.

The "fast SCR" (Equation (8)), first proposed in 1986 [33], proceeds at a much higher reaction rate than "standard SCR" reactions (Equations (6) and (7)) was developed to improve deNO$_x$ efficiency at lower temperatures [34,35]. As indicated by Figure 2 for the NO oxidation reaction, NO could be oxidized to NO_2 at temperatures higher than 150 °C, but NO_2 was not detected at temperature below 200 °C in the SCR test as seen in Figure 3. This might be due to the fact that NO_2 would react with NO and NH_3 according to the fast SCR reaction (Equation (8)).

After further raising the reaction temperature from 200 to 300 °C, NO consumption in the SCR reaction was slightly decreased from 495 ppm to 455 ppm, and concentrations of N_2O and NO_2 increased from 175 to 270 ppm and 0 to 36 ppm, respectively. When reaction temperature increased, oxidation reactions occurred more quickly as indicated by Figures 1 and 2. Hence the reason for slight decreases in the NO consumption at above 200 °C might be due to the fact that NO was oxidized (Equation (5)) as also demonstrated in the literature [26,36–38]. Moreover, the results of Figure 3 also indicated that products of the SCR reaction gradually changed from N_2 to N_2O when raising the temperature from 100 to 300 °C. Therefore, the N_2-selectivity of SCR reaction was gradually decreased from 93% to 41% after increasing the reaction temperature from 100 °C to 300 °C.

2.3. *SCR Reactions with SO_2*

To clarify the byproducts of SCR reactions with SO_2, NO consumption in the SCR reaction and outlet gas concentrations in the NH_3-SCR reaction over MnFe/TiO_2 catalyst at 150 and 300 °C are shown in Figure 4a,b, respectively. It can be seen in Figure 4a that there were four stages during the SCR test period. At stage I where no SO_2 was introduced, the NO consumption in the SCR reaction was very high (495 ppm out of 500 ppm). After 150 ppm SO_2 was introduced, the NO consumption in the SCR reaction decreased rapidly within 60 min SO_2 poisoning (stage II), and they remained roughly stable for another 60 min (stage III), then decreased gradually with time again (stage IV).

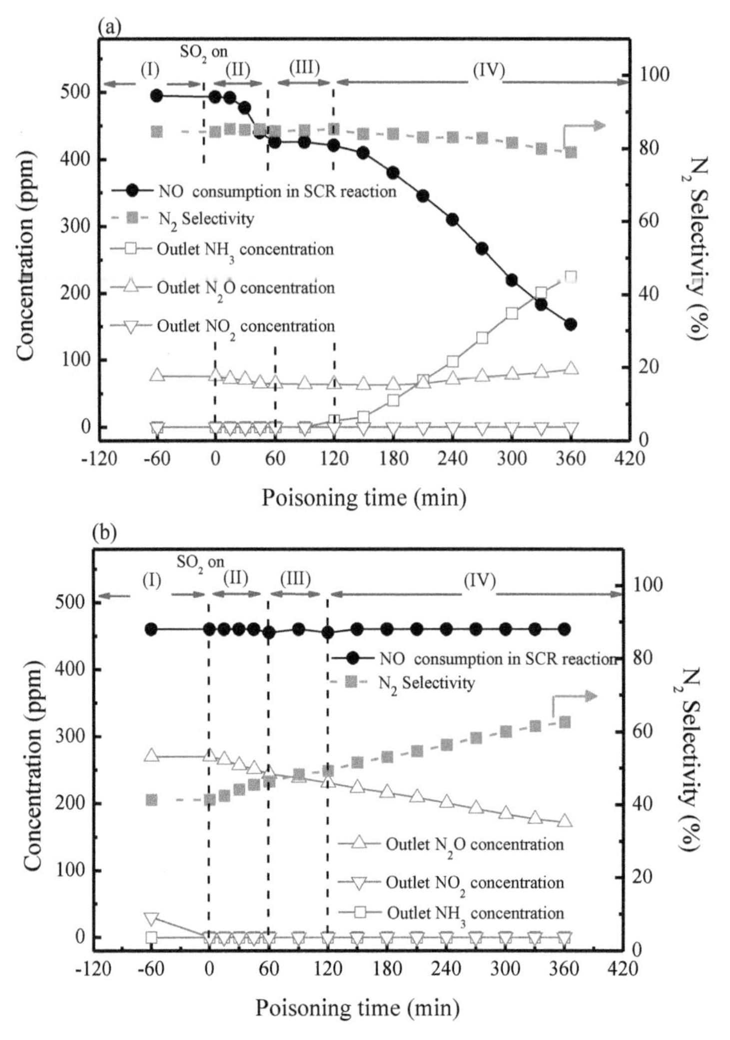

Figure 4. NO consumption and outlet gas concentrations in the NH_3-SCR reaction with SO_2 over MnFe/TiO_2 catalyst at (**a**) 150 °C and (**b**) 300 °C. Reaction conditions: [NO] = 500 ppm, [NH_3] = 500 ppm, [SO_2] = 150 ppm, [O_2] = 10%, balanced with N_2, and GHSV = 50,000 h^{-1}.

In our previous work [32], it was found that the gradual decrease (stage II) of NO consumption in the SCR reaction was probably due to the fact that SO_2 competed with NO to react with NH_3 and form ammonium salts. However, the MnFe/TiO_2 catalyst still had good activity at stage III. Thus, during this stage, there was no sufficient NH_3 to react with NO, which caused NO consumption remained relatively lower but stable at around 425 ppm from 60 to 120 min SO_2 poisoning time as compared to stages I and II. At stage IV, NO consumption in the SCR reaction decreased significantly from 420 ppm to 153 ppm for poisoning time from 120 to 360 min, which was due to accumulation of metal sulfates and ammonium salts which blocked the active sites of catalyst.

On the other hand, no effluent NH_3 was detected from stage I to stage III. Then NH_3 slip occurred at stage IV as seen in Figure 4a, which was due to less active sites and decreasing NO consumption in the SCR reaction. One can also see that concentrations of N_2O slightly decreased from 76 to 63 ppm within 120 min SO_2 poisoning time (stage II and III), then increased gradually to 86 ppm after

360 min SO_2 poisoning time (stage IV). This result indicated that SO_2 would react with NH_3 and form ammonium salts, which cause concentrations of N_2O slightly decreased at stages II and III. Then NH_3 slip occurred at stage IV, which caused increasing N_2O concentrations.

For the SCR test with SO_2 at 300 °C as shown in Figure 4b, it can be seen that the SO_2 poisoning effect was negligible as compared to the poisoning test at low temperature. When the reaction temperature was at 300 °C, the NO consumption in the SCR reaction remained around 460 ppm, which was almost the same as those without SO_2 poisoning. Moreover, during the SCR activity tests no SO_2 concentration was detected, which indicated that all the gas phase SO_2 molecules might be adsorbed and/or reacted with metal catalysts.

The main reason for the inhibition of SO_2 poisoning might be attributed to different SCR reaction mechanisms at different temperatures. Literature results showed that at lower temperatures (<200 °C), SCR reactions would follow the Langmuir−Hinshelwood mechanism. In the Langmuir−Hinshelwood mechanism, SO_2 would compete with NO to be adsorbed on the active sites, which cause the decreases of SCR efficiencies. However, when reaction temperature increased, the SCR reaction mechanism would transform from Langmuir−Hinshelwood mechanism to Eley–Rideal mechanism. In the high temperature range (>200 °C), the SCR reaction mainly followed the Eley–Rideal mechanism, over which the gaseous NO could directly react with an activated ammonia [30,39,40].

One can also see from Figure 4b that at reaction temperature of 300 °C, the SCR system did not have NH_3 slip. The concentrations of NO_2 and N_2O decreased from 30 to 0 ppm and from 270 to 172 ppm, respectively; and the N_2-selectivity increased from 41% to 62% for the SCR reaction from stage II to stage IV. The result indicated that by increasing the reaction temperature to 300 °C the SO_2 poisoning effect could be inhibited, and the N_2 selectivity can be enhanced due to the decreased N_2O concentrations.

The amount of sulfate species on the catalysts was estimated by thermo-gravimetric analysis (TGA) analysis and the results are shown in Figure 5 in terms of differential thermogram (DTG) spectra. The weight loss profiles of all samples showed three distinct decomposition steps: (1). the weight loss at low temperature (<200 °C) was assigned to the water desorption on the catalyst surface. (2). the weight loss at 200–400 °C could be attributed to decomposition of ammonium salts [41–43]. (3). the weight loss at high temperature (>670 °C) was originated from metal sulfates [44,45].

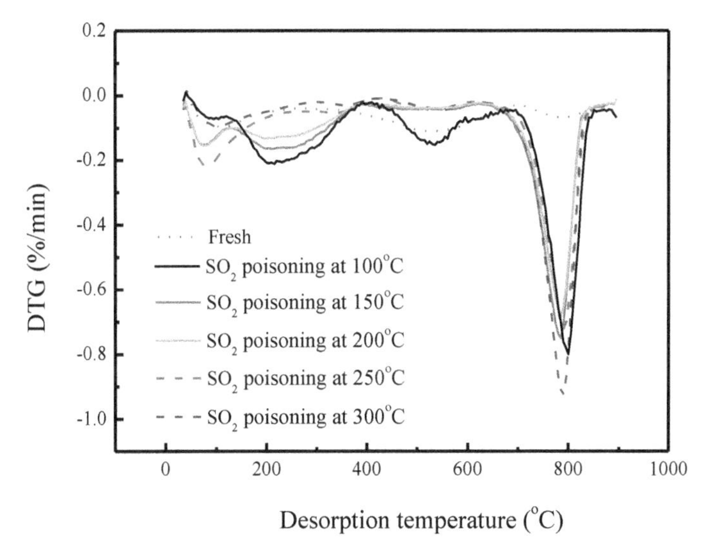

Figure 5. DTG spectra of fresh catalyst and catalysts poisoned at different reaction temperatures after 6 h poisoning time.

It can be seen from Figure 5 that the fresh catalyst had only one major weight loss peak, which appeared at 50–150 °C and corresponded to H_2O desorption on the catalyst surface. On the other hand, decomposition peaks of both ammonium salts (200–400 °C) and metal sulfates (670–900 °C) were observed on all MnFe/TiO_2 catalysts poisoned at temperature ranges of 100–200 °C. When reaction temperatures were 250 °C and 300 °C, there were no decomposition peak of ammonium salts. This reveals that ammonium salts were not formed on the catalyst surface at temperatures above 250 °C. As also noted in Figure 5, there was another peak in the range of temperature of 400–600 °C for the fresh catalyst and poisoned catalyst at 100 °C. However, since we cannot find related literature discussed on this, so the reason for this peak is not clear.

The amounts of sulfate species of fresh and poisoned catalysts are listed in Table 1. It is observed that the amounts of ammonium salts on the catalysts decreased from 2.3 wt.% to negligible amounts by increasing the SCR reaction temperature from 100 to 300 °C. This indicated that the reaction temperature would directly affect the formation of ammonium salts. It is noted that rigid quantification of metal sulfates accumulated on the poisoned catalyst was not possible via the DTG data because some ammonium salts could also be transformed into metal sulfates during the continuous TGA heating process [46]. Hence we can only confirm that all the poisoned catalysts had roughly similar amounts of metal sulfates during the temperature from 100 °C to 250 °C (within experimental error of TGA instrument, ±0.3 wt.%), except for the case at 300 °C.

Table 1. Amounts of ammonium salts and metal sulfates deposited on the catalyst surfaces.

MnFe/TiO_2 Poisoned with Different Reaction Temperatures	Ammonium Salts [a] (by weight) %	Metal Sulfates [b] (by Weight) %
Fresh	0.0	0.0
100 °C	2.3	4.21 [c]
150 °C	1.8	4.11 [c]
200 °C	1.1	4.03 [c]
250 °C	0.1	4.08 [c]
300 °C	0(−0.1)	4.52 [c]

[a] The amount of ammonium salts was calculated by weight difference between the fresh and poisoned catalysts from the TGA spectrum of 200–400 °C. [b] The amount of metal sulfates was calculated by weight difference between the fresh and poisoned catalysts from the TGA spectrum of 670–900 °C. [c] Rigid quantification of the amounts of metal sulfates is not possible via the DTG data because it was possible that some of the ammonium salts could transform into metal sulfates during the heating process of TGA.

At high temperature of 300 °C, the NO consumption was not affected by SO_2. Thus, it is easy to predict that adding an excessive amount of NH_3 tends to be oxidized and forming N_2O and NO as observed in Figure 1. However, at low temperature of 150 °C, the NO consumption was significantly affected by SO_2. Thus, to ensure the gradual decrease of NO consumption at stage II of Figure 4a was due to the competition between SO_2 and NO to react with NH_3, different inlet amounts of NH_3 were tested to study the SO_2 poisoning mechanism at temperature of 150 °C. The results are shown in Figure 6. One can see that adding different amounts of NH_3 had a similar effect on NO consumption at stages I and IV; but it had different NO consumptions at stages II and III. On the other hand, the NH_3 outlet concentrations were different at all stages. At stage I, NH_3 only reacted with a certain amount of NO and thus extra NH_3 slip was detected in the outlet gas. When SO_2 was introduced, NH_3 not only reacted with NO but also reacted with SO_2. Therefore, both the outlet concentrations of NH_3 and NO consumption decreased at stages II and III. Besides, when increasing NH_3 amount to above 550 ppm (i.e., NH_3/NO molar ratio of 1.1), the first decay of NO consumption at stage II could be inhibited. This indicated that NH_3 concentration was sufficient to react with NO and SO_2. The result shown in Figure 6 indicated that if a sufficient amount of NH_3 can be provided to react with both SO_2 and NO, then the first decay of NO consumption could be inhibited at early time of stage II. However, after 120 min of SO_2 poisoning (stage IV), the NO consumption could not be affected by different amounts of NH_3. Thus, the more injection amount of NH_3 led to eventually the more amount of NH_3 slip to the atmosphere.

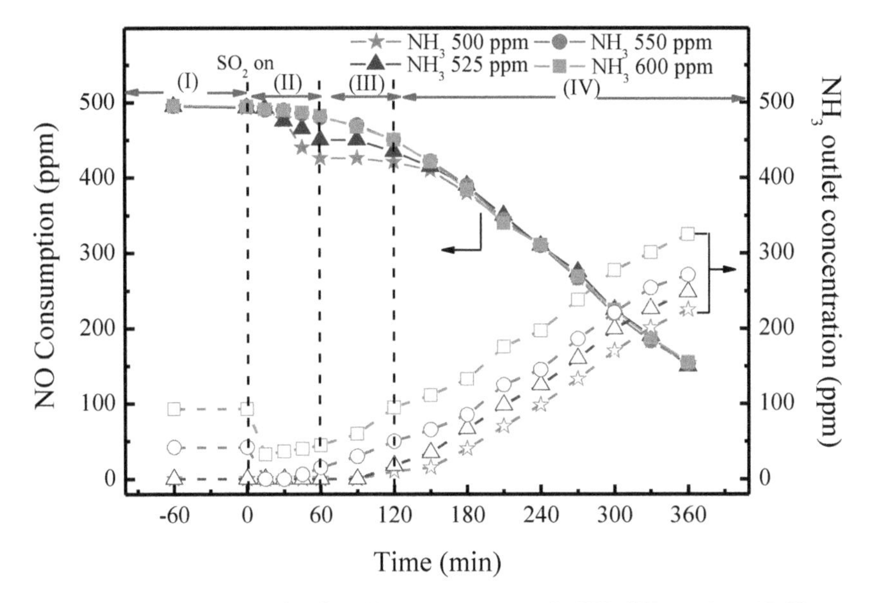

Figure 6. NO consumption and outlet NH_3 concentrations in the NH_3-SCR reaction with SO_2 over MnFe/TiO_2 catalyst at different ammonium amounts. Reaction conditions: Reaction temperature = 150 °C, [NO] = 500 ppm, [NH_3] = 500~650 ppm, [SO_2] = 150 ppm, [O_2] = 10%, balanced with N_2, and GHSV = 50,000 h^{-1}.

2.4. Product and Byproduct Analysis

Since different reactions occurred during the SCR process, i.e., Equations (3)–(12), were considered both with and without SO_2, thus the percentages of product and byproducts can be calculated. Table 2 lists formulas for calculating the percentages of N-containing product (N_2) as well as gaseous and solid byproducts (NO, N_2O and salts) formed during the SCR process. The byproducts could be formed by the SCR reaction as well as by the NH_3 oxidation or the NO oxidation reactions. Because it is difficult to clarify that the outlet NO was from NH_3 oxidation or from the unreacted NO, therefore the NO formation from NH_3 oxidation (Equation (4)) is neglected and all the effluent NO was assumed to be only from the unreacted NO. This is an acceptable assumption since the formation of NO from NH_3 oxidation was very minor (0~7%) at reaction temperatures of 100~300 °C as observed from Figure 1. Besides, it was assumed that the fast SCR reaction (Equation (8)) only served as the intermediate reaction at the low temperature SCR process [35]. Thus, the fast SCR reaction was not considered in the calculation of the percentages of all N-containing product and byproducts.

Based on formulas shown in Table 3, the results on percentages of all N-containing species during the NH_3-SCR process tested at 150 °C are shown in Figure 7a; and those tested at 300 °C are shown in Figure 7b. It can be seen from Figure 7a that at SCR operation temperature of 150 °C and without the presence of SO_2, the major product of SCR process appeared to be N_2. This indicated that MnFe/TiO_2 catalyst can serve as a good catalyst and achieve high N_2 selectivity at low temperature of 150 °C when SO_2 was not presented in the system. The minor presence of N_2O byproduct in the exhaust was majorly from the SCR reaction rather than from the NH_3 oxidation reaction.

Table 2. Formulas for calculating the N-containing product and byproduct percentages during the SCR process.

Product and Byproduct	Reaction Equation	Percentage Calculation Equation
NO_2	**NO oxidation to NO_2 (Equation (5)):** $2NO + O_2 \rightarrow 2NO_2$	**The percentage of NO_2 from NO oxidation** $[NO_2]_{NO}\,(\%) = [\frac{[NO_2]_{out}}{[NO]_{in}+[NH_3]_{in}}] \times 100\%$
Salts $(NH_4)_2SO_4$ NH_4HSO_4	**Ammonium salts** **Equations (10) and (11):** $SO_3 + 2NH_3 + H_2O\ (NH_4)_2SO_4$ $SO_3 + NH_3 + H_2O \rightarrow NH_4HSO_4$	**The percentage of salts** $[Salt](\%) = [\frac{[NH_3]_{Salt}{}^{\#}}{[NO]_{in}+[NH_3]_{in}}] \times 100\%$ $^{\#}$the NH_3 consumption of salts reaction ($[NH_3]_{Salt}$) were indicated by result of Figure 6
	NH_3 slip oxidation to form N_2O **$[N_2O]_{slip}$ (%)**	**The percentage of N_2O from NH_3 slip oxidation** $[N_2O]_{slip}\,(\%) = [\frac{[NH_3]_{out} \times \frac{[N_2O]}{[NH_3]}^{*}}{[NO]_{in}+[NH_3]_{in}}] \times 100\%$ $^{*}\ \frac{[N_2O]}{[NH_3]}$: from results of NH_3 oxidation test (Figure 1)
N_2O	**NH_3 oxidation to form N_2O** **Equation (3):** $2NH_3 + 2O_2 \rightarrow N_2O + 3H_2O$	**The percentage of N_2O from NH_3 oxidation** $[N_2O]NH_3\,(\%) = 1 - [\frac{2[NO]_{SCR}}{[NO]_{in}+[NH_3]_{in}}] \times 100\% - [NO_2]_{NO}(\%) - [Salt](\%) + [N_2O]_{Slip}(\%)$
	SCR reaction to form N_2O **Equation (7):** $4NO + 4NH_3 + 3O_2 \rightarrow 4N_2O + 6H_2O$	**The percentage of N_2O from SCR reaction** $[N_2O]SCR\,(\%) = [\frac{2[N_2O]_{out}}{[NO]_{in}+[NH_3]_{in}}] \times 100\% - [N_2O]_{NH3}(\%)$
N_2	**SCR reaction to form N_2** **Equation (6):** $4NO + 4NH_3 + O_2 \rightarrow 4N_2 + 6H_2O$	**The percentage of N_2 from SCR reaction** $[N_2]SCR\,(\%) = [\frac{2[NO]_{SCR}}{[NO]_{in}+[NH_3]_{in}}] \times 100\% - [N_2O]_{SCR}(\%)$

On the other hand, when SO_2 was introduced at 150 °C, the percentages of N_2 and N_2O decreased after increasing the poisoning time as observed in Figure 7a. Because NH_3 could not be reacted with NO due to decreased availability of active sites, it tends to increase the NO and NH_3 slip to the atmosphere. In addition, the portion of N_2O from NH_3 oxidation was also gradually increased as increasing the SO_2 poisoning time. After 360 min of SO_2 poisoning, the exhaust N_2O from NH_3 oxidation was more than from SCR reaction. One can also see that percentages of ammonium salts were similar at different SO_2 poisoning times. This indicated that certain amounts of NH_3 would preferentially be reacted with SO_2. In addition, the remaining NH_3 would then be reacted with NO or O_2. Moreover, it can be seen that percentages of NO slip were higher than percentages of NH_3 slip at different SO_2 poisoning times, which is due to the fact that NH_3 not only reacted with NO but also reacted with SO_2 and O_2.

For results at 300 °C as seen in Figure 7b, it is observed that NO_2 from NO oxidation disappeared during the SO_2 poisoning. However, the unreacted NO (NO slip) appeared during the SO_2 poisoning. This indicated that the formation of metal sulfates at 300 °C might inhibit the NO oxidation reaction. At high temperature, the SCR reaction mainly followed the Eley–Rideal mechanism, so the continuous decreasing of N_2O concentration as poisoning time increases and no SO_2 concentration in the outlet gases might be related to the reaction between SO_2 and activated NH_3 instead of the NH_3 oxidation reaction at 300 °C. Moreover, the total percentages of N_2 and N_2O from SCR reactions remained almost the same no matter SO_2 was presented in the system or not. The low N_2 selectivity revealed that $MnFe/TiO_2$ catalyst may not be a good candidate for SCR process at 300 °C unless the space velocity can be reduced to further enhance the more complete reduction of NO to N_2 instead of forming the N_2O byproduct. However, it is interested to note that the N_2 selectivity gradually increased and the percentages of N_2O from SCR reaction gradually decreased after increasing SO_2 poisoning time. This indicated that SO_2 promotion phenomenon might exist at 300 °C, which was attributed to the formation of $SO_4{}^{2-}$ on the catalyst surface. This increased NH_3 adsorption and promoted NH_3 reaction with NO via Eley−Rideal mechanism.

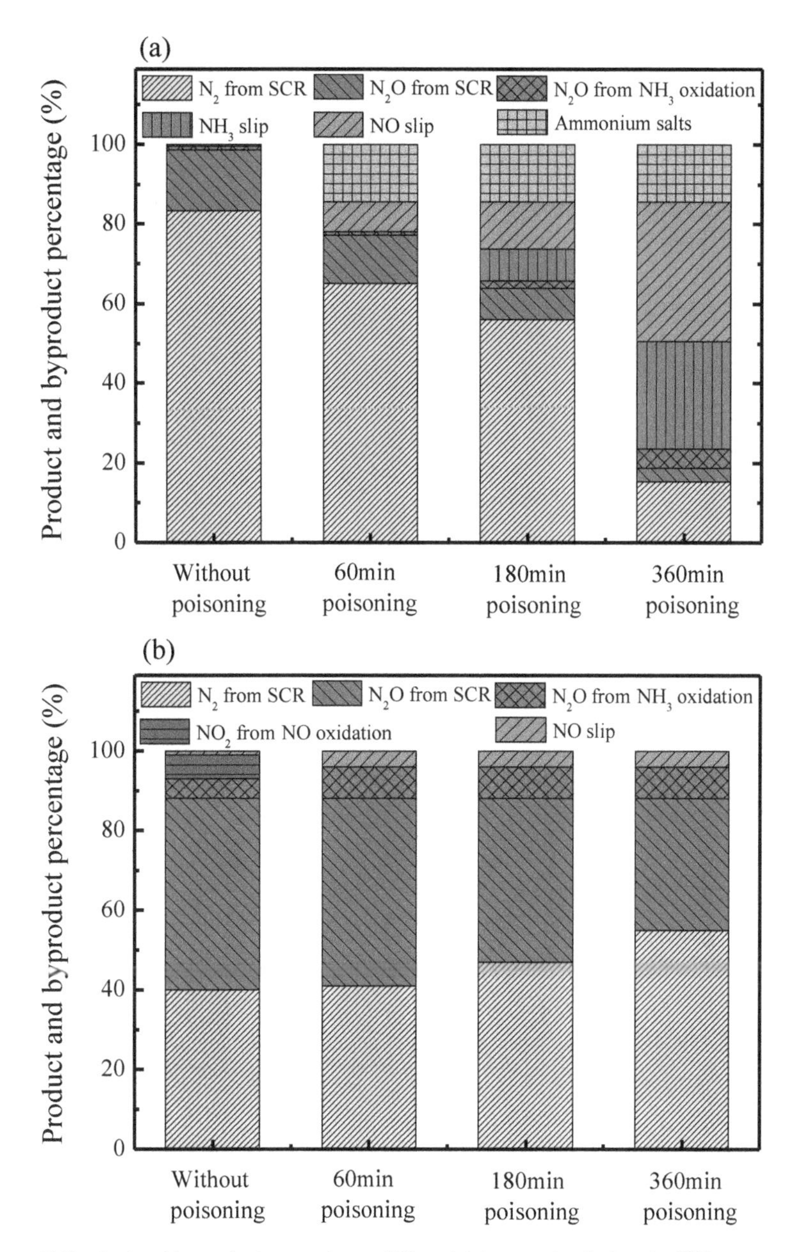

Figure 7. Product and byproduct percentages of N-containing species during the SCR process over MnFe/TiO_2 catalyst with SO_2 poisoning at (**a**) 150 °C and (**b**) 300 °C. The inlet NH_3 and NO molar concentration ratio was 1:1; and the percentage calculation formulas were based on those listed in Table 3.

2.5. Reaction Pathways

From the above results, one could surmise the reaction pathway in the NH_3-SCR system with/without SO_2 at 150 °C and 300 °C as shown in Figure 8. When the SCR system was at 150 °C without SO_2, it can be seen from the top left plot that a fraction of NH_3 and NO would be oxidized to

N_2O and NO_2, respectively. In addition, the major reaction product of SCR reaction was N_2 instead of N_2O. On the other hand, when increasing temperature to 300 °C, it can be seen from the bottom left plot that NH_3 would be oxidized to both N_2O and NO. Moreover, the major reaction product of SCR reaction was N_2O instead of N_2, which was revealed by the low N_2-selectivity as seen in Figure 7b.

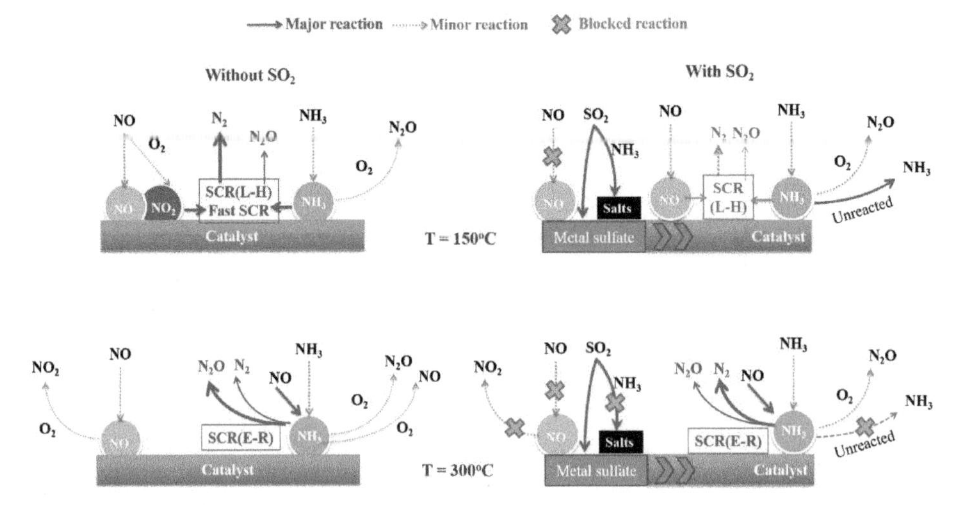

Figure 8. Proposed reaction pathway in the NH_3-SCR system with/without SO_2 at 150 °C and 300 °C. The reaction pathway was based on inlet NH_3 and NO molar concentration ratio of 1:1.

When SO_2 was introduced at 150 °C, it can be seen from the top right plot of Figure 8 that SO_2 would be reacted with both NH_3 and metal catalyst, which resulted in the formation of ammonium salts and metal sulfates, respectively. In addition, because NO could not be adsorbed on metal sulfates, therefore it could not be reacted with NH_3 at low temperature. As a result, unreacted NH_3 (NH_3 slip) turned out to be the major N-containing species in addition to the unreacted NO.

When increasing temperature to 300 °C, it can be seen from the bottom right plot that ammonium salts would not be formed in the presence of SO_2, but SO_2 would react with metal catalyst to form metal sulfates. In addition, gaseous NO could directly react with adsorbed ammonia via Eley−Rideal mechanism [29]. The major reaction product of SCR reaction gradually changed from N_2O to N_2 after increasing poisoning time.

3. Materials and Method

3.1. Reactions in SCR System

In the SCR system, it may contain oxidation reactions, SCR reactions and SO_2 poisoning reactions [46–52].

NH_3 oxidation:

$$2NH_3 + 2O_2 \rightarrow N_2O + 3H_2O \quad (3)$$

$$4NH_3 + 5O_2 \rightarrow 4NO + 6H_2O \quad (4)$$

NO oxidation:

$$2NO + O_2 \rightarrow 2NO_2 \quad (5)$$

SCR reactions:

$$4NO + 4NH_3 + O_2 \rightarrow 4N_2 + 6H_2O \quad (6)$$

$$4NO + 4NH_3 + 3O_2 \rightarrow 4N_2O + 6H_2O \quad (7)$$

$$NO + NO_2 + 2NH_3 \rightarrow 2N_2 + 3H_2O \text{ (fast SCR)} \quad (8)$$

SO_2 oxidation and poisoning:

$$SO_2 + 1/2\, O_2 \rightarrow SO_3 \quad (9)$$

$$SO_3 + 2NH_3 + H_2O \rightarrow (NH_4)_2SO_4 \quad (10)$$

$$SO_3 + NH_3 + H_2O \rightarrow NH_4HSO_4 \quad (11)$$

$$SO_2 + \text{metal} \rightarrow \text{metal sulfates (e.g., } MnSO_4 \text{ and } Ti(SO_4)_2) \quad (12)$$

In this study, experimental tests were designed to clarify the products and byproducts of the above reactions.

3.2. Synthesis of MnFe/TiO2 Catalysts

Mn and Fe metal oxides were supported on TiO_2 (in the form of $TiO(OH)_2$) by the co-precipitation method. In a typical procedure, 8 g of TiO_2 (China Steel Corp., Kaohsiung, Taiwan), 11.57 g of ferric nitrate 9-hydrate (99%,J.T. Baker, Radnor, PA, USA), 7.13 g of manganese (II) acetate tetrahydrate (99%, Merck, Kenilworth, NJ, USA) and D.I. water (76 g) were mixed then adjusted to pH = 10 with 25 wt.% ammonia solution to form a precipitate. It was filtered and washed thoroughly with D.I. water, then dried at 120 °C for 12 h. Finally, the material was calcined at 350 °C for 6 h in air.

3.3. Catalyst Reaction

The NH_3 oxidation, NO oxidation and SCR activity tests were carried out at atmospheric pressure in a fixed-bed reactor loaded with sieved pelletized (16–30 mesh) catalysts. The operation conditions of inlet gas concentrations, reaction temperatures, and gas hourly space velocity for different tests are shown in Table 3. Under typical NH_3 oxidation, NO oxidation, and SCR reaction tests, the concentrations of NH_3 and/or NO were the same at 500 ppmv, and the SO_2 concentration was 150 ppmv if SO_2 poisoning effect was considered. The feed gases were mixed in a gas mixer. Then the catalysts were preheated in the reactor for 30 min to ensure that an isothermal reaction temperature was reached. During the oxidation and SCR test, the NO and SO_2 concentrations at the inlet and outlet of the reactor were monitored by a NO/SO_2 analyzer (Ultramat 23, SIEMENS, Munich, Germany). In addition, the concentrations of NH_3, N_2O, and NO_2 at the inlet and outlet of the reactor were monitored by a FTIR Spectrophotometer (Bomem MB 104,San Jose, CA, USA and ITRI, Hsinchu, Taiwan).

Table 3. Operation conditions of experimental tests in this study.

	NH_3 (ppmv)	NO (ppmv)	SO_2 (ppmv)	O_2 (%)	Temperature (°C)	GHSV (hr^{-1})
NH_3 oxidation test	500	0	0	10%	100–300	50,000
NO oxidation test	0	500	0	10%	100–300	50,000
SCR test without SO_2	500	500	0	10%	100–300	50,000
SCR test with SO_2	500	500	150	10%	150 & 300	50,000
SCR test with SO_2 at different NH_3 amounts	500–600	500	150	10%	150	50,000

The NO consumption due to SCR reactions (Equations (6)–(8)) must be subtracted by the NO oxidation to NO_2 (Equation (5)). Thus, the NO consumption of SCR, $[NO]_{SCR}$ is defined by:

$$[NO]_{SCR} = [NO]_{in} - ([NO]_{out} - [NO_2]_{out}) \quad (13)$$

Since the N-containing product and byproduct of SCR reaction are N_2 and N_2O from Equations (6)–(8), thus the N_2-selectivity of SCR reactions was calculated by

$$N_2 \text{ selectivity of SCR reaction } = \left[1 - \frac{[NO]_{out} + [NH_3]_{out} + [N_2O]_{out} + [NO_2]_{out}}{[NO]_{in} + [NH_3]_{in}}\right] \times 100\% \quad (14)$$

The TGA was conducted to determine the sulfates species forming on the surface of the catalysts with a NETZSCH TG 209 F1 apparatus. The heating program was carried out under airflow of 10 mL/min with a heating rate of 10 °C/min from room temperature to 900 °C.

4. Conclusions

This study employed $MnFe/TiO_2$ catalyst to study the product/byproducts for the oxidation of NH_3, the oxidation of NO and the NH_3-SCR reaction with/without SO_2 to understand the reaction pathway of medium to low-temperature SCR process. For SCR operation temperature of 150 °C without the presence of SO_2, the major product of SCR process appeared to be N_2. The minor presence of N_2O byproduct was majorly from the SCR reaction instead of from the NH_3 oxidation reaction. Moreover, the result indicated that products of the SCR reaction gradually changed from N_2 to N_2O when raising the temperature from 100 to 300 °C. Therefore, the N_2-selectivity of SCR reaction was gradually decreased. One the other hand when SO_2 was introduced at 150 °C, the percentages of N_2 and N_2O decreased after increased poisoning time. However, when increasing temperature to 300 °C, the percentages of N_2 increased while that of N_2O decreased after increasing the poisoning time. This indicated the existence of SO_2 promotion effect on the NH_3-SCR at 300 °C.

Author Contributions: Conceptualization, T.L. and H.B.; methodology, T.L. and H.B.; validation, H.B.; formal analysis, T.L.; investigation, T.L.; resources, H.B.; data curation, T.L. and H.B.; writing—original draft preparation, T.L.; writing—review and editing, H.B.; visualization, T.L. and H.B.; supervision, H.B.; project administration, H.B.; funding acquisition, H.B.

Funding: This research received no external funding.

Acknowledgments: The authors gratefully acknowledge the financial support from the Ministry of Science and Technology, Taiwan through grant No: MOST 105-3113-E-009-003.

Conflicts of Interest: The authors declare no conflict of interest.

References

1. Busca, G.; Lietti, L.; Ramis, G.; Berti, F. Chemical and mechanistic aspects of the selective catalytic reduction of NO_x by ammonia over oxide catalysts: A review. *Appl. Catal. B Environ.* **1998**, *18*, 1–36. [CrossRef]
2. Chang, Y.Y.; Yan, Y.L.; Tseng, C.H.; Syu, J.Y.; Lin, W.Y.; Yuan, Y.C. Development of an Innovative Circulating Fluidized-Bed with Microwave System for Controlling NO_x. *Aerosol Air Qual. Res.* **2012**, *12*, 379–386. [CrossRef]
3. Garcia-Bordeje, E.; Pinilla, J.L.; Lazaro, M.J.; Moliner, R.; Fierro, J.L.G. Role of sulphates on the mechanism of NH_3-SCR of NO at low temperatures over presulphated vanadium supported on carbon-coated monoliths. *J. Catal.* **2005**, *233*, 166–175. [CrossRef]
4. Gan, L.N.; Guo, F.; Yu, J.; Xu, G.W. Improved Low-Temperature Activity of V_2O_5-WO_3/TiO_2 for Denitration Using Different Vanadium Precursors. *Catalysts* **2016**, *6*, 25. [CrossRef]
5. Kompio, P.G.W.A.; Bruckner, A.; Hipler, F.; Auer, G.; Loffler, E.; Grunert, W. A new view on the relations between tungsten and vanadium in V_2O_5-WO_3/TiO_2 catalysts for the selective reduction of NO with NH_3. *J. Catal.* **2012**, *286*, 237–247. [CrossRef]
6. Krocher, O. Selective Catalytic Reduction of NO_x. *Catalysts* **2018**, *8*, 459. [CrossRef]
7. Balle, P.; Geiger, B.; Kureti, S. Selective catalytic reduction of NO_x by NH_3 on Fe/HBEA zeolite catalysts in oxygen-rich exhaust. *Appl. Catal. B Environ.* **2009**, *85*, 109–119. [CrossRef]
8. Kim, M.H.; Yang, K.H. The Role of Fe_2O_3 Species in Depressing the Formation of N_2O in the Selective Reduction of NO by NH_3 over V_2O_5/TiO_2-Based Catalysts. *Catalysts* **2018**, *8*, 134.

9. Lee, S.M.; Kim, S.S.; Hong, S.C. Systematic mechanism study of the high temperature SCR of NO_x by NH_3 over a W/TiO_2 catalyst. *Chem. Eng. Sci.* **2012**, *79*, 177–185.
10. Ning, R.; Chen, L.; Li, E.; Liu, X.; Zhu, T. Applicability of V_2O_5-WO_3/TiO_2 Catalysts for the SCR Denitrification of Alumina Calcining Flue Gas. *Catalysts* **2019**, *9*, 220. [CrossRef]
11. Gao, R.H.; Zhang, D.S.; Liu, X.G.; Shi, L.Y.; Maitarad, P.; Li, H.R.; Zhang, J.P.; Cao, W.G. Enhanced catalytic performance of V_2O_5-$WO_3/Fe_2O_3/TiO_2$ microspheres for selective catalytic reduction of NO by NH_3. *Catal. Sci. Technol.* **2013**, *3*, 191–199. [CrossRef]
12. Xu, W.Q.; He, H.; Yu, Y.B. Deactivation of a Ce/TiO_2 Catalyst by SO_2 in the Selective Catalytic Reduction of NO by NH_3. *J. Phys. Chem. C* **2009**, *113*, 4426–4432. [CrossRef]
13. Yang, R.; Huang, H.F.; Chen, Y.J.; Zhang, X.X.; Lu, H.F. Performance of Cr-doped vanadia/titania catalysts for low-temperature selective catalytic reduction of NO_x with NH_3. *Chin. J. Catal.* **2015**, *36*, 1256–1262. [CrossRef]
14. Kong, Z.J.; Wang, C.; Ding, Z.N.; Chen, Y.F.; Zhang, Z.K. Enhanced activity of $Mn_xW_{0.05}Ti_{0.95-x}O_{2-\delta}$ for selective catalytic reduction of NO_x with ammonia by self-propagating high-temperature synthesis. *Catal. Commun.* **2015**, *64*, 27–31. [CrossRef]
15. Lian, Z.H.; Liu, F.D.; He, H.; Shi, X.Y.; Mo, J.S.; Wu, Z.B. Manganese-niobium mixed oxide catalyst for the selective catalytic reduction of NO_x with NH_3 at low temperatures. *Chem. Eng. J.* **2014**, *250*, 390–398. [CrossRef]
16. Gao, C.; Shi, J.W.; Fan, Z.Y.; Gao, G.; Niu, C.M. Sulfur and Water Resistance of Mn-Based Catalysts for Low-Temperature Selective Catalytic Reduction of NO_x: A Review. *Catalysts* **2018**, *8*, 11. [CrossRef]
17. Putluru, S.S.R.; Schill, L.; Godiksen, A.; Poreddy, R.; Mossin, S.; Jensen, A.D.; Fehrmann, R. Promoted V_2O_5/TiO_2 catalysts for selective catalytic reduction of NO with NH_3 at low temperatures. *Appl. Catal. B Environ.* **2016**, *183*, 282–290. [CrossRef]
18. Gil, S.; Garcia-Vargas, J.M.; Liotta, L.F.; Pantaleo, G.; Ousmane, M.; Retailleau, L.; Giroir-Fendler, A. Catalytic Oxidation of Propene over Pd Catalysts Supported on CeO_2, TiO_2, Al_2O_3 and M/Al_2O_3 Oxides (M = Ce, Ti, Fe, Mn). *Catalysts* **2015**, *5*, 671–689. [CrossRef]
19. Liu, L.; Gao, X.; Song, H.; Zheng, C.H.; Zhu, X.B.; Luo, Z.Y.; Ni, M.J.; Cen, K.F. Study of the Promotion Effect of Iron on Supported Manganese Catalysts for No Oxidation. *Aerosol Air Qual. Res.* **2014**, *14*, 1038–1046. [CrossRef]
20. Shi, J.; Zhang, Z.H.; Chen, M.X.; Zhang, Z.X.; Shangguan, W.F. Promotion effect of tungsten and iron co-addition on the catalytic performance of MnO_x/TiO_2 for NH_3-SCR of NOx. *Fuel* **2017**, *210*, 783–789. [CrossRef]
21. Zhu, L.; Zhong, Z.P.; Yang, H.; Wang, C.H. NH_3-SCR Performance of Mn-Fe/TiO_2 Catalysts at Low Temperature in the Absence and Presence of Water Vapor. *Water Air Soil Poll.* **2016**, *227*, 476. [CrossRef]
22. Deng, S.C.; Zhuang, K.; Xu, B.L.; Ding, Y.H.; Yu, L.; Fan, Y.N. Promotional effect of iron oxide on the catalytic properties of Fe-MnO_x/TiO_2 (anatase) catalysts for the SCR reaction at low temperatures. *Catal. Sci. Technol.* **2016**, *6*, 1772–1778. [CrossRef]
23. Dong, L.F.; Fan, Y.M.; Ling, W.; Yang, C.; Huang, B.C. Effect of Ce/Y Addition on Low-Temperature SCR Activity and SO_2 and H_2O Resistance of MnO_x/ZrO_2/MWCNTs Catalysts. *Catalysts* **2017**, *7*, 181. [CrossRef]
24. Mu, W.T.; Zhu, J.; Zhang, S.; Guo, Y.Y.; Su, L.Q.; Li, X.Y.; Li, Z. Novel proposition on mechanism aspects over Fe-Mn/ZSM-5 catalyst for NH_3-SCR of NO_x at low temperature: Rate and direction of multifunctional electron-transfer-bridge and in situ DRIFTs analysis. *Catal. Sci. Technol.* **2016**, *6*, 7532–7548. [CrossRef]
25. Chen, T.; Guan, B.; Lin, H.; Zhu, L. In situ DRIFTS study of the mechanism of low temperature selective catalytic reduction over manganese-iron oxides. *Chin. J. Catal.* **2014**, *35*, 294–301. [CrossRef]
26. Zhou, G.Y.; Zhong, B.C.; Wang, W.H.; Guan, X.J.; Huang, B.C.; Ye, D.Q.; Wu, H.J. In situ DRIFTS study of NO reduction by NH_3 over Fe-Ce-Mn/ZSM-5 catalysts. *Catal. Today* **2011**, *175*, 157–163. [CrossRef]
27. Lin, Q.C.; Li, J.H.; Ma, L.; Hao, J.M. Selective catalytic reduction of NO with NH_3 over Mn-Fe/USY under lean burn conditions. *Catal. Today* **2010**, *151*, 251–256. [CrossRef]
28. Zhang, K.; Yu, F.; Zhu, M.Y.; Dan, J.M.; Wang, X.G.; Zhang, J.L.; Dai, B. Enhanced Low Temperature NO Reduction Performance via MnO_x-Fe_2O_3/Vermiculite Monolithic Honeycomb Catalysts. *Catalysts* **2018**, *8*, 100. [CrossRef]

29. Liu, F.D.; He, H.; Ding, Y.; Zhang, C.B. Effect of manganese substitution on the structure and activity of iron titanate catalyst for the selective catalytic reduction of NO with NH_3. *Appl. Catal. B Environ.* **2009**, *93*, 194–204. [CrossRef]
30. Shu, Y.; Sun, H.; Quan, X.; Chen, S.O. Enhancement of Catalytic Activity Over the Iron-Modified Ce/TiO_2 Catalyst for Selective Catalytic Reduction of NO_x with Ammonia. *J. Phys. Chem. C* **2012**, *116*, 25319–25327. [CrossRef]
31. Yang, S.J.; Xiong, S.C.; Liao, Y.; Xiao, X.; Qi, F.H.; Peng, Y.; Fu, Y.W.; Shan, W.P.; Li, J.H. Mechanism of N_2O Formation during the Low-Temperature Selective Catalytic Reduction of NO with NH_3 over Mn-Fe Spinel. *Environ. Sci. Technol.* **2014**, *48*, 10354–10362. [CrossRef] [PubMed]
32. Lee, T.Y.; Liou, S.Y.; Bai, H.L. Comparison of titania nanotubes and titanium dioxide as supports of low-temperature selective catalytic reduction catalysts under sulfur dioxide poisoning. *J. Air Waste Manag.* **2017**, *67*, 292–305. [CrossRef] [PubMed]
33. Tuenter, G.; Vanleeuwen, W.F.; Snepvangers, L.J.M. Kinetics and Mechanism of the NO_x Reduction with NH_3 on V_2O_5-WO_3-TiO_2 Catalyst. *Ind. Eng. Chem. Prod. Res. Dev.* **1986**, *25*, 633–636. [CrossRef]
34. Kang, M.; Park, E.D.; Kim, J.M.; Yie, J.E. Manganese oxide catalysts for NO_x reduction with NH_3 at low temperatures. *Appl. Catal. A Gen.* **2007**, *327*, 261–269. [CrossRef]
35. Ruggeri, M.P.; Grossale, A.; Nova, I.; Tronconi, E.; Jirglova, H.; Sobalik, Z. FTIR in situ mechanistic study of the NH_3-NO/NO_2 "Fast SCR" reaction over a commercial Fe-ZSM-5 catalyst. *Catal. Today* **2012**, *184*, 107–114. [CrossRef]
36. Park, K.H.; Lee, S.M.; Kim, S.S.; Kwon, D.W.; Hong, S.C. Reversibility of Mn Valence State in MnO_x/TiO_2 Catalysts for Low-temperature Selective Catalytic Reduction for NO with NH_3. *Catal. Lett.* **2013**, *143*, 246–253. [CrossRef]
37. Cao, F.; Xiang, J.; Su, S.; Wang, P.Y.; Hu, S.; Sun, L.S. Ag modified Mn-Ce/gamma-Al_2O_3 catalyst for selective catalytic reduction of NO with NH_3 at low-temperature. *Fuel Process. Technol.* **2015**, *135*, 66–72. [CrossRef]
38. Xu, H.D.; Fang, Z.T.; Cao, Y.; Kong, S.; Lin, T.; Gong, M.C.; Chen, Y.Q. Influence of Mn/(Mn plus Ce) Ratio of MnO_x-CeO_2/WO_3-ZrO_2 Monolith Catalyst on Selective Catalytic Reduction of NO_x with Ammonia. *Chin. J. Catal.* **2012**, *33*, 1927–1937. [CrossRef]
39. Liu, F.D.; Asakura, K.; He, H.; Shan, W.P.; Shi, X.Y.; Zhang, C.B. Influence of sulfation on iron titanate catalyst for the selective catalytic reduction of NO_x with NH_3. *Appl. Catal. B Environ.* **2011**, *103*, 369–377. [CrossRef]
40. Shu, Y.; Aikebaier, T.; Quan, X.; Chen, S.; Yu, H.T. Selective catalytic reaction of NO_x with NH_3 over Ce-Fe/TiO_2-loaded wire-mesh honeycomb: Resistance to SO_2 poisoning. *Appl. Catal. B Environ.* **2014**, *150*, 630–635. [CrossRef]
41. Zhang, L.F.; Zhang, X.L.; Lv, S.S.; Wu, X.P.; Wang, P.M. Promoted performance of a MnO_x/PG catalyst for low-temperature SCR against SO_2 poisoning by addition of cerium oxide. *RSC Adv.* **2015**, *5*, 82952–82959. [CrossRef]
42. Pourkhalil, M.; Moghaddam, A.Z.; Rashidi, A.; Towfighi, J.; Mortazavi, Y. Preparation of highly active manganese oxides supported on functionalized MWNTs for low temperature NO_x reduction with NH_3. *Appl. Surf. Sci.* **2013**, *279*, 250–259. [CrossRef]
43. Wang, X.B.; Gui, K.T. Fe_2O_3 particles as superior catalysts for low temperature selective catalytic reduction of NO with NH_3. *J. Environ. Sci.* **2013**, *25*, 2469–2475. [CrossRef]
44. Jin, R.B.; Liu, Y.; Wang, Y.; Cen, W.L.; Wu, Z.B.; Wang, H.Q.; Weng, X.L. The role of cerium in the improved SO_2 tolerance for NO reduction with NH_3 over Mn-Ce/TiO_2 catalyst at low temperature. *Appl. Catal. B Environ.* **2014**, *148*, 582–588. [CrossRef]
45. Shen, B.X.; Zhang, X.P.; Ma, H.Q.; Yao, Y.; Liu, T. A comparative study of Mn/CeO_2, Mn/ZrO_2 and Mn/Ce-ZrO_2 for low temperature selective catalytic reduction of NO with NH_3 in the presence of SO_2 and H_2O. *J. Environ. Sci. China* **2013**, *25*, 791–800. [CrossRef]
46. Lee, T.; Bai, H. Metal Sulfate Poisoning Effects over MnFe/TiO_2 for Selective Catalytic Reduction of NO by NH_3 at Low Temperature. *Ind. Eng. Chem. Res.* **2018**, *57*, 4848–4858. [CrossRef]
47. Busca, G.; Larrubia, M.A.; Arrighi, L.; Ramis, G. Catalytic abatement of NO_x: Chemical and mechanistic aspects. *Catal. Today* **2005**, *107–108*, 139–148. [CrossRef]
48. Kapteijn, F.; Singoredjo, L.; Andreini, A.; Moulijn, J.A. Activity and Selectivity of Pure Manganese Oxides in the Selective Catalytic Reduction of Nitric-Oxide with Ammonia. *Appl. Catal. B Environ.* **1994**, *3*, 173–189. [CrossRef]

49. Wang, Y.L.; Li, X.X.; Zhan, L.; Li, C.; Qiao, W.M.; Ling, L.C. Effect of SO_2 on Activated Carbon Honeycomb Supported CeO_2-MnO_x Catalyst for NO Removal at Low Temperature. *Ind. Eng. Chem. Res.* **2015**, *54*, 2274–2278. [CrossRef]
50. Yang, W.W.; Liu, F.D.; Xie, L.J.; Lan, Z.H.; He, H. Effect of V_2O_5 Additive on the SO_2 Resistance of a Fe_2O_3/AC Catalyst for NH_3-SCR of NO_x at Low Temperatures. *Ind. Eng. Chem. Res.* **2016**, *55*, 2677–2685. [CrossRef]
51. Qiu, L.; Wang, Y.; Pang, D.D.; Ouyang, F.; Zhang, C.L.; Cao, G. Characterization and Catalytic Activity of Mn-Co/TiO_2 Catalysts for NO Oxidation to NO_2 at Low Temperature. *Catalysts* **2016**, *6*, 9. [CrossRef]
52. Lippits, M.J.; Gluhoi, A.C.; Nieuwenhuys, B.E. A comparative study of the selective oxidation of NH_3 to N_2 over gold, silver and copper catalysts and the effect of addition of Li_2O and CeO_x. *Catal. Today* **2008**, *137*, 446–452. [CrossRef]

Article

Microstructural Characteristics of Vehicle-Aged Heavy-Duty Diesel Oxidation Catalyst and Natural Gas Three-Way Catalyst

Tomi Kanerva [1,†], Mari Honkanen [1], Tanja Kolli [2], Olli Heikkinen [3,‡], Kauko Kallinen [4], Tuomo Saarinen [5,§], Jouko Lahtinen [3], Eva Olsson [6], Riitta L. Keiski [2] and Minnamari Vippola [1,*]

1. Faculty of Engineering and Natural Sciences, Tampere University, P.O. Box 589, FI-33014 Tampere, Finland; tomi.kanerva@ttl.fi (T.K.); mari.honkanen@tuni.fi (M.H.)
2. Faculty of Technology, University of Oulu, P.O. Box 4300, FI-90014 Oulu, Finland; tanja.kolli@oulu.fi (T.K.); riitta.keiski@oulu.fi (R.L.K.)
3. Department of Applied Physics, Aalto University, P.O. Box 110, FI-00076 Aalto, Finland; olli.heikkinen@gmail.com (O.H.); jouko.lahtinen@aalto.fi (J.L.)
4. Dinex Finland Oy, Vihtavuorentie 162, FI-41330 Vihtavuori, Finland; kki@dinex.fi
5. SSAB Europe Oy, Rautaruukintie 155, FI-92101 Raahe, Finland; tuomo.saarinen@sandvik.com
6. Department of Physics, Chalmers University of Technology, SE-412 96 Göteborg, Sweden; eva.olsson@chalmers.se

* Correspondence: minnamari.vippola@tuni.fi; Tel.: +358-40-8490148

† Currently Finnish Institute of Occupational Health, P.O. Box 40, FI-00032 Työterveyslaitos, Finland.

‡ Currently Murata Electronics Oy, Myllynkivenkuja 6, FI-01621 Vantaa, Finland.

Received: 3 December 2018; Accepted: 18 January 2019; Published: 1 February 2019

Abstract: Techniques to control vehicle engine emissions have been under increasing need for development during the last few years in the more and more strictly regulated society. In this study, vehicle-aged heavy-duty catalysts from diesel and natural gas engines were analyzed using a cross-sectional electron microscopy method with both a scanning electron microscope and a transmission electron microscope. Also, additional supporting characterization methods including X-ray diffractometry, X-ray photoelectron spectroscopy, Fourier-transform infrared spectroscopy and catalytic performance analyses were used to reveal the ageing effects. Structural and elemental investigations were performed on these samples, and the effect of real-life ageing of the catalyst was studied in comparison with fresh catalyst samples. In the real-life use of two different catalysts, the poison penetration varied greatly depending on the engine and fuel at hand: the diesel oxidation catalyst appeared to suffer more thorough changes than the natural gas catalyst, which was affected only in the inlet part of the catalyst. The most common poison, sulphur, in the diesel oxidation catalyst was connected to cerium-rich areas. On the other hand, the severities of the ageing effects were more pronounced in the natural gas catalyst, with heavy structural changes in the washcoat and high concentrations of poisons, mainly zinc, phosphorus and silicon, on the surface of the inlet part.

Keywords: catalyst deactivation; diesel; natural gas; SEM; TEM; poisoning

---

1. Introduction

Since the introduction of the first catalytic converter for automobile exhaust gas catalysis, the development and research for cleaner and lower emissions have been attractive challenges, due to ever-tightening emission standards. There have been various directions in the development of more efficient systems for exhaust gas after-treatment, from engine design to catalytic converter and fuel technology development. In the recent decades, while the reduction of petroleum car emissions has improved and promising results from diesel engine emissions have arisen, the aspect of low-emission fuels has gained more and more interest. For example, biodiesel, ethanol and natural gas are challenging the traditional fuels in transportation use for the future. Although there have been advances in the lowering of emissions of these more efficient combustion processes, the deactivation and ageing problems of the catalysts remain at a certain level [1–12]. To meet the demands for more efficient and clean engines, there is a great need to study deactivation and ageing effects of these different types of fuels with catalytic materials and other phenomena.

Catalyst ageing effects and deactivation during its lifecycle in vehicles is a very complex phenomenon. There are two main problems typically described in the studies of those catalysts: poisoning and thermal ageing of the catalyst, both of them greatly depending on the used catalyst materials and conditions of the use. Poisoning refers to blocking of active sites for catalytic reactions in the catalyst, resulting from the chemisorption of fuel and lubricant impurities onto the catalytically active sites. Thermal ageing due to high temperature conditions in the vehicle engine can cause loss of the active surface via, for example, sintering and phase transformations. In heavy-duty vehicle engines, the roles and characteristics of these problems are still quite seldom addressed in comparison to studies from smaller vehicles and/or gasoline-fueled cars [7–14].

It is well known that sulphur and phosphorus originating from fuel and/or lubricating oil decrease efficiency of diesel and natural gas oxidation catalysts, for example, [11,15–18]. For example, in the Pd-rich catalysts for methane oxidation in natural gas engine exhausts, aluminum sulphate species formed under SO_2 exposure. This inhibited the CH_4 oxidation reaction in a broad temperature range [18]. In the Pt-based diesel oxidation catalysts, the formation of phosphates decreased the specific surface area of the catalyst and decreased catalyst efficiency [16].

In this study, vehicle-aged catalysts were analyzed, providing insight into ageing effects on real-life catalysts. According to the authors' knowledge, this information is mostly lacking [8]. The vehicle-aged and fresh heavy-duty catalysts from diesel and natural gas engines were analyzed using a cross-sectional electron microscopy method for both a scanning electron microscope (SEM) and (scanning) transmission electron microscope ((S)TEM) equipped with an energy dispersive spectrometer (EDS). Also, additional supporting characterization methods including X-ray diffractometry (XRD), X-ray photoelectron spectroscopy (XPS), Fourier-transform infrared (FTIR) spectroscopy and catalytic performance analyses were used to reveal the ageing effects. The fresh catalyst samples were studied in a similar manner as the vehicle-aged samples for finding out the effects of the real engine use on the structures and properties of the catalysts. Alumina was the main support material in the metallic support sheet for both catalysts, and platinum was the precious metal in the diesel oxidation catalyst together with palladium in the natural gas catalyst.

2. Results and Discussion

2.1. Diesel Oxidation Catalyst

2.1.1. Electron Microscopy

The cross-sectional SEM images of the diesel oxidation catalyst in the used form are presented in Figure 1. The diesel oxidation catalyst washcoat is an alumina-based mix of metal oxides, mainly zirconium, cerium and lanthanum, supporting the platinum catalysts. Platinum is spread evenly across the layers.

The SEM-EDS line analyses of selected elements in the inlet catalyst sheet are shown in Figure 2. Both the SEM-EDS line analyses and spot analyses indicate a higher sulphur content in the inner layer of the washcoat. As can be seen from Figure 2, the sulphur content follows the content of ceria in the washcoat, indicating an attachment between S and Ce. Sulphur is a known poison for ceria [19–23].

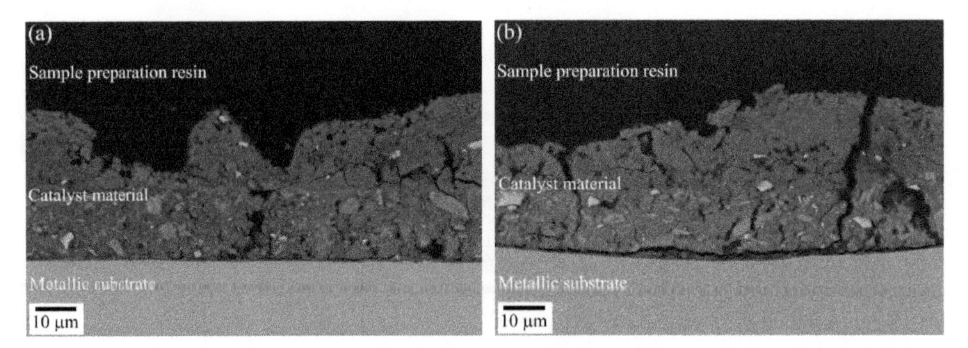

Figure 1. The cross-sectional SEM image of the diesel oxidation catalyst, (**a**) the inlet part of the vehicle-aged catalyst and (**b**) the outlet part of the vehicle-aged catalyst.

Phosphorus content follows alumina (Al and O) near the surface and decreases towards the bottom layer of the washcoat. The SEM-EDS spot analyses of layers of the inlet catalyst washcoat give sulphur contents maxima of 10 wt % for the ceria-rich inner layer and 5 wt % for the outer layer containing less ceria. Phosphorus was detected, but there are challenges in the calculations due to the P K_{α}-peak overlapping with the Pt M_{α}-peak and the Zr L_{α}-peak in the spectrum. However, the amount of phosphorus is significantly higher in the outer rather than in the inner layer. The EDS results from the used inlet and outlet diesel oxidation catalyst layers were similar.

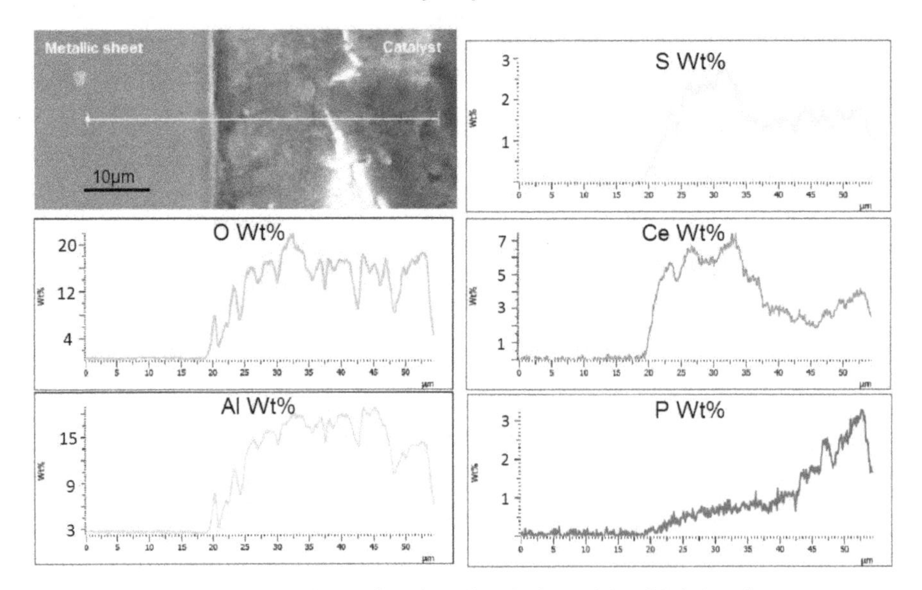

Figure 2. SEM-EDS line analysis from the vehicle-aged diesel (inlet) catalyst.

The EDS analyses from the TEM samples supported the SEM-EDS analyses (Figure 2). Detected poisons in the aged diesel oxidation catalyst samples, from both the inlet and outlet, were S, P, K, Ca and Zn, sulphur clearly with the highest percentage. These are typical poisons in diesel engines originating from fuels, lubricants and detergents [4]. TEM images from the fresh diesel oxidation catalyst sample (Figure 3a,b) show the fine structure of the washcoat with small Pt particles. The size of the Pt particles was <5 nm measured from the STEM images. In the used inlet sample (Figure 4a,c), the size of the platinum particles was 7.0 ± 1.9 nm (100 particles measured). Thus, the particle size increased slightly in the vehicle ageing. In addition, carbon (soot) and phosphorus were detected regularly (Figure 4b,c). Phosphorus exposure is known to cause such growth and shape effects in the noble metals [11,12].

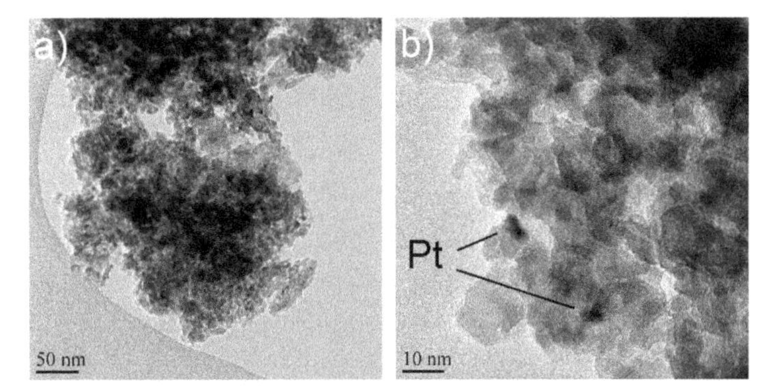

Figure 3. TEM images from the fresh diesel oxidation catalyst with (**a**) lower and (**b**) higher magnification.

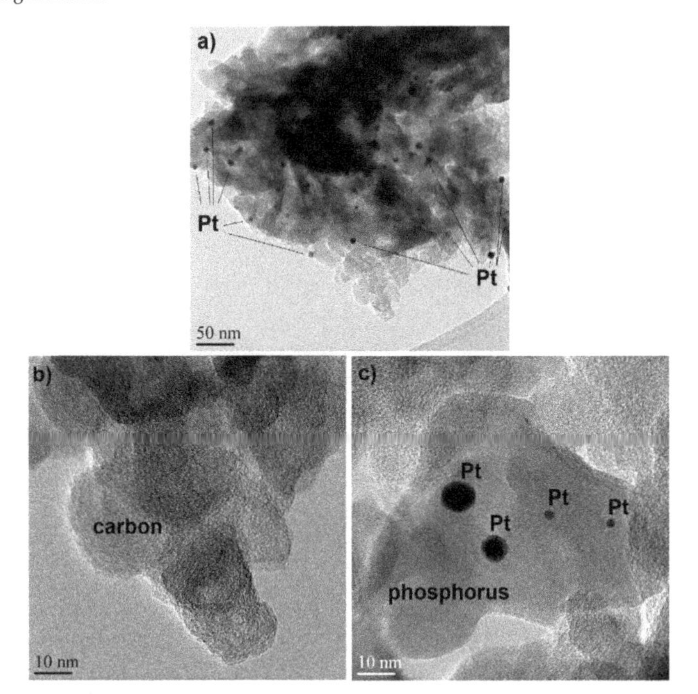

Figure 4. TEM images from (**a**) the vehicle-aged diesel inlet catalyst. (**b**) and (**c**) showing the carbon and phosphorus poisoned areas, respectively.

The cross-sectional TEM images from the fresh and vehicle-aged diesel oxidation catalysts are presented in Figure 5. The cross-sectional TEM–EDS map images from selected elements for the used inlet diesel sample are shown in Figure 6. In the studied sample, sulphur and phosphorus were the only poisons with a sufficient amount to be included in the EDS map. As seen in Figure 6, ceria and sulphur cover the same area in the analysed cross-section, as also indicated in the SEM-EDS line analyses (Figure 2). These Ce- and S-rich areas are clearly detectable as darker areas in the TEM images in Figure 5b. Phosphorus covers the washcoat more evenly, following the contents of aluminium and oxygen (Figure 6), as seen in the SEM-EDS line analyses (Figure 2). Overall, the platinum particles have grown in use, and there seem to be larger Pt particles outside the Ce/S-rich area than inside this area. A possibility of the Ce/S effect on the Pt particle sintering could be an interesting topic for future investigations, while such a phenomenon has not been extensively reported.

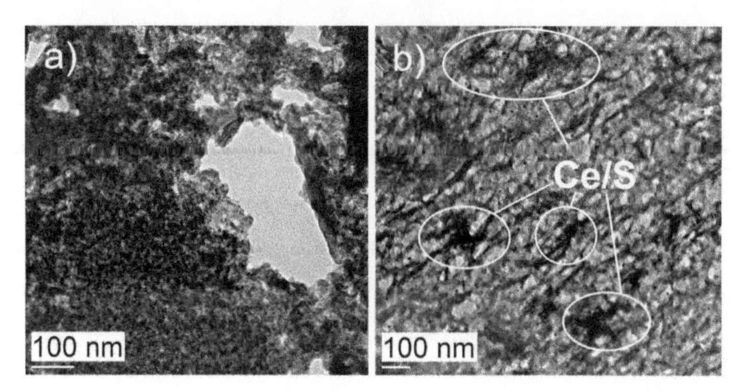

Figure 5. The cross-sectional TEM images of (**a**) the fresh diesel oxidation catalyst and (**b**) the vehicle-aged inlet diesel oxidation catalyst.

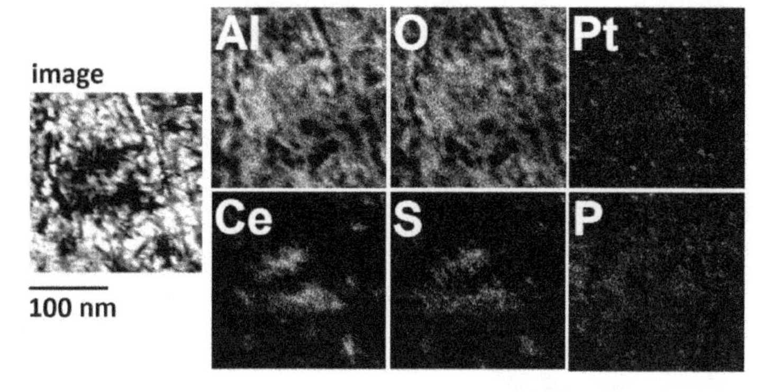

Figure 6. Cross-sectional TEM–EDS map from the vehicle-aged diesel (inlet) catalyst.

2.1.2. XRD and XPS Analysis, FTIR-ATR Measurements and Catalyst Performance Experiments

The X-ray diffraction analysis of diesel oxidation catalyst sheets is presented in Figure 7, revealing also an indication of an increase in the platinum particle size. The platinum peak was higher and sharper in the used samples. Differences between inlet and outlet samples were not observed, while both included obscure peaks arising from poisons and their compounds. This indicates that the deactivation has been effective throughout the catalyst from the inlet to the outlet.

The XPS analysis indicated that the binding energy of the platinum 3d5/2 line was 316.4 eV in the fresh sample and 312.5 eV in the aged one (Figure 8). This suggests that significant platinum reduction takes place during ageing [24]. As for catalyst poisons, survey scans showed prominent spectral lines for sulphur and phosphorus, and thus they were measured more thoroughly. The S 2p line was observed at 169.1 eV, whereas the P 2s line was located at 190.8 eV. The binding energies indicate that the poisons appear on the catalyst surface as sulphate and phosphate compounds [24]. The sulphur content was found to be 5.3 wt % and the phosphorus content 7.0 wt %.

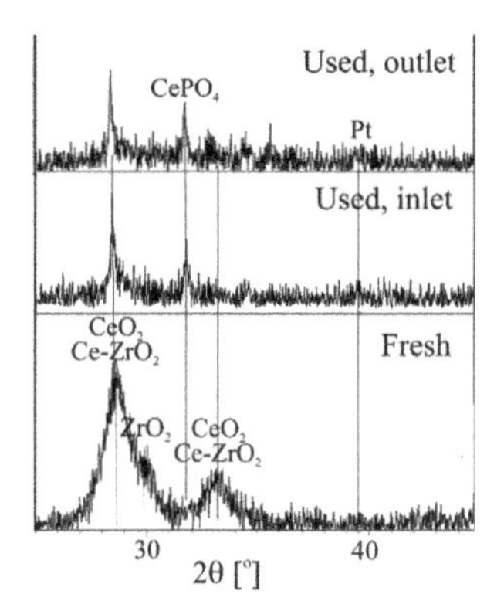

Figure 7. XRD patterns of the fresh and used diesel oxidation catalysts.

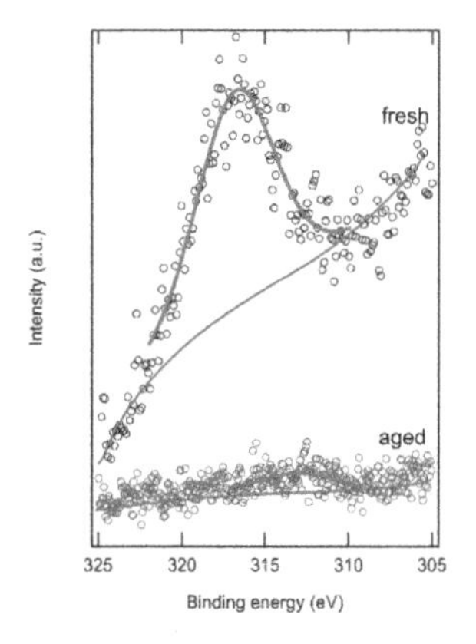

Figure 8. XPS spectra on platinum 4d5/2 lines measured from the diesel oxidation catalyst. While Shirley background subtraction has been used in the analysis, the background lines in the images have been drawn to guide the eye.

The FTIR-ATR spectra of the vehicle-aged and fresh diesel oxidation catalysts are presented in Figure 9. Only some differences can be found between the fresh and the aged catalyst spectra, from which the three most obvious peaks (1073, 1365 and 1625 cm^{-1}) are indicated.

The first broad peak at around 1100 cm^{-1} can be assigned to the sulphur species; for example, CeO_2 sulphates are known to give a broad peak at ~1200 cm^{-1} and bulk sulphates near to 1160 cm^{-1}. For the gas-phase SO_2, the major peaks are 1151 cm^{-1} and 1363 cm^{-1}, and for the gas phase, the SO_3 major peaks are 1061 cm^{-1} and 1391 cm^{-1} [19]. An indication of aluminium sulphates can also be found at 1360 cm^{-1}, 1290 cm^{-1}, 1170 cm^{-1} and 1045 cm^{-1} [25]. The presence of phosphorus can also be considered, since the group frequency for the functional group of the phosphate ion is 1100–1000 cm^{-1}. The broad peak at 1626 cm^{-1} was not clearly assigned to any of the poisons in the used diesel oxidation catalyst. According to the literature, there is a characteristic group frequency of organic nitrates at 1620–1640 cm^{-1} [26].

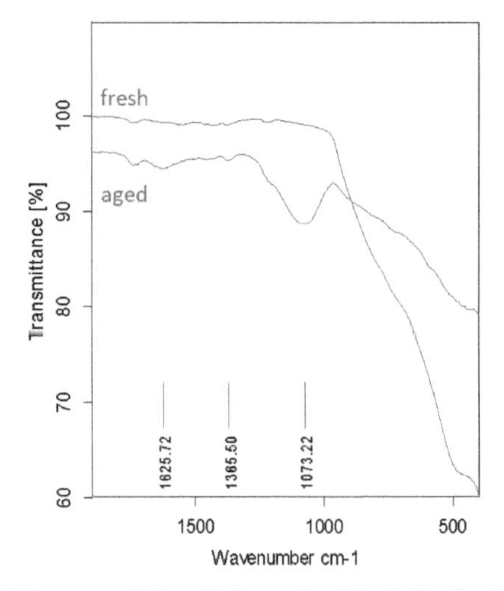

Figure 9. FTIR-ATR spectra of the vehicle-aged and fresh diesel oxidation catalysts.

The studied gases in the catalyst performance tests were those typical for lean diesel exhaust gas. CO and C_3H_6 were used to compare the performance of diesel and natural gas catalyst materials. The results of the performance test are presented in Figure 10. As can be observed from the light-off curves of CO and C_3H_6, the aged diesel oxidation catalyst was surprisingly more active than the fresh diesel oxidation catalyst. With the aged diesel oxidation catalyst, after the very lean reaction gas mixture experiment, the colour of the catalyst was changed from black to light grey, as shown in Figure 10. The reason for this phenomenon is that the oxygen has "cleaned" the catalyst surface, by burning carbon, and therefore more active sites for CO adsorption are exposed. This can to some extent explain why the aged catalyst was more active than the fresh one. As seen in Figure 11, a lot of carbon can be found in the aged sheet. The SEM-EDS analyses from these sheets indicate a decrease in the amount of carbon in the sheet surface (from around 30 wt % of carbon in the used sheet to around 15 wt % in the performance-tested light-grey sheet).

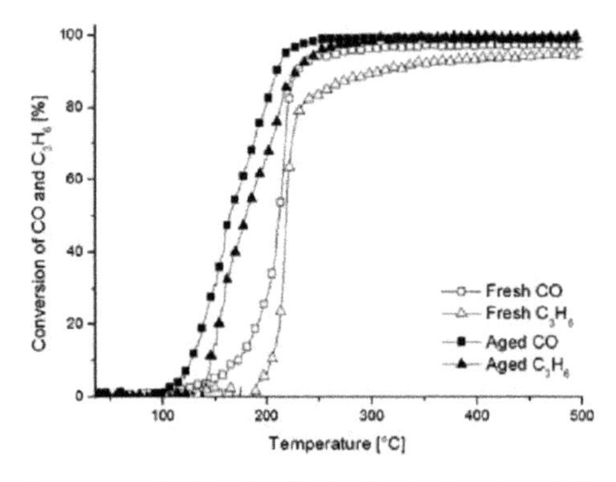

Figure 10. The conversion curves of CO and C_3H_6 for the fresh and aged diesel oxidation catalysts. Conditions: m = 530–550 mg, 500 ppm CO, 300 ppm C_3H_6, and 12 vol % air, balanced with N_2.

Figure 11. The sections of the diesel oxidation catalyst sheets: fresh, vehicle-aged and performance-tested.

2.2. Natural Gas Catalyst

2.2.1. Electron Microscopy

The cross-sectional SEM images of natural gas catalysts in the used form are presented in Figure 12. The NG catalyst washcoat is an alumina-based mixture of metal oxides, mainly zirconium and cerium, supporting palladium and platinum catalysts.

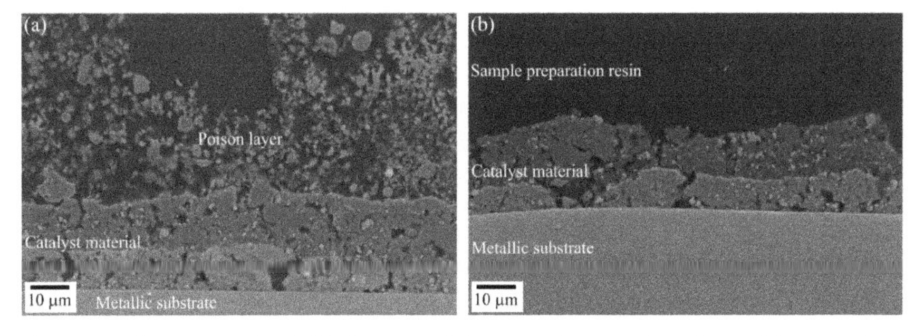

Figure 12. The cross-sectional SEM images of the natural gas catalyst, (**a**) the inlet part of the vehicle-aged catalyst and (**b**) the outlet part of the vehicle-aged catalyst.

In Figure 13, the SEM-EDS line analyses of the used inlet natural gas catalyst sheet are presented. On the top of the used inlet sample there is a porous layer containing mostly high concentrations of carbon and other poisoning elements, and this is excluded from the line analyses. Contents of poisons in the line analyses showed the presence of zinc, phosphorus and silicon throughout the washcoat. Zinc seems to attach to zirconia with contents up to 10 wt %, while silicon mainly follows the alumina content throughout the washcoat with contents of 5–12 wt %, and phosphorus is mostly in the outer layer of the catalyst with the content of 1–9 wt %. In the SEM-EDS spot analyses, a number of additional poisons were observed, mostly in the outer layer of the catalyst. These included potassium, chlorine, calcium and copper in variable contents depending on the analysed area. The natural gas outlet part did not have any poison layer, and EDS results did not show significant amounts of poisons in the outlet catalyst.

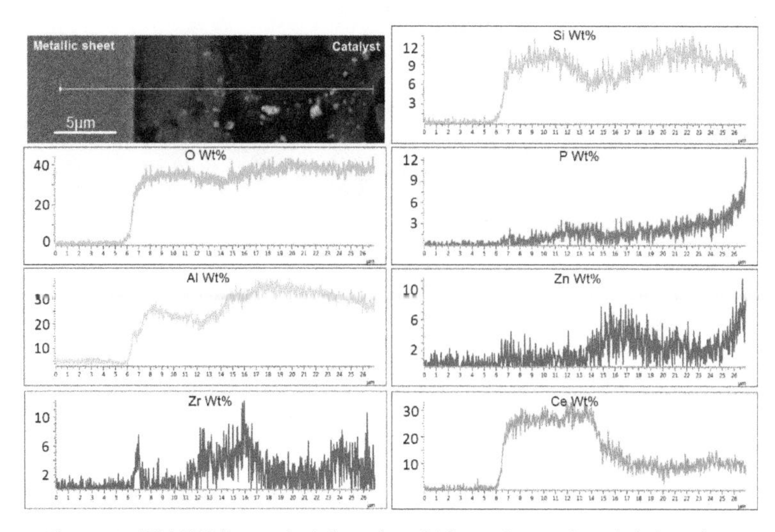

Figure 13. SEM-EDS line analysis from the vehicle-aged natural gas (inlet) catalyst.

The fine structure of the fresh NG catalyst washcoat is presented in Figure 14. Detected poisons in the TEM-EDS spot analyses from aged natural gas catalyst samples (Figure 15) were S, P, K, Zn, Si, Ca and Cu in various amounts. Copper was not calculated in these analyses, since the powder sample holder is made of copper and thus gives an additional copper signal to the spectrum. The used outlet catalyst showed few or no structural changes or poisonous impurities after use (Figure 15b).

Figure 14. TEM image from the fresh natural gas catalyst.

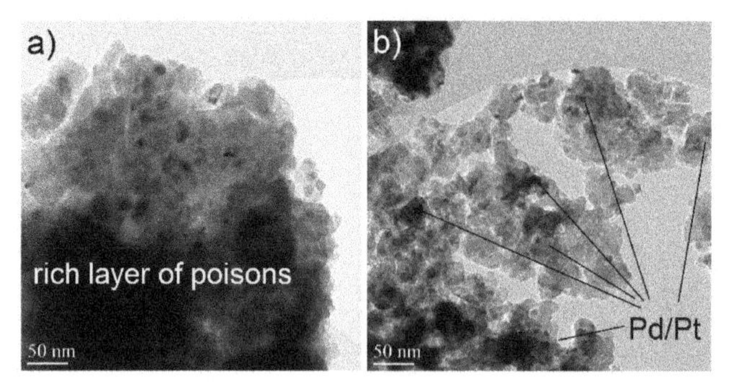

Figure 15. TEM images from the vehicle-aged natural gas catalyst (**a**) inlet, (**b**) outlet.

In the NG catalyst inlet part, significant noble metal particle growth was observed after the use, as seen in the cross-sectional TEM images in Figure 16a,b. In the fresh catalyst, the metal particle size was <5 nm, and after use it was 54 ± 23 nm as measured from the STEM images (100 particles measured). The structure of the used inlet catalyst had become more heterogeneous and dense compared to the fresh structure due to heavy sintering during use.

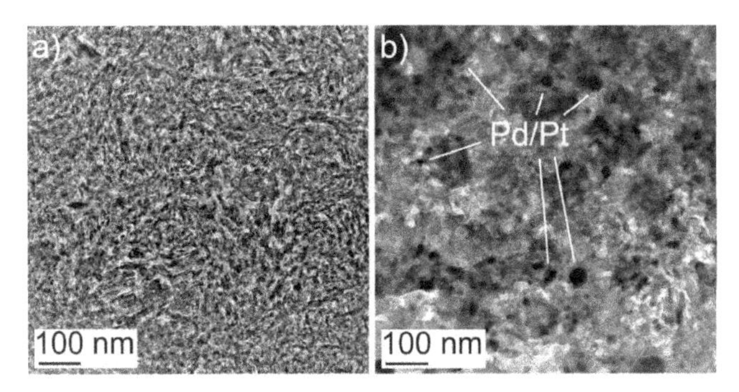

Figure 16. The cross-sectional TEM image of (**a**) the fresh natural gas catalyst and (**b**) the vehicle-aged inlet natural gas catalyst.

Three main poisoning elements, Ca, Cu and P, were included in the cross-sectional TEM-EDS map (Figure 17) of the inlet NG sample. One possible origin for copper can be the lubricant oil of the natural gas engine. In addition, copper seems to be mainly found together with noble metal particles (Figure 17). The presence of copper in the used catalysts needs further investigations in order to verify the occurrence of the phenomenon. High local concentrations of calcium were detected in both the EDS spot and map analyses. Pt was not included in the EDS map of the used sample due to its low amount. Again, phosphorus was spread quite evenly across the sample. The phosphorus detection has some uncertainties, due to the overlapping of Pt and Zr peaks.

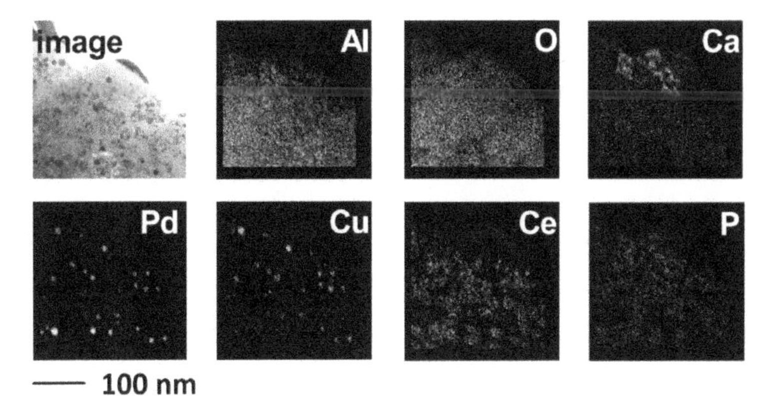

Figure 17. The cross-sectional TEM-EDS map of the inlet natural gas sample.

2.2.2. XRD and XPS Analysis, FTIR-ATR Measurements and Catalyst Performance Experiments

The XRD patterns of the vehicle-aged and fresh natural gas catalysts are presented in Figure 18. The main components of the catalyst are cerium zirconium oxide, cerium oxide, alumina and palladium. The amount of Pt is too low to be detected by XRD. The composition of the catalyst is complex and therefore the analysis of the patterns is challenging, especially for the used samples, where several peaks remained unidentified. However, noble metal crystal growth can be detected, as indicated with the Pd peak in Figure 18 in the spectra from the inlet and outlet parts of the catalyst. The presence of poisons is obvious in the pattern of the inlet part covering the actual catalyst structure. One of the identified poisons was ZnO, and zinc was also detected by the EDS line analysis (Figure 13).

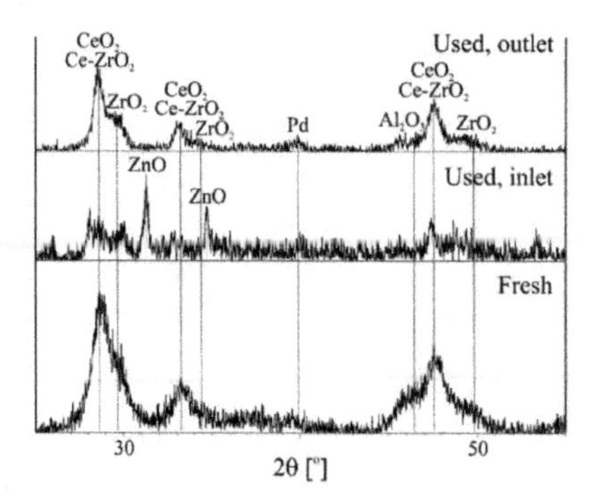

Figure 18. XRD patterns of the fresh and used natural gas catalysts.

The XPS analysis indicated several catalytic poisons in the inlet end of the aged catalyst, most notably silicon (14 wt %), calcium (4.7 wt %) and phosphorus (2.6 wt %), but also potassium (0.8 wt %) and sulphur (0.14 wt %). Silicon appeared to be as SiO_2, and calcium was also found to be oxidized [23]. Potassium was metallic, whereas sulphur and phosphorus were found as sulphates and phosphates, respectively [24]. The surveys on the outlet end showed almost no poisons. Only small traces of calcium (0.5 wt %) and sulphur (0.1 wt %) were found. The FTIR-ATR spectra of the vehicle-aged and fresh natural gas catalysts are presented in Figure 19. The clear differences between the fresh and the aged catalyst spectra are indicated by peaks at 1092, 1025, 725 and 662 cm^{-1}. Peaks at around 1100–1000 cm^{-1} can be assigned, for example, to sulphur species of gas-phase SO_3 and aluminium sulphates. Again, the phosphorus presence is possible, and also the C–S functional group frequencies can be assigned to peaks at 725 cm^{-1} and 662 cm^{-1} [19,25,26].

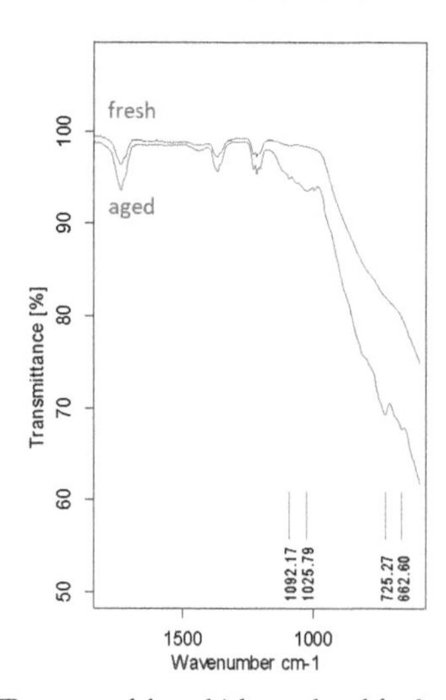

Figure 19. FTIR-ATR spectra of the vehicle-aged and fresh natural gas catalysts.

The studied gases in the catalyst performance tests were CO and C_3H_6. The results of the performance test are presented in Figure 20. For the natural gas catalyst, the fresh one was more active than the aged catalyst, as expected. Poisons and palladium crystal growth had affected the active sites of the catalyst, increasing the light-off temperature.

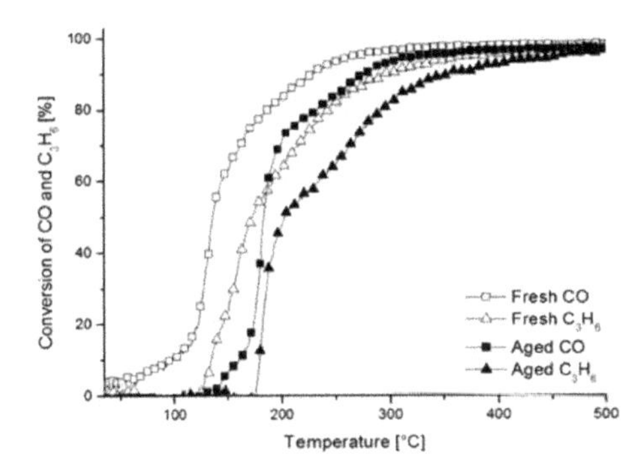

Figure 20. The conversion curves of CO and C_3H_6 for the fresh and aged natural gas catalysts.

3. Materials and Methods

In this study, two different catalysts were studied after the vehicle ageing. Catalysts were designed and manufactured by Dinex Finland Oy. Both of the catalysts were compared to similar fresh catalysts in order to find out ageing-induced effects after use. The diesel oxidation catalyst was from a heavy-duty vehicle and it had been in service for 80,000 km. The natural gas vehicle catalyst was used for 85,000 km in a bus application, which had suffered some engine problems and thus, according to the catalyst manufacturer, had been severely poisoned during use. The catalyst support materials were mainly alumina with oxides of zirconium, lanthanum and cerium on a metallic substrate sheet. Precious metals were platinum in the diesel oxidation catalyst and palladium with platinum in the natural gas vehicle catalyst.

3.1. Electron Microscopy Characterization

Catalyst samples were studied with a transmission electron microscope and scanning electron microscope. The SEM analyses were carried out with field-emission SEMs (FESEMs) equipped with EDSs from Oxford Instruments (Oxford, UK). FESEMs used were: Zeiss ULTRAplus (Carl Zeiss, Oberkochen, Germany) together with EDS with INCAx-act silicon-drift detector (SDD) and Jeol JEM-7000F (Tokyo, Japan) together with EDS with XMaxN SDD. Samples for SEM-EDS analyses were prepared from the cross-sections of the metallic support sheets by mechanical grinding and polishing. Prior to SEM-EDS analyses, the polished samples were carbon-coated to avoid sample charging during SEM studies. The TEM analyses were carried out with analytical (scanning) transmission electron microscopes, JEM 2010 TEM from Jeol, equipped with an energy-dispersive spectrometer (EDS), ThermoNoran from Thermo Scientific (Waltham, MA, USA), F200 STEM from Jeol and CM200 FEG–TEM from FEI (Eindhoven, The Netherlands) equipped with an Oxford Instruments EDS system, all operated at 200 kV. Samples from catalyst powders separated from metallic catalyst sheets were prepared for TEM by crushing the powder between glass slides. The crushed powder was dispersed onto a holey carbon-covered Cu grid with ethanol.

The cross-sectional samples for TEM analyses were prepared by an in-situ lift-out technique with focused ion beam milling in an FEI Strata 235 Dual Beam. Due to the poor electrical conductivity of the ceramic matrix, the materials had to be coated with a thin carbon film to reduce any charging from the beam in the FIB system. The lift-out procedure started with depositing a $25 \times 5 \times 1$ μm^3 Pt layer to protect the material during milling. Trenches, about 10 µm deep were then milled on each side of the Pt strip (step 2, Figure 21). Membranes, with a size of approximately 20×10 μm^2 (step 3, Figure 21), were lifted out from the materials and mounted on Cu grids (step 5, Figure 21). The membranes were then thinned to electron transparency with ion beam currents down to 50 pA. Due to the porosity of the materials, it was difficult to achieve a homogeneous thickness over large areas. Figure 21 presents SEM images of different steps of the FIB sample preparation. Image of the final stage of the preparation (step 6) shows the sample used in the TEM analyses.

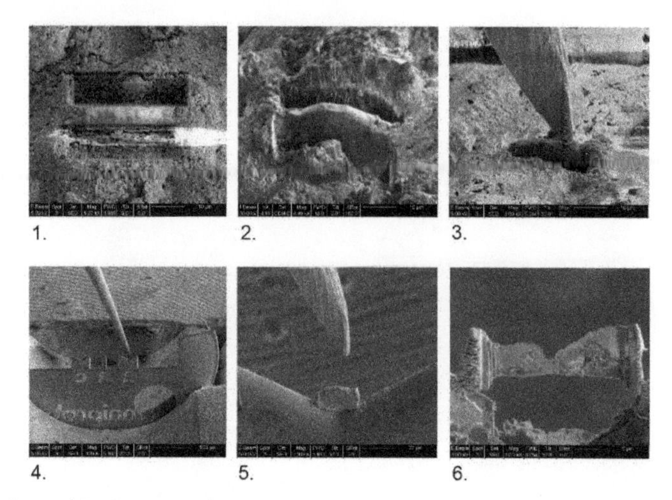

Figure 21. Focused Ion Beam sample preparation steps 1–6. Steps 1 and 2 are from the milled trenches on the surface of the catalyst. Steps 3–5 are from the lift-out and mounting procedure. Step 6 is the final thinned sample.

3.2. Catalyst Characterization

Phase structures of catalysts were studied with an X-ray diffractometer (XRD), the Empyrean multipurpose diffractometer with PIXcel3D solid-state detector from PANalytical (Almelo, The Netherlands) using copper Kα-radiation. Phases were identified by using the database (PDF-4+ 2015) from the International Centre for Diffraction Data (Newtown Square, PA, USA). The elemental concentrations and chemical states of the studied samples were analysed by using X-ray photoelectron spectroscopy (XPS, SSX-100, Surface Science Instruments, Mountain View, CA, USA) using monochromatic aluminium Kα radiation. The samples were pretreated by vacuum for a few hours before the measurements. The equipment for the FTIR-ATR (Fourier-transform infrared–attenuated total reflectance) analyses of surface compositions was a Bruker Optics Tensor 27 (Billerica, MA, USA). The ATR diamond accessory method was used to measure the spectra in the range from 4000 cm^{-1} to 400 cm^{-1} with the resolution of 4 cm^{-1} of both the used inlet and fresh catalyst sheet samples.

Catalytic performance for the fresh and aged diesel as well as natural gas catalysts was measured by using laboratory-scale light-off experiments. A dry lean gas mixture was used to reveal differences between samples. The gas mixture contained 500 ppm of CO, 300 ppm of C_3H_6, and 12 vol % of air, balanced with N_2. The measurements were carried out at atmospheric pressure in a tubular furnace with a quartz reactor. A metal sheet sample (0.53–0.55 g) was placed in the reactor tube. The gas flows were controlled by mass flow controllers (Brooks 5850TR, Brooks Instrument, Hatfield, PA, USA) and the total flow during the experiment was 1 dm^3/min. The temperature of the reactor was increased from room temperature up to 500 °C with a linear heating rate of 10 °C/min. The compound concentrations in the feed and product gases were measured as a function of temperature every 5 s by a GasmetTM FTIR gas analyser (Helsinki, Finland). Oxygen concentration was determined by using a paramagnetic oxygen analyser (ABB Advanced Optima, Zürich, Switzerland).

4. Conclusions

Catalysts from a diesel engine and natural gas engine were studied both fresh and after their use in heavy-duty vehicles. Both of the catalysts had suffered poisoning- and ageing-induced microstructural changes in the support material and in the noble metals. The diesel oxidation catalyst poisons of sulphur, phosphorus and additional oil lubricant-based impurities were detected throughout the diesel oxidation catalyst, in both inlet and outlet parts. Slight platinum particle growth and poisoning-induced shape transformation to a more spherical form were observed especially in the areas of phosphorus poisoning, typically in the alumina-rich areas and on the surface of the washcoat. In addition, platinum reduction took place during ageing. The most common poison in the diesel oxidation catalyst, sulphur, was connected to cerium-rich areas in the diesel oxidation catalyst samples, forming sulphate compounds. For the natural gas oxidation catalyst, the poisoning and structural effects were only present in the inlet part of the catalyst. Significant palladium and platinum particle sintering was observed in the inlet part of the catalyst. In addition, the washcoat structure had changed severely due to the ageing. Also, the poison variety and amounts were larger in the natural gas oxidation catalyst samples, partly due to the failure history of the engine. The catalyst performance studies indicated this very well. In the real-life use of two different catalysts, the poison penetration through the catalyst varied greatly depending on the engine and fuel at hand. The diesel oxidation catalyst appeared to suffer from more thorough changes than the natural gas catalyst, which was affected only in the inlet part of the catalyst. On the other hand, the severity of the effects could be more pronounced in the natural gas oxidation catalyst with heavy structural changes, that is, densification of the washcoat and sintering of noble metals, and high concentrations of poisons in the inlet catalyst surfaces. In this study, it was highlighted that the experimental setup for investigations should be as comprehensive as possible to reveal numerous aspects of vehicle ageing of catalysts for different fuel varieties.

Author Contributions: Conceptualization, T.K. (Tomi Kanerva), M.H. and M.V.; methodology and investigation T.K. (Tomi Kanerva), M.H., T.K. (Tanja Kolli), O.H. and T.S.; writing—original draft preparation, T.K. (Tomi Kanerva), M.H. and M.V.; writing—review and editing, all; supervision, K.K., J.L., E.O., R.L.K. and M.V.

Funding: Authors acknowledge financial support from the 100th Anniversary Foundation of the Federation of Finnish Technology Industries and the Academy of Finland is thanked for funding (Decision numbers 138798 and 139187).

Acknowledgments: Gustaf Östberg is acknowledged for performing FIB sample preparations and associated TEM analyses. Maija Hoikkanen and Juha-Pekka Nikkanen are acknowledged for carrying out the FTIR-ATR measurements. This work made use of Tampere Microscopy Center facilities at Tampere University.

Conflicts of Interest: The authors declare no conflict of interest.

References

1. DieselNet. Available online: https://www.dieselnet.com/ (accessed on 16 January 2019).
2. Heck, R.M.; Farrauto, R.J. Automobile Exhaust catalysts. *Appl. Catal. A* **2001**, *221*, 443–457. [CrossRef]
3. Gandhi, H.S.; Graham, G.W.; McCabe, R.W. Automotive exhaust catalysis. *J. Catal.* **2003**, *216*, 433–442. [CrossRef]
4. Andersson, J.; Antonsson, M.; Eurenius, L.; Olsson, E.; Skoglundh, M. Deactivation of diesel oxidation catalysts: Vehicle- and synthetic ageing correlations. *Appl. Catal. B* **2007**, *72*, 71–81. [CrossRef]
5. Argyle, M.D.; Bartholomew, C.H. Heterogeneous Catalyst Deactivation and Regeneration: A Review. *Catalysts* **2015**, *5*, 145–269. [CrossRef]
6. Air Topics. Available online: https://www.epa.gov/environmental-topics/air-topics (accessed on 20 February 2018).
7. Wang, D.; Epling, B.; Nova, I.; Szanyi, J. Advances in Automobile Emissions Control Catalysis. *Catal. Today* **2016**, *267*, 1–2. [CrossRef]

8. Wiebenga, M.H.; Kim, C.H.; Schmieg, S.J.; Oh, S.H.; Brown, D.B.; Kim, D.H.; Lee, J.-H.; Peden, C.H.F. Deactivation mechanisms of Pt/Pd-based diesel oxidation catalysts. *Catal. Today* **2012**, *184*, 197–204. [CrossRef]
9. Matam, S.K.; Otal, E.H.; Aguirre, M.H.; Winkler, A.; Ulrich, A.; Rentsch, D.; Weidenkaff, A.; Ferri, D. Thermal and chemical aging of model three-way catalyst Pd/Al_2O_3 and its impact on the conversion of CNG vehicle exhaust. *Catal. Today* **2012**, *184*, 237–244. [CrossRef]
10. Honkanen, M.; Kärkkäinen, M.; Viitanen, V.; Jiang, H.; Kallinen, K.; Huuhtanen, M.; Vippola, M.; Lahtinen, J.; Keiski, R.L.; Lepistö, T. Structural characteristics of natural-gas-vehicle-aged oxidation catalyst. *Top. Catal.* **2013**, *56*, 576–585. [CrossRef]
11. Kärkkäinen, M.; Kolli, T.; Honkanen, M.; Heikkinen, O.; Huuhtanen, M.; Kallinen, K.; Lepistö, T.; Lahtinen, J.; Vippola, M.; Keiski, R.L. The effect of phosphorus exposure on diesel oxidation catalysts Part I: Activity measurements, elementary and surface analysis. *Top. Catal.* **2015**, *58*, 961–970. [CrossRef]
12. Honkanen, M.; Kärkkäinen, M.; Heikkinen, O.; Kallinen, K.; Kolli, T.; Huuhtanen, M.; Lahtinen, J.; Keiski, R.L.; Lepistö, T.; Vippola, M. The effect of phosphorus exposure on diesel oxidation catalysts—Part II: Characterization of structural changes by transmission electron microscopy. *Top. Catal.* **2015**, *58*, 971–976. [CrossRef]
13. Wilburn, M.S.; Ebling, W.S. Sulfur deactivation and regeneration of mono- and bimetallic Pd-Pt methane oxidation catalysts. *Appl. Catal. B* **2017**, *206*, 589–598. [CrossRef]
14. Kanerva, T.; Rahkamaa-Tolonen, K.; Vippola, M.; Lepistö, T. Preparation of cross-sectional transmission electron microscopy samples from vehicle-aged and fresh diesel catalysts. In Proceedings of the EUROPACAT VIII, Turku, Finland, 26–31 August 2007.
15. Millo, F.; Rafigh, M.; Andreata, M.; Vlachos, T.; Arya, P.; Miceli, P. Impact of high sulfur fuel and de-sulfaction process on a close-coupled diesel oxidation catalyst and diesel particulate filter. *Fuel* **2017**, *198*, 58–67. [CrossRef]
16. Väliheikki, A.; Kärkkäinen, M.; Honkanen, M.; Heikkinen, O.; Kolli, T.; Kallinen, K.; Huuhtanen, M.; Vippola, M.; Lahtinen, J.; Keiski, R.L. Deactivation of Pt/SiO_2-ZrO_2 diesel oxidation catalysts by sulphur, phosphorus and their combinations. *Appl. Catal. B* **2017**, *218*, 409–419. [CrossRef]
17. Gremminger, A.; Lott, P.; Menno, M.; Casapu, M.; Grunwaldt, J.-D.; Deutschmann, O. Sulfur poisoning and regeneration of bimetallic Pd-Pt methane oxidation catalysts. *Appl. Catal. B* **2017**, *218*, 833–843. [CrossRef]
18. Sadokhina, N.; Smedler, G.; Nylén, U.; Olofsson, M.; Olsson, L. Deceleration of SO_2 poisoning on $PtPd/Al_2O_3$ catalyst during complete methane oxidation. *Appl. Catal. B* **2018**, *236*, 384–395. [CrossRef]
19. Luo, T.; Vohs, J.M.; Gorte, R.J. An Examination of Sulfur Poisoning on Pd/Ceria Catalysts. *J. Catal.* **2002**, *210*, 397–404. [CrossRef]
20. Majumdar, S.S.; Alexander, A.-M.; Gawade, P.; Celik, G.; Ozkan, U.S. Effect of alumina incorporation on the sulfur tolerance of the dual-catalyst aftertreatment system for reduction of nitrogen oxides under lean conditions. *Catal. Today* **2019**, *320*, 204–213. [CrossRef]
21. Beck, D.D.; Sommers, J.W.; Dimaggio, C.L. Axial characterization of oxygen storage capacity in close-coupled lightoff and underfloor catalytic converters and impact of sulfur. *Appl. Catal. B* **1997**, *11*, 273–290. [CrossRef]
22. Beck, D.D.; Sommers, J.W. Impact of sulfur on the performance of vehicle-aged palladium monoliths. *Appl. Catal. B* **1994**, *6*, 185–200. [CrossRef]
23. Beck, D.D. Impact of sulfur on three-way automotive catalyst performance and catalyst diagnostics. *Catal. Deactiv.* **1997**, *111*, 21–38. [CrossRef]
24. *NIST X-ray Photoelectron Spectroscopy Database*; version 4; National Institute of Standards and Technology: Gaithersburg, MD, USA, 2012. Available online: http://srdata.nist.gov/xps/ (accessed on 1 July 2016).
25. Gracia, F.J.; Guerrero, S.; Wolf, E.E.; Miller, J.T.; Kropf, A.J. Kinetics, operando FTIR, and controlled atmosphere EXAFS study of the effect of sulfur on Pt-supported catalysts during CO oxidation. *J. Catal.* **2005**, *233*, 372–387. [CrossRef]
26. Coates, J. *Encyclopedia of Analytical Chemistry*; Meyers, R.A., Ed.; John Wiley & Sons Ltd.: Chichester, UK, 2000.

 catalysts

Article

In Situ Regeneration and Deactivation of Co-Zn/H-Beta Catalysts in Catalytic Reduction of NO_x with Propane

Hua Pan [1], Dongmei Xu [1], Chi He [2,*] and Chao Shen [1,*]

[1] College of Biology and Environment Engineering, Zhejiang Shuren University, Hangzhou 310015, China; panhua.7@163.com (H.P.); dm25xu@163.com (D.X.)
[2] School of Energy and Power Engineering, Xi'an Jiaotong University, Xi'an 710049, China
* Correspondence: chi_he@xjtu.edu.cn (C.H.); shenchaozju@hotmail.com (C.S.); Tel.: +86-029-82668572 (C.H.); +86-0571-88297172 (C.S.)

Received: 23 October 2018; Accepted: 24 December 2018; Published: 30 December 2018

Abstract: Regeneration and deactivation behaviors of Co-Zn/H-Beta catalysts were investigated in NO_x reduction with C_3H_8. Co-Zn/H-Beta exhibited a good water resistance in the presence of 10 vol.% H_2O. However, there was a significant drop off in N_2 yield in the presence of SO_2. The formation of surface sulfate and coke decreased the surface area, blocked the pore structure, and reduced the availability of active sites of Co-Zn/H-Beta during the reaction of NO reduction by C_3H_8. The activity of catalyst regenerated by air oxidation followed by H_2 reduction was higher than that of catalyst regenerated by H_2 reduction followed by air oxidation. Among the catalysts regenerated by air oxidation followed by H_2 reduction with different regeneration temperatures, the optimal regeneration temperature was 550 °C. The textural properties of poisoned catalysts could be restored to the levels of fresh catalysts by the optimized regeneration process. The regeneration process of air oxidation followed by H_2 reduction could recover the active sites of cobalt and zinc species from sulfate species, as well as eliminate coke deposition on poisoned catalysts. The regeneration pathway of air oxidation followed by H_2 reduction is summarized as initial removal of coke by air oxidation and final reduction of the sulfate species by H_2.

Keywords: sulfur poisoning; coke deposition; *in situ* regeneration; Co-Zn/H-Beta; NO_x reduction by C_3H_8

---

1. Introduction

Selective catalytic reduction of NO_x by hydrocarbons (HC-SCR) was considered a promising technology for NO_x removal [1]. Among the HC-SCR catalysts, Co/Beta has attracted much attention due to its excellent activity and N_2 selectivity [2–7]. The long-term stability [3], nature of active sites [4], influences of Co loading and precursor [5], preparation method [6], and mechanism of HC-SCR [7] have been widely investigated on Co/Beta, especially for C_3H_8-SCR [2–6]. To improve the stability and activity of Co/zeolites in HC-SCR, many studies also focused on the modification of Co/zeolites by adding promoters, such as Co-Pd/HBeta [8], Co-In/ferrierite [9], and Co-Zn/HZSM-5 [10].

Sulfur tolerance is a great challenge for deNO_x catalysts. Unfortunately, Co-based zeolites exhibited unsatisfactory activity in the presence of SO_2 [11,12]. On the other hand, coke formation, originating from hydrocarbons, also resulted in the deactivation of deNO_x catalysts for HC-SCR [13]. H_2 reduction is widely used for the regeneration treatment of catalysts deactivated by SO_2, such as NO_x storage-reduction (NSR) catalysts [14–18], and catalytic reduction of NO by NH_3 (NH_3-SCR) catalysts [19,20]. In the case of catalysts deactivated by coke deposition, air calcination was considered as an efficient regeneration method [21–25]. Hence, a regeneration process that combines H_2 reduction with air oxidation may be a potential technology for *in situ* regeneration of $deNO_x$ catalysts in HC-SCR. To our knowledge, no reports focused on the regeneration of HC-SCR catalysts deactivated by dual impacts of SO_2 and coke deposition.

In the present paper, Co-Zn/H-Beta was chosen as a $deNO_x$ catalyst for C_3H_8-SCR, because it showed good catalytic activity [26]. The regeneration of Co-Zn/H-Beta catalysts deactivated by SO_2 and coke deposition was performed in a combined *in situ* process of air oxidation and H_2 reduction. The effects of the regeneration sequence and regeneration temperature on the regeneration efficiency were investigated.

2. Results and Discussion

2.1. Stability of Catalyst

Figure 1 illustrates the influences of SO_2 and H_2O on the activity of Co-Zn/H-Beta at 450 °C for 80 h. Co-Zn/H-Beta had excellent catalytic activity, with 95% N_2 yield obtained for C_3H_8-SCR at 450 °C without addition of SO_2 and H_2O. The catalytic activity decreased slightly in the presence of 10 vol.% H_2O. Upon removing H_2O from the feeding gas, N_2 yield almost returned to its original level. This demonstrates that Co-Zn/H-Beta displays a good water resistance. However, N_2 yield declined significantly during 50–200 ppm SO_2 co-feeding for 15 h. Only 64% N_2 yield was achieved when 200 ppm SO_2 was added into the feeding gas. Upon switching off the SO_2, N_2 yield increased from 64% to 71.5%, which was far away from the initial activity of 95%. When both 10 vol.% H_2O and 200 ppm SO_2 were added simultaneously for 7 h, N_2 yield further dropped from 71.5% to 62%. Upon switching off the SO_2 and H_2O, a partial recovery of catalytic activity was observed. However, N_2 yield decreased gradually after aging for 40 h without adding SO_2 and H_2O. After the stability experiment, the color of the catalyst turned from light gray to black. This demonstrates that carbon deposition occurs on Co-Zn/H-Beta catalysts during the reaction of NO reduction by C_3H_8.

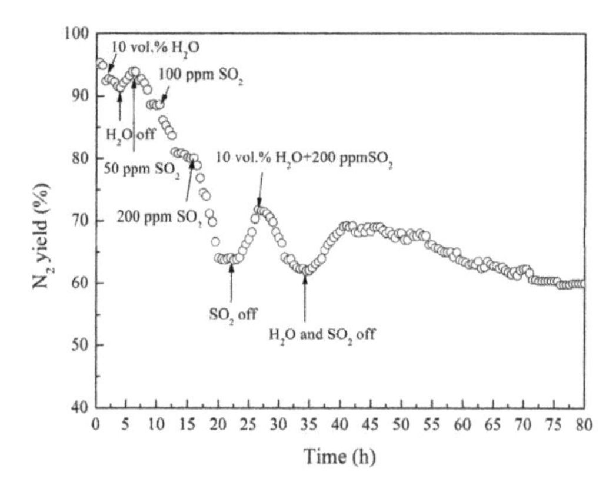

Figure 1. The stability of Co-Zn/H-Beta in C_3H_8-SCR under the dual effects of SO_2 and H_2O.

2.2. Regeneration Performance

Figure 2 presents the influence of regeneration sequence on the activity of the poisoned catalyst at a fixed regeneration temperature of 450 °C. Compared with the poisoned Co-Zn/H-Beta-D catalysts, the regenerated catalysts exhibited higher activity. The activity of regenerated catalysts decreased in the order of: Co-Zn/H-Beta-R ($O_2 + H_2$, 450 °C) > Co-Zn/H-Beta-R ($H_2 + O_2$, 450 °C) > Co-Zn/H-Beta-R (H_2, 450 °C) > Co-Zn/H-Beta-R (O_2, 450 °C). Interestingly, Co-Zn/H-Beta-R ($O_2 + H_2$, 450 °C) displayed higher activity than Co-Zn/H-Beta-R ($H_2 + O_2$, 450 °C). This suggests that combined regeneration is better than single regeneration, and air oxidation followed by H_2 reduction is an optimal regeneration sequence for the deactivated Co-Zn/H-Beta catalyst in C_3H_8-SCR.

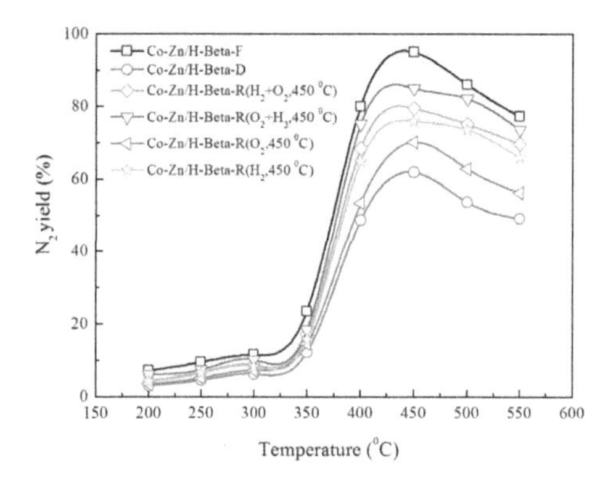

Figure 2. Optimization of regeneration sequence.

Figure 3 exhibits the effect of regeneration temperatures on the activity of the catalysts regenerated by the combined regeneration process. The catalytic activity of both Co-Zn/H-Beta-R (O_2 + H_2) and Co-Zn/H-Beta-R (H_2 + O_2) catalysts increased with the regeneration temperature from 450 to 550 °C. The regenerated Co-Zn/H-Beta-R (O_2 + H_2, 550 °C) displayed similar activity to the fresh catalyst. This indicates that the optimal regeneration temperature was 550 °C. Compared to off-site treatment of solution washing [27] and *in situ* regeneration by H_2 reduction for deactivated deNO$_x$ catalysts [28], the *in situ* regeneration process of air oxidation followed by H_2 reduction showed more convenient operation and higher regeneration efficiency, respectively. Thus, although this comparison may be taken with caution because different reaction conditions were employed, the *in situ* regeneration process of air oxidation followed by H_2 reduction is a promising technology for the regeneration of deactivated Co-Zn/H-Beta catalyst.

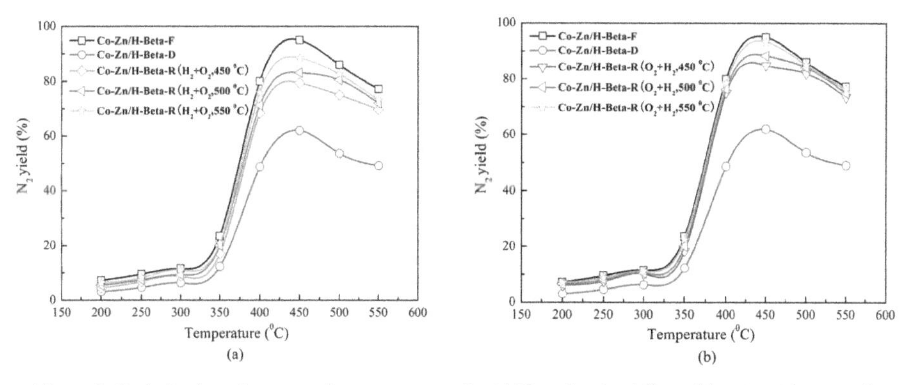

Figure 3. Optimization of regeneration temperature for (**a**) H_2 reduction followed by air oxidation and (**b**) air oxidation followed by H_2 reduction.

2.3. Structural and Textural Properties of Catalysts

Table 1 illustrates the textural properties of the samples. Compared with the Co-Zn/H-Beta-F sample, a significant decrease in surface area, microporous area, and pore volume was detected for the Co-Zn/H-Beta-D catalysts. This implies that sulfate species and coke were deposited both on the surface and in the micropores of the deactivated catalysts. For the regenerated catalysts, surface area, microporous area, and pore volume greatly increased after both the combined regeneration process and single regeneration process. Co-Zn/H-Beta-R (O_2 + H_2, 550 °C) even showed similar values of textural properties to the fresh sample. This implies that air oxidation followed by H_2 reduction at 550 °C could eliminate the sulfates and coke deposited over the deactivated Co-Zn/H-Beta.

Table 1. Textural properties and X-ray photoelectron spectroscopy (XPS) results of samples.

Sample	S_{BET} [a] (m^2 g^{-1})	S_{micro} [b] (m^2 g^{-1})	V_{total} [a] (cm^3 g^{-1})	Binding Energy of $Co2p_{3/2}$ (eV)	Interval between $Co2p_{3/2}$ and $Co2p_{1/2}$ (eV)
Co-Zn/H-Beta-F	428.42	338.88	0.17	780.5	15.8
Co-Zn/H-Beta-D	333.65	264.14	0.13	778.6	17.2
Co-Zn/H-Beta-R (O_2, 550 °C)	384.56	298.22	0.15	/	/
Co-Zn/H-Beta-R (H_2, 550 °C)	360.12	282.36	0.14	/	/
Co-Zn/H-Beta-R (H_2 + O_2, 450 °C)	388.50	305.97	0.15	782.7	14.6
Co-Zn/H-Beta-R (O_2 + H_2, 450 °C)	418.49	323.97	0.16	781.5	15.2
Co-Zn/H-Beta-R (O_2 + H_2, 500 °C)	419.91	336.47	0.17	781.4	15.1
Co-Zn/H-Beta-R (O_2 + H_2, 550 °C)	426.54	337.37	0.17	782.5	15.1

[a] BET method. [b] t Plot method.

The X-ray diffraction patterns (XRD) of Co-Zn/H-Beta catalysts are shown in Figure 4. The position of the main diffraction peak around 2θ = 22.4° is generally taken as evidence of lattice contraction/expansion of the Beta structure [29,30]. The peaks at 21.8°, 25.1°, 28.4°, 29.3°, and 43.6° were assigned to Beta-type zeolite. The deactivated and regenerated catalysts preserved the typical Beta crystal structure. The diffraction peaks of Co_3O_4 (PDF#73-1701), CoO (PDF#71-1178), and $Zn(OH)_2$ (PDF#74-0094 and PDF#71-2115) were detected for all samples. However, the diffraction peak intensity of $Zn(OH)_2$ in the deactivated catalyst was weaker than in the fresh and regenerated samples. No new peak related to sulfate species was observed on any sample. This implies that sulfate species might exist as amorphous bulk species, which could not be detected by XRD.

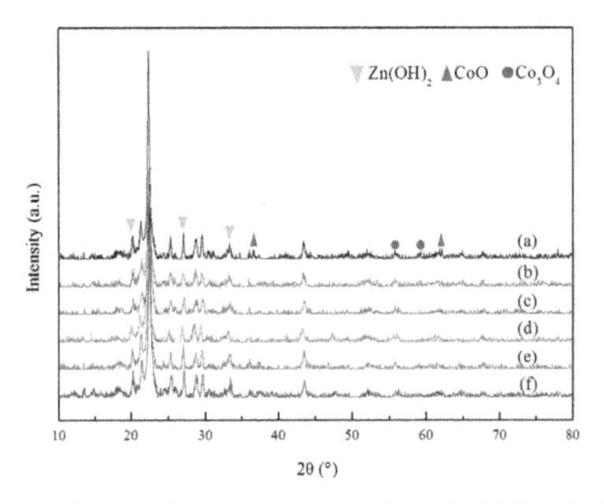

Figure 4. XRD patterns of Co-Zn/H-Beta catalysts. (**a**) Co-Zn/H-Beta-F, (**b**) Co-Zn/H-Beta-D, (**c**) Co-Zn/H-Beta-R (H_2 + O_2, 450 °C), (**d**) Co-Zn/H-Beta-R (O_2 + H_2, 450 °C), (**e**) Co-Zn/H-Beta-R (O_2 + H_2, 500 °C), (**f**) Co-Zn/H-Beta-R (O_2 + H_2, 550 °C).

To identify the state of surface species on various catalysts, the samples were measured by XPS. In Figure 5, the significant movement of the binding energy value of Co $2p_{3/2}$ was observed for both deactivated and regenerated samples, compared to the fresh sample. The binding energy of Co $2p_{3/2}$ shifted toward a lower value (778.6 eV) in the deactivated catalyst. However, the Co $2p_{3/2}$ peaks of regenerated catalysts were shifted to a higher binding energy (781.4–782.7 eV). In Co 2p spectra, shake-up peaks were observed for all samples. This means that metallic cobalt was absent in all samples, because the spectrum of metallic cobalt does not contain shake-up satellite structure at all [31]. Table 1 illustrates the results of Co 2p in the fresh, deactivated, and regenerated catalysts. According to the position of the Co $2p_{3/2}$ peak and interval between Co $2p_{3/2}$ and Co $2p_{1/2}$, CoO, Co_3O_4, $Co(OH)_2$, and $ZnCo_2O_4$ were present in the fresh sample [32,33]. For the deactivated catalyst, the lowest binding energy of Co $2p_{3/2}$ (778.6 eV) and highest interval between Co $2p_{3/2}$ and Co $2p_{1/2}$ (17.2 eV) were observed, which was quite different from other catalysts in Table 1. This implies that sulfate species were generated on the surface of the deactivated catalyst. For Co-Zn/H-Beta-R

($H_2 + O_2$, 450 °C), the highest binding energy of Co $2p_{3/2}$ (782.7 eV) and lowest interval between Co $2p_{3/2}$ and Co $2p_{1/2}$ (14.6 eV) were detected among all catalysts. Cobalt species mainly exist as a $CoAl_2O_4$ state on Co-Zn/H-Beta-R ($H_2 + O_2$, 450 °C) [33]. $CoAl_2O_4$ was recognized to be inactive for NO-SCR [34,35]. Thus, the catalytic activity of Co-Zn/H-Beta-R ($H_2 + O_2$, 450 °C) was lower than that of Co-Zn/H-Beta-R ($O_2 + H_2$) catalysts (Figure 2). In the case of Co-Zn/H-Beta-R ($O_2 + H_2$) catalysts, the binding energy value of Co $2p_{3/2}$ was from 781.4 to 782.5 eV, and the interval between Co $2p_{3/2}$ and Co $2p_{1/2}$ was about 15.1 eV. This indicates that CoO, Co_3O_4, $Co(OH)_2$, and $ZnCo_2O_4$ were present in Co-Zn/H-Beta-R ($O_2 + H_2$) catalysts [32,33]. Therefore, the regeneration process of air oxidation followed by H_2 reduction could promote the recovery of the deactivated cobalt species.

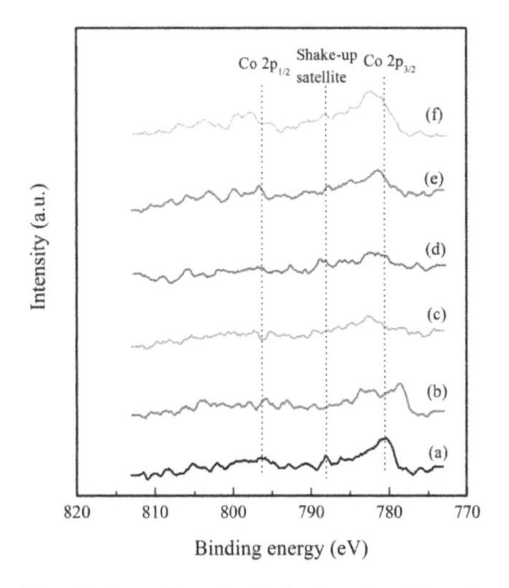

Figure 5. XPS spectra of the Co 2p regions for fresh, deactivated, and regenerated catalysts. (**a**) Co-Zn/H-Beta-F, (**b**) Co-Zn/H-Beta-D, (**c**) Co-Zn/H-Beta-R ($H_2 + O_2$, 450 °C), (**d**) Co-Zn/H-Beta-R ($O_2 + H_2$, 450 °C), (**e**) Co-Zn/H-Beta-R ($O_2 + H_2$, 500 °C), (**f**) Co-Zn/H-Beta-R ($O_2 + H_2$, 550 °C).

Figure 6 presents XPS spectra of the Zn 2p regions for the fresh, deactivated, and regenerated catalysts. For all samples, binding energy of Zn $2p_{3/2}$ and interval between Zn $2p_{3/2}$ and Zn $2p_{1/2}$ were 1022–1023 eV and 23 eV, respectively, which could be attributed to $Zn(OH)^+$ [36] or ZnO [37]. Compared to the fresh sample, a lower binding energy of Zn $2p_{3/2}$ was observed for the deactivated sample, which was similar with the variation trend of Co $2p_{3/2}$ lines. The binding energy of Zn $2p_{3/2}$ shifting to a lower value may be due to the formation of sulfate species on the surface of the deactivated catalysts. For Co-Zn/H-Beta-R ($O_2 + H_2$) catalysts, the position of the binding energy of Zn $2p_{3/2}$ and Zn $2p_{1/2}$ was the same as the fresh sample. However, a lower binding energy of Zn $2p_{3/2}$ and Zn $2p_{1/2}$ was also observed for Co-Zn/H-Beta-R ($H_2 + O_2$, 450 °C), which was similar with the deactivated sample. Thus, regeneration sequence was important to the regeneration of poisoned catalysts. The peak intensity of both Co $2p_{3/2}$ and Zn $2p_{3/2}$ was enhanced with the increase of regeneration temperature from 450 to 550 °C, meaning that a regeneration temperature of 550 °C is optimal.

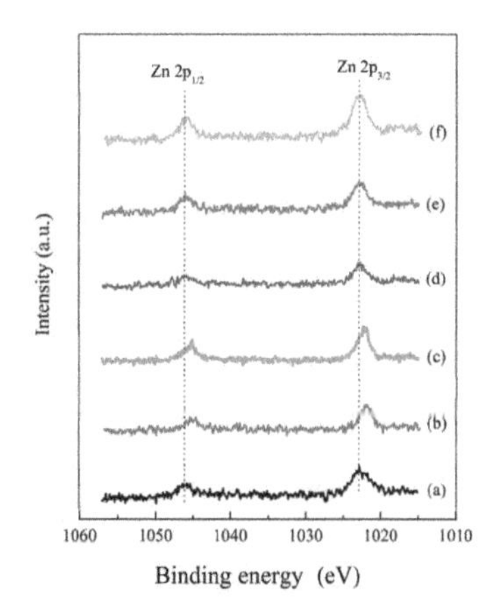

Figure 6. XPS spectra of the Zn 2p regions for fresh, deactivated, and regenerated catalysts. (**a**) Co-Zn/H-Beta-F, (**b**) Co-Zn/H-Beta-D, (**c**) Co-Zn/H-Beta-R (H_2 + O_2, 450 °C), (**d**) Co-Zn/H-Beta-R (O_2 + H_2, 450 °C), (**e**) Co-Zn/H-Beta-R (O_2 + H_2, 500 °C), (**f**) Co-Zn/H-Beta-R (O_2 + H_2, 550 °C).

Figure 7 presents the H_2-temperature programmed reduction (H_2-TPR) of Co-Zn/H-Beta catalysts. The TPR peak centered at around 345 °C is ascribed to the reduction of Co_3O_4 [38]. The peaks centered at 423 °C and 512 °C could correspond to the reduction of CoO_x on the catalyst surface and in the catalyst pore, respectively [38]. The broad peaks centered at 550 °C and 565 °C could be ascribed to the reduction of sulfate and $CoAl_2O_4$, respectively. The high temperature reduction peaks of 620 °C and 800 °C may be assigned to the reduction peaks of $Zn(OH)^+$ and ZnO, respectively. The reduction peak of sulfate (550 °C) is clearly detected for Co-Zn/H-Beta-D and Co-Zn/H-Beta-R (O_2, 450 °C) catalysts. Thus, air oxidation could not remove sulfate on deactivated catalysts. The reduction peak of $CoAl_2O_4$ was observed for Co-Zn/H-Beta-R (H_2 + O_2, 450 °C), but not for Co-Zn/H-Beta-R (O_2 + H_2, 450 °C), Co-Zn/H-Beta-R (O_2 + H_2, 550 °C), and Co-Zn/H-Beta-R (H_2, 450 °C). This may be due to the diffusion of Co species on the surface into the pore of zeolite, and interaction with extra-framework Al^{3+} cations at high temperature during the SCR reaction and regeneration process, resulting in the formation of $CoAl_2O_4$ [39]. This means that air oxidation can promote the formation of $CoAl_2O_4$, while H_2 reduction could inhibit $CoAl_2O_4$ formation. $CoAl_2O_4$ was recognized to be inactive for HC-SCR [34,35]. Therefore, air oxidation followed by H_2 reduction is an optimal regeneration sequence for the deactivated Co-Zn/H-Beta catalysts in C_3H_8-SCR.

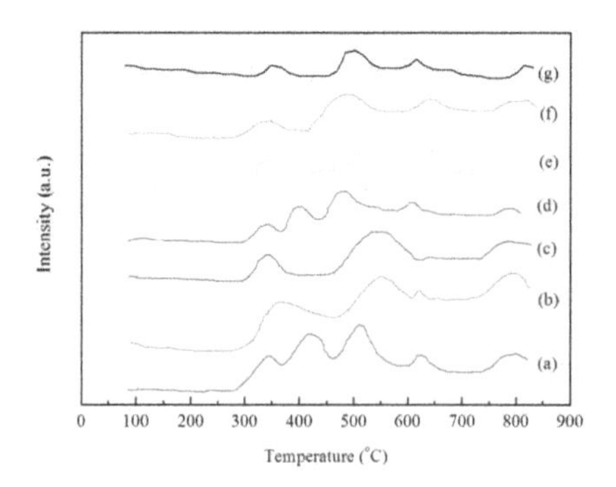

Figure 7. H_2-TPR of fresh, deactivated, and regenerated catalysts. (**a**) Co-Zn/H-Beta-F, (**b**) Co-Zn/H-Beta-D, (**c**) Co-Zn/H-Beta-R (O_2, 550 °C), (**d**) Co-Zn/H-Beta-R (H_2, 550 °C), (**e**) Co-Zn/H-Beta-R (H_2 + O_2, 450 °C), (**f**) Co-Zn/H-Beta-R (O_2 + H_2, 450 °C), (**g**) Co-Zn/H-Beta-R (O_2 + H_2, 550 °C).

Figure 8 shows thermo gravimetric analysis (TGA) curves of various catalysts. The weight loss of all samples in the temperature range of 50–200 °C can be attributed to the desorption of H_2O. However, the deactivated catalyst displayed another stage of weight losses at temperatures above 650 °C, which corresponded to the combustion of coke and sulfate deposited on the catalyst. The regenerated (Co-Zn/H-Beta-R (H_2 + O_2, 450 °C)) sample showed slighter weight losses than the deactivated sample did at temperatures above 650 °C. Interestingly, no significant weight losses were observed for the regenerated (Co-Zn/H-Beta-R (O_2 + H_2, 550 °C)) sample. This further indicates that a combined *in situ* regeneration process of air oxidation followed by H_2 reduction at 550 °C is efficient for regeneration of Co-Zn/H-Beta in C_3H_8-SCR. In summary, the regeneration pathway is illustrated in Scheme 1. Coke on the deactivated catalyst was initially removed by air oxidation at 550 °C. Finally, the sulfate species were reduced to the active cobalt and zinc species by H_2 at 550 °C.

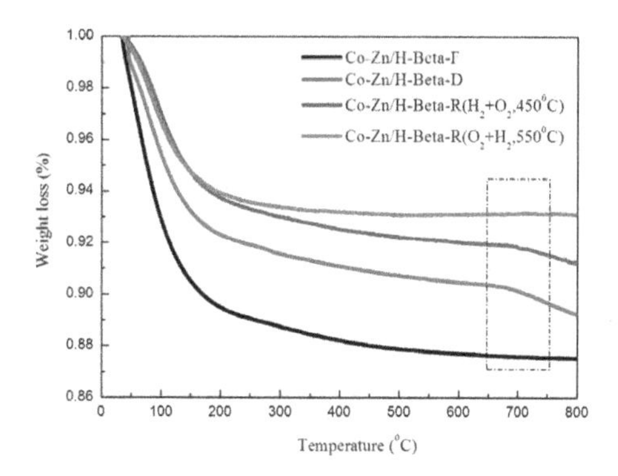

Figure 8. TGA curves of fresh, deactivated, and regenerated catalysts.

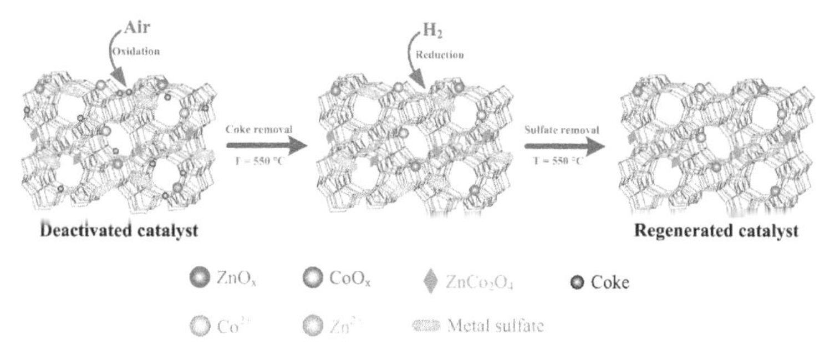

Scheme 1. Regeneration pathway for the combined polluted Co-Zn/H-Beta catalyst.

3. Materials and Methods

3.1. Catalyst Preparation

Beta zeolite with an atomic ratio Si/Al = 25 was purchased commercially in H-form from Nankai University (Tianjin, China). Co-Zn/H-Beta catalysts were synthesized according to the method described elsewhere [40], using $Co(NO_3)_2$ and $Zn(NO_3)_2$ as precursors. All the chemicals were purchased from Macklin Inc. (Shanghai, China). The cobalt and zinc content of Co-Zn/H-Beta was 2 wt.%. The fresh samples were noted as Co-Zn/H-Beta-F.

3.2. Deactivation and Regeneration of Catalysts

In the deactivation process, the samples were exposed to a mixture of 50–200 ppm SO_2, 600 ppm NO, 25 ppm NO_2, 600 ppm C_3H_8, 6 vol.% O_2, 10 vol.% H_2O, and balance of N_2 at 450 °C for 45 h.

The deactivated catalysts were denoted as Co-Zn/H-Beta-D. All the gases were purchased from Hangzhou Jingong Special Gas Co., Ltd. (Hangzhou, China).

The deactivated catalysts were regenerated by 60 mL min^{-1} air and 400 mL min^{-1} 5 vol.% H_2/Ar at regeneration temperatures from 450 to 550 °C for 60 min. The regenerated catalysts were denoted as Co-Zn/H-Beta-R (O_2, H_2, O_2 + H_2, and H_2 + O_2, T), where O_2 means that the deactivated catalysts were regenerated by air oxidation, and H_2 means that the deactivated catalysts were regenerated by H_2 reduction. O_2 + H_2 means that the deactivated catalysts were regenerated by air oxidation followed by H_2 reduction. H_2 + O_2 means that the deactivated catalysts were regenerated by H_2 reduction followed by air oxidation. T is the regeneration temperature.

3.3. Catalytic Activity Measurement

Catalytic activity tests were performed in a self-made packed-bed flow micro-reactor (1 cm i.d.), operating at atmospheric pressure. A 1.5 mL (0.73 g) volume of catalyst powder was held on a quartz frit at the center of the reactor. A typical feeding gas composition was 600 ppm NO, 25 ppm NO_2, 600 ppm C_3H_8, and 6 vol.% O_2, with N_2 as the balance gas. Each feeding gas flow rate was controlled independently by a mass flow controller (D07, Beijing Sevenstar Electronics Co.,Ltd., Beijing, China). The overall flow rate was 460 mL min^{-1}, which was equal to a space velocity of 18,400 h^{-1}.

NO and NO_2 concentrations were continuously monitored by an infrared gas analyzer (Xi'an Juneng Corporation, Xi'an, China). The products were analyzed using a gas chromatograph (Linghua GC9890, Shanghai Ling-Hua Instrument Co., Ltd, Shanghai, China) equipped with a thermal conductivity detector (TCD). A porapak Q column (Shanghai Ling-Hua Instrument Co., Ltd., Shanghai, China) was used for separation of CO_2, N_2O and NO. The data was collected in the steady state reaction. In this work, N_2 selectivity is over 95%, because N_2O concentration is below 10 ppm. Thus, the catalytic activity was assessed in terms of the following equation:

$$N_2 \text{ yield } (\%) = \frac{(NO_{in} + NO_{2,in}) - (NO_{out} + NO_{2,out} + 2N_2O_{out})}{(NO_{in} + NO_{2,in})} \times 100\% \quad (1)$$

3.4. Catalyst Characterizations

N_2 adsorption isotherms (ASAP2460, Micromeritics Instrument (Shanghai) Ltd., Shanghai, China) and XRD (Rigaku, Tokyo, Japan) and XPS (PHI-5000C ESCA system, Perkin–Elmer, Waltham, MA, USA) measurements were carried out according to the method described in our previous work [41]. TGA was conducted on an HCR-3 analyzer (Beijing Henven Scientific Instrument Co., Ltd., Beijing, China) from 30 to 800 °C under air, with a heating rate of 5 °C/min.

4. Conclusions

Co-Zn/H-Beta catalysts exhibited a poor stability in the presence of SO_2 and H_2O for C_3H_8-SCR, due to SO_2 poisoning and carbon deposition. The formation of surface sulfate and coke reduced the availability of surface active sites and textural properties of the catalysts. Coke deposition and surface sulfate could be reduced by the combined processes of air oxidation followed by H_2 reduction *in situ*. The combined process of air oxidation followed by H_2 reduction is an available technology for the regeneration of Co-Zn/H-Beta catalysts in C_3H_8-SCR, for removing NO_x from lean-burn and diesel exhausts.

Author Contributions: Conceptualization, H.P. and C.H.; Investigation, H.P. and D.X.; Supervision, C.S.; Writing-original draft, H.P.

Funding: This work was financially supported by the National Natural Science Foundation of China (21677114), Natural Science Foundation of Zhejiang Province of China (LY19E080023), and Young and middle-aged academic team project of Zhejiang Shuren University. We also acknowledge the support of Key Laboratory of Microbial Technology for Industrial Pollution Control of Zhejiang Province.

Conflicts of Interest: The authors declare no conflict of interest.

References

1. Mendes, A.N.; Ribeiro, M.F.; Henriques, C.; Da Costa, P. On the Effect of Preparation Methods of PdCe-MOR Catalysts as NO_x CH_4-SCR System for Natural Gas Vehicles Application. *Catalysts* **2015**, *5*, 1815–1830. [CrossRef]
2. Tabata, T.; Ohtsuka, H.; Sabatino, L.M.F.; Bellussi, G. Selective catalytic reduction of NO_x by propane on Co-loaded zeolites. *Microporous Mesoporous Mater.* **1998**, *21*, 517–524. [CrossRef]
3. Tabata, T.; Kokitsu, M.; Ohtsuka, H.; Okada, O.; Sabatino, L.M.F.; Bellussi, G. Study on catalysts of selective reduction of NO_x using hydrocarbons for natural gas engines. *Catal. Today* **1996**, *27*, 91–98. [CrossRef]
4. Ohtsuka, H.; Tabata, T.; Okada, O.; Sabatino, L.M.F.; Bellussi, G. A study on selective reduction of NO_x by propane on Co-Beta. *Catal. Lett.* **1997**, *44*, 265–270. [CrossRef]
5. Chen, H.H.; Shen, S.C.; Chen, X.Y.; Kawi, S. Selective catalytic reduction of NO over Co/beta-zeolite: Effects of synthesis condition of beta-zeolites, Co precursor, Co loading method and reductant. *Appl. Catal. B Environ.* **2004**, *50*, 37–47. [CrossRef]
6. Čapek, L.; Dědeček, J.; Sazama, P.; Wichterlová, B. The decisive role of the distribution of Al in the framework of beta zeolites on the structure and activity of Co ion species in propane-SCR–NO_x in the presence of water vapour. *J. Catal.* **2010**, *272*, 44–54. [CrossRef]
7. Pietrzyk, P.; Dujardin, C.; Gora-Marek, K.; Granger, P.; Sojka, Z. Spectroscopic IR, EPR, and operando DRIFT insights into surface reaction pathways of selective reduction of NO by propene over the Co–BEA zeolite. *Phys. Chem. Chem. Phys.* **2012**, *14*, 2203–2215. [CrossRef]
8. Ferreira, A.P.; Henriques, C.; Ribeiro, M.F.; Ribeiro, F.R. SCR of NO with methane over Co-HBEA and PdCo-HBEA catalysts: The promoting effect of steaming over bimetallic catalyst. *Catal. Today* **2005**, *107–108*, 181–191. [CrossRef]
9. Kubacka, A.; Janas, J.; Sulikowski, B. In/Co-ferrierite: A highly active catalyst for the CH_4-SCR NO process under presence of steam. *Appl. Catal. B Environ.* **2006**, *69*, 43–48. [CrossRef]
10. Ren, L.L.; Zhang, T.; Liang, D.B.; Xu, C.H.; Tang, J.W.; Lin, L.W. Effect of addition of Zn on the catalytic activity of a Co/HZSM-5 catalyst for the SCR of NO_x with CH_4. *Appl. Catal. B Environ.* **2002**, *35*, 317–321. [CrossRef]
11. Zhang, J.Q.; Liu, Y.Y.; Fan, W.B.; He, Y.; Li, R.F. Effect of SO_2 on Catalytic Performance of CoH-FBZ for Selective Catalytic Reduction of NO by CH_4 in The Presence of O_2. *Environ. Eng. Sci.* **2007**, *24*, 292–300. [CrossRef]
12. Chen, S.W.; Yan, X.L.; Wang, Y.; Chen, J.Q.; Pan, D.H.; Ma, J.H.; Li, R.F. Effect of SO_2 on Co sites for NO-SCR by CH_4 over Co-Beta. *Catal. Today* **2011**, *175*, 12–17. [CrossRef]
13. Krishna, K.; Makkee, M. Coke formation over zeolites and CeO_2-zeolites and its influence on selective catalytic reduction of NO_x. *Appl. Catal. B Environ.* **2005**, *59*, 35–44. [CrossRef]
14. Corbos, E.C.; Courtois, X.; Bion, N.; Marecot, P.; Duprez, D. Impact of the support oxide and Ba loading on the sulfur resistance and regeneration of Pt/Ba/support catalysts. *Appl. Catal. B Environ.* **2008**, *80*, 62–71. [CrossRef]
15. Wang, Q.; Zhu, J.H.; Wei, S.Y.; Chung, J.S.; Guo, Z.H. Sulfur Poisoning and Regeneration of NO_x Storage-Reduction Cu/$K_2Ti_2O_5$ Catalyst. *Ind. Eng. Chem. Res.* **2010**, *49*, 7330–7335. [CrossRef]
16. Liu, Z.Q.; Anderson, J.A. Influence of reductant on the regeneration of SO_2-poisoned Pt/Ba/Al_2O_3 NO_x storage and reduction catalyst. *J. Catal.* **2004**, *228*, 243–253. [CrossRef]
17. Tanaka, T.; Amano, K.; Dohmae, K.; Takahashi, N.; Shinjoh, H. Studies on the regeneration of sulfur-poisoned NO_x storage and reduction catalysts, including a Ba composite oxide. *Appl. Catal. A Gen.* **2013**, *455*, 16–24. [CrossRef]
18. Le Phuc, N.; Corbos, E.C.; Courtois, X.; Can, F.; Marecot, P.; Duprez, D. NO_x storage and reduction properties of Pt/$Ce_xZr_{1-x}O_2$ mixed oxides: Sulfur resistance and regeneration, and ammonia formation. *Appl. Catal. B Environ.* **2009**, *93*, 12–21. [CrossRef]
19. Doronkin, D.E.; Khan, T.S.; Bligaard, T.; Fogel, S.; Gabrielsson, P.; Dahl, S. Sulfur poisoning and regeneration of the Ag/γ-Al_2O_3 catalyst for H_2-assisted SCR of NO_x by ammonia. *Appl. Catal. B Environ.* **2012**, *117*, 49–58. [CrossRef]
20. Chang, H.Z.; Li, J.H.; Yuan, J.; Chen, L.; Dai, Y.; Arandiyan, H.; Xu, J.Y.; Hao, J.M. Ge, Mn-doped CeO_2–WO_3 catalysts for NH_3–SCR of NO_x: Effects of SO_2 and H_2 regeneration. *Catal. Today* **2013**, *201*, 139–144. [CrossRef]

21. Serrano, D.P.; Aguado, J.; Rodríguez, J.M.; Peral, A. Catalytic cracking of polyethylene over nanocrystalline HZSM-5: Catalyst deactivation and regeneration study. *J. Anal. Appl. Pyrol.* **2007**, *79*, 456–464. [CrossRef]
22. Aguado, J.; Serrano, D.P.; Escola, J.M.; Briones, L. Deactivation and regeneration of a Ni supported hierarchical Beta zeolite catalyst used in the hydroreforming of the oil produced by LDPE thermal cracking. *Fuel* **2013**, *109*, 679–686. [CrossRef]
23. Madaan, N.; Gatla, S.; Kalevaru, V.N.; Radnik, J.; Lücke, B.; Brückner, A.; Martin, A. Deactivation and regeneration studies of a $PdSb/TiO_2$ catalyst used in the gas-phase acetoxylation of toluene. *J. Catal.* **2011**, *282*, 103–111. [CrossRef]
24. Villegas, J.I.; Kumar, N.; Heikkilä, T.; Lehto, V.P.; Salmi, T.; Murzin, D.Y. Isomerization of n-butane to isobutane over Pt-modified Beta and ZSM-5 zeolite catalysts: Catalyst deactivation and regeneration. *Chem. Eng. J.* **2006**, *120*, 83–89. [CrossRef]
25. Mazzieri, V.A.; Pieck, C.L.; Vera, C.R.; Yori, J.C.; Grau, J.M. Analysis of coke deposition and study of the variables of regeneration and rejuvenation of naphtha reforming trimetallic catalysts. *Catal. Today* **2018**, *133–135*, 870–878. [CrossRef]
26. Zhang, Y.T.; Pan, H.; Li, W.; Shi, Y. SCR of NO_x with C_3H_8 over Co/H-beta Modified by Zn at High GHSV. *J. Chem. Eng. Chin. Univ.* **2009**, *23*, 236–239.
27. Khodayari, R. Regeneration of commercial TiO_2-V_2O_5-WO_3 SCR catalysts used in bio fuel plants. *Appl. Catal. B Environ.* **2001**, *30*, 87–99. [CrossRef]
28. Pan, H.; Jian, Y.F.; Yu, Y.K.; He, C.; Shen, Z.X.; Liu, H.X. Regeneration and sulfur poisoning behavior of In/H-BEA catalyst for NO_x reduction by CH_4. *Appl. Surf. Sci.* **2017**, *401*, 120–126. [CrossRef]
29. Dzwigaj, S.; Janas, J.; Machej, T.; Che, M. Selective catalytic reduction of NO by alcohols on Co- and Fe-Siβ catalysts. *Catal. Today* **2007**, *119*, 133–136. [CrossRef]
30. Reddy, J.S.; Sayari, A. A simple method for the preparation of active Ti beta zeolite catalysts. *Stud. Surf. Sci. Catal.* **1995**, *94*, 309–316.
31. Chung, K.S.; Massoth, F.E. Studies on molybdena-alumina catalysts: VII. Effect of cobalt on catalyst states and reducibility. *J. Catal.* **1980**, *64*, 320–331. [CrossRef]
32. Stranick, M.S.; Houalla, M.; Hercules, D.M. Spectroscopic characterization of $TiO_2Al_2O_3$ and $CoAl_2O_3TiO_2$ catalysts. *J. Catal.* **1987**, *106*, 362–368. [CrossRef]
33. Zsoldos, Z.; Guczi, L. Structure and Catalytic Activity of Alumina Supported Platinum Cobalt Bimetallic Catalysts. 3. Effect of Treatment on the Interface Layer. *J. Phys. Chem.* **1992**, *96*, 9393–9400. [CrossRef]
34. Gomez-Garcia, M.A.; Pitchon, V.; Kiennemann, A. Pollution by nitrogen oxides: An approach to NO_x abatement by using sorbing catalytic materials. *Environ. Int.* **2005**, *31*, 445–467. [CrossRef] [PubMed]
35. Yan, J.; Kung, M.C.; Sachtler, W.M.H.; Kung, H.H. Co/Al_2O_3 Lean NO_x Reduction Catalyst. *J. Catal.* **1997**, *172*, 178–186. [CrossRef]
36. Gong, T.; Qin, L.J.; Lu, J.; Feng, H. ZnO modified ZSM-5 and Y zeolites fabricated by atomic layer deposition for propane conversion. *Phys. Chem. Chem. Phys.* **2016**, *18*, 601–614. [CrossRef] [PubMed]
37. Kicir, N.; Tuken, T.; Erken, O.; Gumus, C.; Ufuktepe, Y. Nanostructured ZnO films in forms of rod, plate and flower: Electrodeposition mechanisms and characterization. *Appl. Surf. Sci.* **2016**, *377*, 191–199. [CrossRef]
38. Wang, X.; Chen, H.Y.; Sachtler, W.M.H. Catalytic reduction of NO_x by hydrocarbons over Co/ZSM-5 catalysts prepared with different methods. *Appl. Catal. B Environ.* **2000**, *26*, L227–L239. [CrossRef]
39. Zhang, Q.; Wang, X.D. Charaterization of the phase and valency state of Co on Co/HZSM-5 catalysts. *Chem. J. Chin. Univ.* **2002**, *23*, 129–131.
40. Shi, Y.; Su, Q.F.; Chen, J.; Wei, J.W.; Yang, J.T.; Pan, H. Combination non-thermal plasma and low temperature-C_3H_8-SCR over Co-In/H-beta catalyst for NO_x abatement. *Environ. Eng. Sci.* **2009**, *26*, 1107–1113. [CrossRef]
41. Pan, H.; Guo, Y.H.; Bi, H.T. NO_x adsorption and reduction with C_3H_6 over Fe/zeolite catalysts: Effect of catalyst support. *Chem. Eng. J.* **2015**, *280*, 66–73. [CrossRef]

Article

The Role of Impregnated Sodium Ions in Cu/SSZ-13 NH_3-SCR Catalysts

Chen Wang [1], Jun Wang [2], Jianqiang Wang [2], Zhixin Wang [2], Zexiang Chen [2], Xiaolan Li [1], Meiqing Shen [2,*], Wenjun Yan [3] and Xue Kang [4,*]

1 School of Environment and safety Engineering, North University of China, Taiyuan 030051, China; chenwang87@nuc.edu.cn (C.W.); lxldmu@163.com (X.L.)
2 Key Laboratory for Green Chemical Technology of State Education Ministry, School of Chemical Engineering & Technology, Tianjin University, Tianjin 300072, China; wangjun@tju.edu.cn (J.W.); jianqiangwang@tju.edu.cn (J.W.); wangzhixin@tju.edu.cn (Z.W.); tjuczx@tju.edu.cn (Z.C.)
3 Analytical Instrumentation Center, Institute of Coal Chemistry, Chinese Academy of Sciences, Taiyuan 030001, China; yanwenjun@sxicc.ac.cn
4 School of Chemical Engineering and Technology, North University of China, Taiyuan 030051, China
* Correspondence: mqshen@tju.edu.cn (M.S.); kx19871111@tju.edu.cn (X.K.); Tel.: +86-155-3689-0402 (X.K.)

Received: 30 October 2018; Accepted: 27 November 2018; Published: 30 November 2018

Abstract: To reveal the role of impregnated sodium (Na) ions in Cu/SSZ-13 catalysts, Cu/SSZ-13 catalysts with four Na-loading contents were prepared using an incipient wetness impregnation method and hydrothermally treated at 600 °C for 16 h. The physicochemical property and selective catalytic reduction (SCR) activity of these catalysts were studied to probe the deactivation mechanism. The impregnated Na exists as Na^+ on catalysts and results in the loss of both Brönsted acid sites and Cu^{2+} ions. Moreover, the high loading of Na ions destroy the framework structure of Cu/SSZ-13 and forms new phases (SiO_2/$NaSiO_3$ and amorphous species) when Na loading was higher than 1.0 mmol/g. The decreased Cu^{2+} ions finally transformed into Cu_xO, CuO, and $CuAlO_x$ species. The inferior SCR activity of Na impregnated catalysts was mainly due to the reduced contents of Cu^{2+} ions at kinetic temperature region. The reduction in the amount of acid sites and Cu^{2+} ions, as well as copper oxide species (Cu_xO and CuO) formation, led to low SCR performance at high temperature. Our study also revealed that the existing problem of the Na ions' effect should be well-considered, especially at high hydrothermal aging when diesel particulate filter (DPF) is applied in upstream of the SCR applications.

Keywords: Cu/SSZ-13; NH_3-SCR; sodium ions; deactivation mechanism

---

1. Introduction

So far, selective catalytic reduction of NO_x with ammonia (NH_3-SCR) has been proved to be the most excellent post-processing technique for lean NO_x control in diesel vehicle emissions. Cu/SSZ-13, one of the chabazite (CHA) zeolites with a simple topological structure, receives a lot of attention because it provides the opportunity to understand the SCR mechanism of zeolites [1]. Besides, compared with other commercial zeolites, such as ZSM-5 and BEA, Cu/SSZ-13 catalysts with outstanding NH_3-SCR performance and hydrothermal stability are suitable for selection as a good candidate to meet the China VI emission regulation [2–8].

Sodium (Na) ions, one type of alkali ions, usually comes from urea solution, bio-diesel fuel, and aerosol particulates. Since Na can exchange with acid sites on NH_3-SCR catalysts, the Na effect has been recognized as a non-negligible problem in NH_3-SCR catalysts. For Cu/SAPO-34 catalysts, Ma et al. [9] found the NH_3-SCR activity of Cu/SAPO-34 was greatly reduced at a high content of alkali (>0.5%) mainly due to the decreased amount of isolated Cu^{2+} formed CuO_x clusters. Wang et al. [10]

further found that alkali decreased the number of Brönsted acid sites and NH_3 coverage, and would decrease the SCR reaction rate. In our previous study [11], the different contents of Na impact on Cu/SAPO-34 were also studied. Except for the decrease of active sites and acidity, the results also showed the framework of Cu/SAPO-34 damaged and $CuAlO_x$ species formed at a high content of sodium (>0.8%), and all these factors hindered the SCR activity.

In recent years, some studies have also considered the Na ions' effect on Cu/SSZ-13 catalysts, but most of them focused on co-cation Na ions. Gao et al. [12] investigated the effects of co-cation Na on Cu/SSZ-13 catalyst. They found ≈1.78% Na promoted SCR performance at low temperatures and helped to improve the hydrothermal stability of Cu/SSZ-13 because Na ions modified the redox of active sites and protected the framework of CHA structure. Zhao et al. [13] found a high amount of co-cation Na ions decreased the hydrothermal stability of Al-rich Cu/SSZ-13 at 750 °C for 5 h because the excess amount of Na ions weakened the interaction between Cu ions and the zeolitic framework and formed Cu_xO species. Xie et al. [14] found one-pot-synthesized Cu/SSZ-13 with higher co-cation Na contents showed poorer hydrothermal stability at 750 °C for 16 h, which was attributed to Cu species with poor stability and CHA structure deterioration. Even though some achievements have been made on co-cation Na, the conclusion could not be applied in real-world applications because co-cation Na ions have already existed before Cu exchange and could not represent the deposition of Na in the real-world. As far as we know, few studies have investigated the effect of impregnated Na on Cu/SSZ-13 catalysts. Fan et al. [15] studied the impregnated alkali and alkaline metal effect on Cu/SSZ-13 after hydrothermal aging. They found less of a decrease of SCR activity in 0.5 mmol/g Na-impregnated samples compared with fresh Cu/SSZ-13 when samples were hydrothermally treated at 700 °C for 12 h. However, the tolerance to Na ions and the overall influence could not be obtained because only one impregnated sample was shown. Therefore, the effect of impregnated Na on Cu/SSZ-13 should be further studied.

In combination with conclusions made on Na impregnated Cu/SAPO-34 catalysts, in this study, the following key points about the Na effect on Cu/SSZ-13 should be carefully considered: (1) whether Na ions affect SCR performance, (2) whether Na ions damage the framework of Cu/SSZ-13, and (3) whether other copper species except for Cu_xO will form. In order to probe its nature, four different Na metal loading samples with 0.3, 0.5, 1.0, and 1.5 mmol/g (Na/Al = 0.23, 0.38, 0.77, and 1.15) were obtained using an incipient wetness impregnation method and were hydrothermally treated under the relative lower temperature at 600 °C for 16 h. Through characterizations with Brunauer–Emmett–Teller measurements (BET), X-ray diffraction (XRD), and nuclear magnetic resonance (NMR), the change law of the framework of Cu/SSZ-13 catalysts upon hydrothermal aging was revealed. NH_3-TPD and Ex-situ diffuse reflectance infrared Fourier transform spectra (ex situ DRIFTs) were used to investigate the acidity variation. The nature of copper species changing was acquired using H_2 temperature-programmed reduction (H_2-TPR), electron paramagnetic resonance (EPR), and UV-Visible diffuse reflectance spectra (UV-Vis DRS spectra). Furthermore, the existing state of Na over Cu/SSZ-13 catalysts had been studied using CO_2-DRIFTs measurements.

2. Results

2.1. Structural Characterization

2.1.1. BET and XRD Results

To probe the effect of Na ions on catalysts after 600 °C hydrothermal treatment, all catalysts were first measured using BET and the results are shown in Table 1. The reduction of the surface area of the Na-impregnated samples shows a strong relation with the amount of Na introduced into catalysts. When F-Cu went through a low content Na impregnation treatment, a lower reduction of surface area was found on Na-Cu-0.3 and Na-Cu-1.5. Meanwhile, the high contents of Na impregnation led to an obvious decline in the BET surface area, where a reduction of 25.3% and 65% were observed on Na-Cu-1.0 and Na-Cu-1.5, respectively, compared with F-Cu.

Table 1. Na contents and BET surface area of fresh and Na impregnated catalysts.

Samples' Name	Na Contents (%) [1]	BET Surface Area	ΔS (%) [2]
F-Cu	0	792	—
Na-Cu-0.3	0.7	747	5.6
Na-Cu-0.5	1.2	737	6.9
Na-Cu-1.0	2.3	591	25.3
Na-Cu-1.5	3.5	277	65.0

[1] Na contents measured using ICP-AES; [2] $\Delta S(\%) = \frac{S_{F-Cu} - S_{Na-Cu-x}}{S_{F-Cu}} \times 100\%$.

To further investigate the impact of introduced Na ions on the Cu/SSZ-13 catalyst, XRD experiments were performed. As shown in Figure 1a, the diffraction peaks of the chabazite phase (2θ = 9.6°, 13°, 16.3°, 18.0°, 20.9°, and 25.3°) can be observed in fresh and Na introduced catalysts, illustrating that all catalysts had the typical chabazite (CHA) structure [16,17]. Figure 1b shows the relative crystallinity of different catalysts. The detailed calculation method of the relative crystallinity of catalysts are found in our previous study [11]. The 4%, 10%, 28%, and 56% reduction in crystallinity were found for Na-Cu-0.3, Na-Cu-0.5, Na-Cu-1.0, and Na-Cu-1.5, respectively, compared with F-Cu. The changing trend of the relative crystallinity was consistent with the BET results. It can be concluded that the decrease of BET surface area was due to the declined structure integrity of Cu/SSZ-13. Notably, the diffraction peak centered at 21.63° assigned to SiO_2 or $NaSiO_3$ crystalline appeared on Na-Cu-1.0 and Na-Cu-1.5 [11,18], which also suggests that the CHA structure was no longer intact. The intensity of this new peak became more intense with an increased Na amount. Besides that, some amorphous phases were also observed in Na-Cu-1.5 because the baseline of XRD pattern was non-horizontal between 18° and 25°.

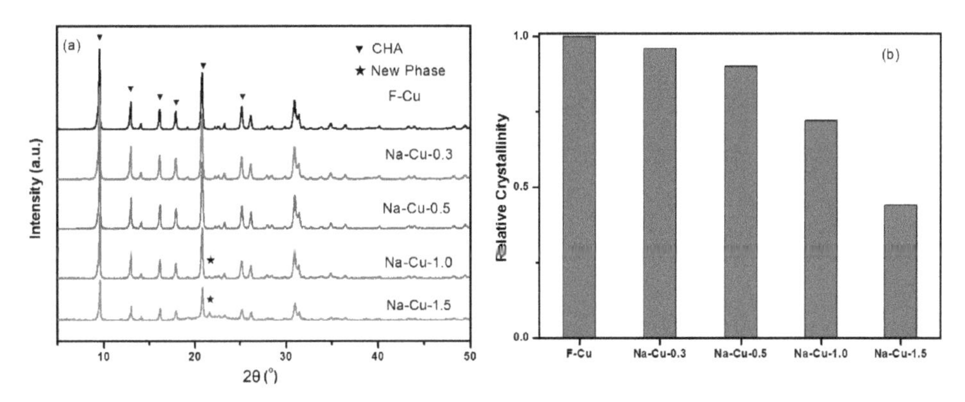

Figure 1. XRD pattern (**a**) and relative crystallinity (**b**) of fresh and Na-impregnated catalysts.

2.1.2. NH_3-TPD Results

Figure 2a shows the NH_3-TPD results of fresh and Na introduced Cu/SSZ-13 catalysts. For fresh Cu/SSZ-13, two peaks centered at ≈250 °C (peak A) and ≈420 °C (peak B) can be observed. In order to identify the different acid sites, DRIFTS experiments were conducted on F-Cu with stepwise NH_3 adsorption with increasing temperature (Figure S1). The desorption peak at ≈250 °C (peak A) was considered to be NH_3 desorbed on Lewis acid sites (Cu^{2+} ions) and the high-temperature desorption peak centered at ≈420 °C (peak B) was assigned to NH_3 released from Brönsted acid sites (Si-O(H)-Al). For Na-impregnated catalysts, another peak centered at 190 °C (peak C) appeared. The new peak should be assigned to NH_3 desorbed from Na^+ sites, which ws consistent with the result of Gao's study [12].

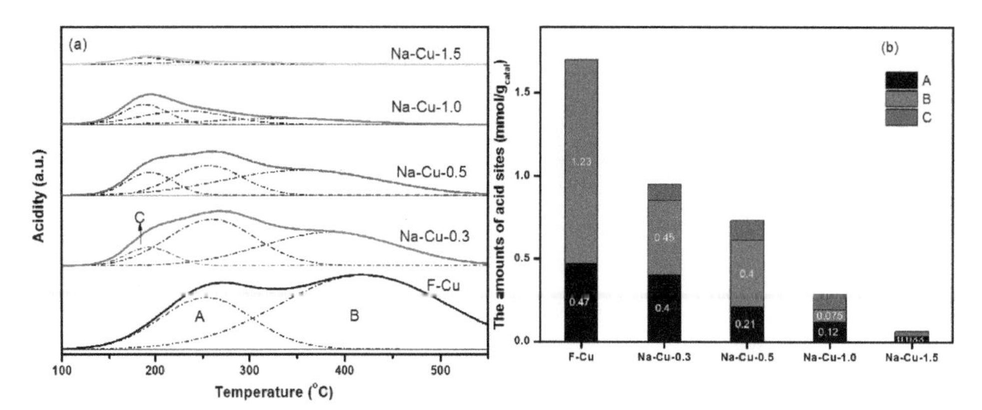

Figure 2. NH_3-TPD results (**a**) and acidity (**b**) of fresh and Na-impregnated catalysts. The reaction was carried out with a feed containing 500 ppm NH_3, balance N_2, and with GHSV = 80,000 h^{-1}.

Figure 2b shows the acid contents of different catalysts. The fresh Cu/SSZ-13 showed the highest acidity among all catalysts. After the Na-impregnation process, the acidity declined and the reduction degree rose with increase amounts of introduced Na. When the Na loading reached 0.3 mmol/g, the desorption peak B assigned to Si-O(H)-Al bonds mainly decreased. Then, peak A, attributed to Cu^{2+} ions, declined when the Na loading was higher than 0.3 mmol/g. The total acidity decreased in the following order: F-Cu (1.7 mmol/g) > Na-Cu-0.3 (0.95 mmol/g) > Na-Cu-0.5 (0.73 mmol/g) > Na-Cu-1.0 (0.29 mmol/g) > Na-Cu-1.5 (0.06 mmol/g).

2.1.3. Ex Situ DRIFTs Results

Figure 3 shows the ex situ DRIFTs results of the fresh sample with those of Na-impregnated ones. The IR bands in the 3500–3800 cm^{-1} region were related to the stretching vibration modes of OH groups (–OH) [2,12,19]. IR bands at 3733 cm^{-1} corresponded to isolated silanols (SiOH) and at 3655 cm^{-1} to $[Cu(OH)]^+$ [2]. Two strong bands at 3604 and 3580 cm^{-1} were assigned to the Brönsted OH groups as (Si-O(H)-Al) [2,12,19]. Moreover, the 1000–800 cm^{-1} region in the IR corresponded to a framework T–O–T vibration that was perturbed by the presence of exchanged Cu ions. The bands at 950 and 900 cm^{-1} were associated with an internal asymmetric framework vibration perturbed by two types of copper cations ($[Cu(OH)]^+$ in 8 MR and Cu^{2+} in 6 MR) [5,20,21].

For the fresh and Na impregnated catalysts, the introduction of Na reduced the intensity of Brönsted OH groups and exchanged Cu ions over Cu/SSZ-13. Furthermore, the IR band intensity of these groups showed a downward trend with the amount of Na increasing (from 0.3 mmol/g to 1.5 mmol/g).

2.1.4. CO_2-DRIFTs Results

To probe the Na existing state, CO_2-DRIFTs tests were performed, and the results are shown in Figure 4. Two IR bands at 2354 and 2346 cm^{-1} were assigned to the vibrations of CO_2 interacting with Na^+ ions on exchange sites on catalysts [11].

When Na was introduced in Cu/SSZ-13 catalysts, two IR bands appeared indicating Na existed as an ionic state on Si-O(Na^+)-Al.

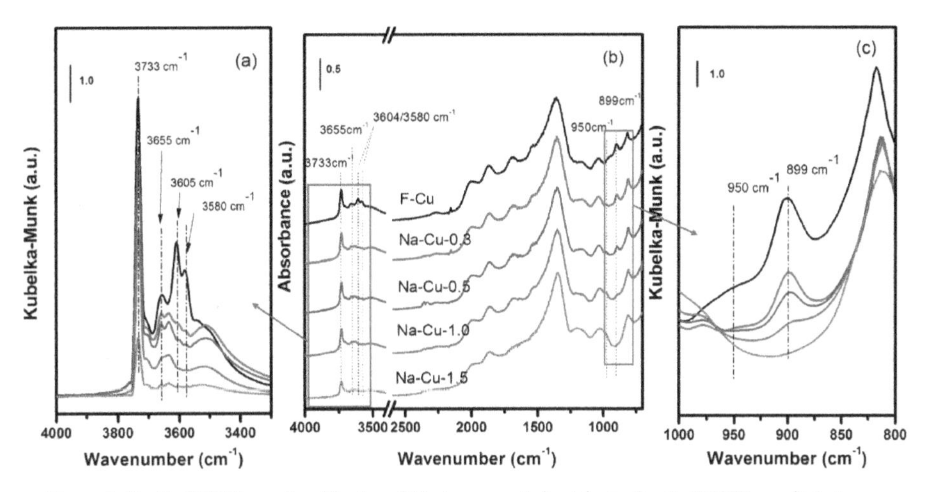

Figure 3. Ex situ DRIFTs results of fresh and Na-impregnated catalysts. Ex situ DRIFTs results between 4000 and 3300 cm^{-1}, units in Kubelka-Munk (**a**); ex situ DRIFTs results between 4000 and 800 cm^{-1}, units in absorbance (**b**); ex situ DRIFT spectra between 1000 and 800 cm^{-1}, units in Kubelka-Munk (**c**).

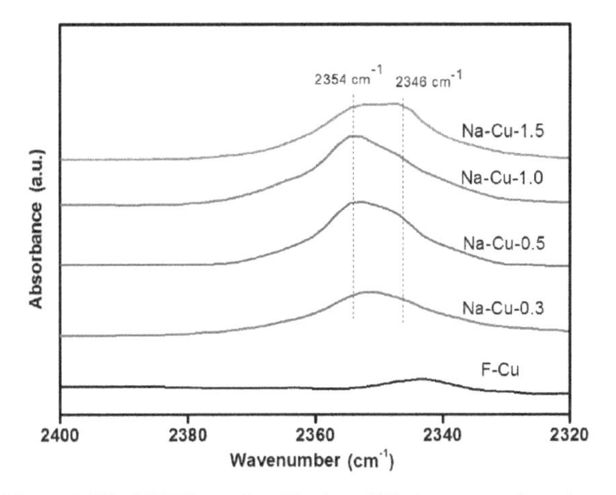

Figure 4. CO_2-DRIFTs results of fresh and Na-impregnated catalysts.

2.2. Copper Species Variation on Cu/SSZ-13

Since Cu species bring high NH_3-SCR performance, the variation of copper species affected by Na introducing should be considered. Thus, in this part, the overall change regulation of Cu species is qualitatively and quantitatively measured.

2.2.1. EPR Results

Since other copper state and oxidation species are EPR silent [22], EPR is a sensitive method to measure the amount of $[Cu(OH)]^+$ in 8 MR and Cu^{2+} in 6 MR using hydrated catalysts [3,22]. Figure 5a shows the EPR spectra of the fresh and Na introduced Cu/SSZ-13 catalysts. All Na-impregnated catalysts showed the same coordination environment as the F-Cu with $g_{//}$ = 2.39 and $A_{//}$ = 136 G, which is consistent with other studies [23]. Meanwhile, another new hyperfine splitting over Na-Cu-1.5 ($g_{//}$ = 2.32 and $A_{//}$ = 147 G) was observed, and a new feature was also found at 3310 G on Na-Cu-1.5, which was attributed to the $CuAlO_x$ species [11,24,25].

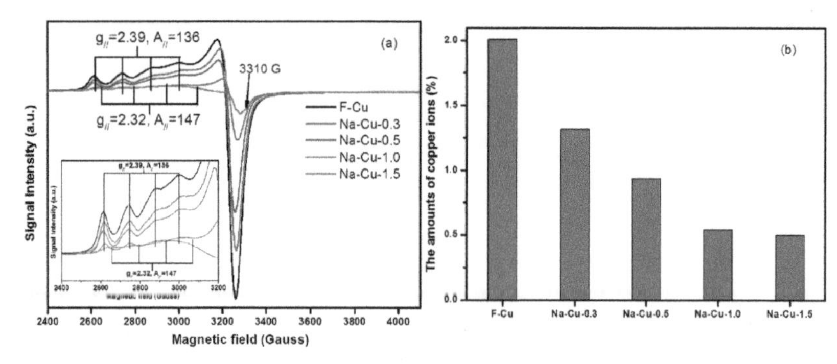

Figure 5. EPR spectra (**a**) and the quantitative results (**b**) of fresh and Na-impregnated catalysts.

Figure 5b compares the number of Cu^{2+} ions on the Na impregnated Cu/SSZ-13 catalysts from EPR spectra using the standard $CuSO_4$ solution as a reference. The number of Cu^{2+} ions decreased in the following order: F-Cu (2.01%) > Na-Cu-0.3 (1.32%) > Na-Cu-0.5 (0.94%) > Na-Cu-1.0 (0.54%) > Na-Cu-1.5 (0.50%). Notably, since Cu^{2+} ions in $CuAlO_x$ species have an EPR signal [11,24,25], the number of Cu^{2+} ions over Na-Cu-1.5 contained both the amounts of Cu^{2+} ions left and Cu^{2+} in $CuAlO_x$ species.

2.2.2. H_2-TPR Results

H_2-TPR measurement is a common method to get the variation of all reducible copper species, thus the fresh and Na impregnated catalysts were measured and the results are shown in Figure 6a. For the fresh sample, two peaks centered at ≈220 °C and ≈350 °C represent the reduction of $[Cu(OH)]^+$ in the CHA cages to Cu^+ and the reduction of Cu^{2+} near 6 MR to Cu^+, respectively [5,26–28]. Another peak centered at 280–300 °C was present for the Na impregnated catalysts, which was assigned to the reduction of copper oxides species to Cu^0 [6,26]. Meanwhile, the reduction peak at ≈410 °C, which was attributed to the reduction of $CuAlO_x$ species, is shown for Na-Cu-1.5 [13,29,30].

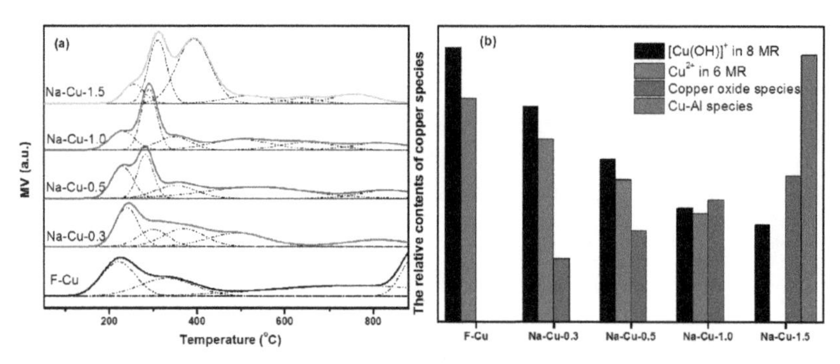

Figure 6. H_2-TPR results (**a**) and qualitative results (**b**) of fresh and Na-impregnated catalysts.

Figure 6b shows the relative contents of the different copper species from the H_2-TPR results. The amounts of Cu^{2+} declined with an increase in Na loading, while copper oxide species contents increased. Besides, the $CuAlO_x$ species appeared on Na-Cu-1.5, which is consistent with our EPR results shown in Figure 5.

2.2.3. UV-Vis DRS Results

To further prove the variation of different Cu species amounts on catalysts, UV-Vis DRS spectra, a qualitative generic analysis method, was used and the results are shown in Figure 7. The band centered at 205 nm was assigned to the oxygen-to-metal charge transfer related to Cu^{2+} ions [9,15]. The 300–600 nm region was attributed to transitions in Cu_xO, such as charge transfer transitions in O–Cu–O and Cu–O–Cu [4,31]. In addition, the region between 600 and 800 nm was assigned to the electron d–d transitions of Cu^{2+} in distorted octahedral surrounding by oxygen in CuO particles [9,15].

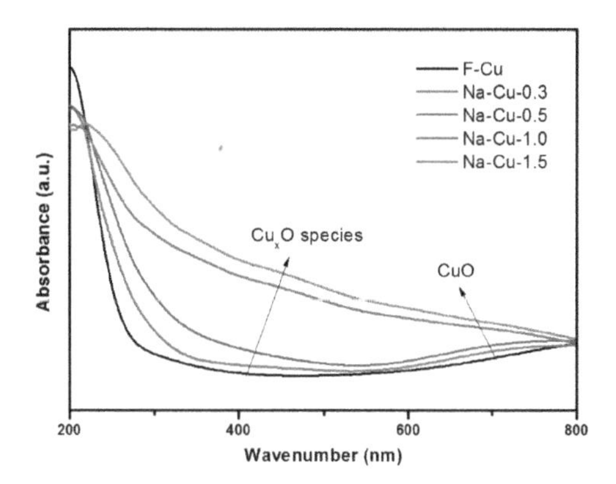

Figure 7. UV-Vis DRS results of fresh and Na-impregnated catalysts.

In general, the intensity of band at 205 nm, assigned to Cu^{2+} ions, declined when Na content increased in the Cu/SSZ-13. Meanwhile, the intensity of the regions attributed to Cu_xO (300–600 nm) and CuO (600–800 nm) increased as Na contents rose.

2.3. NH_3-SCR Reactions over Cu/SSZ-13

2.3.1. NH_3-SCR Activity

Figure 8a shows the NH_3-SCR reaction of fresh and Na-impregnated Cu/SSZ-13 catalysts. Compared to F-Cu, the NO_x conversion of Na-impregnated catalysts declined with the increase in Na loading. Na-Cu-1.5 was the least active catalyst among all the samples over the whole temperature region. In addition, the N_2O formation should also be considered. The highest N_2O formation was found on the fresh catalysts at 250 °C (5 ppm). While, compared with F-Cu, the Na-introduced catalysts decreases the N_2O formation by 1 to 4 ppm at 250 °C and increased the N_2O concentration by 1 to 3 ppm at high temperatures (Figure 8b).

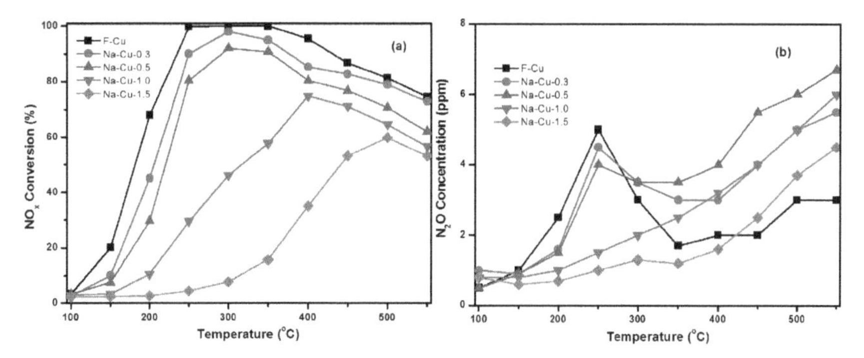

Figure 8. (**a**) NOx conversion as a function of reaction temperature over catalysts; (**b**) N_2O formation during NH_3-SCR reaction on the catalysts (**b**). The reaction was carried out with a feed containing 500 ppm NO_x, 500 ppm NH_3, 10% O_2, 3% H_2O, balance N_2, and with GHSV = 80,000 h^{-1}.

2.3.2. Activation Energy (Ea) Measurements

Figure 9 shows the Arrhenius plots for the SCR reaction over the fresh and Na-impregnated catalysts and all kinetic measurements were taken at <20% NO_x conversion. It is clearly seen that the SCR reaction rate decreased with an increase of Na intake. Moreover, the apparent activation energies (Ea) change with different catalysts. All samples, except for Na-Cu-1.5, had similar Ea (Ea = 62 ± 2 kJ/mol); meanwhile, the Ea value of Na-Cu-1.5 was considerably lower (Ea = 28.5 kJ/mol).

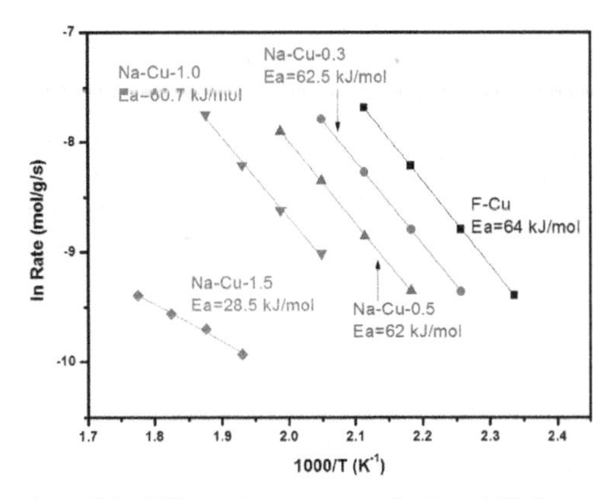

Figure 9. Arrhenius plots of the SCR reaction rates over fresh and Na-impregnated catalysts. The reaction was carried out with a feed containing 500 ppm NO_x, 500 ppm NH_3, 10% O_2, 3% H_2O, balance N_2, and with GHSV = 316,800 h^{-1}.

3. Discussion

3.1. Effect of Na Ions on CHA Structure over Cu/SSZ-13

For Na impregnated catalysts with lower Na loadings (Na-Cu-0.3 and Na-Cu-0.5), the Na content had little impact on the CHA structure integrity because of the slight reduction of surface area and relative F-Cu crystallinity according to our BET and XRD results (Table 1 and Figure 1). Even though the CHA structure was intact, the acid sites, including Brönsted acid sites and Cu^{2+} ions, declined with an increase of Na amounts, as shown in our NH_3-TPD and ex situ DRIFTs results (Figures 2 and 3). This was induced by impregnated Na, which decreased acidity and exchanged the acid sites on Cu/SSZ-13 during post-treatment, as proved via CO_2-DRIFTs results (Figure 4). The surface area and relative crystallinity decreased significantly when Na loadings were greater than 2% in catalysts (Na-Cu-1.0 and Na-Cu-1.5). Moreover, SiO_2/$NaSiO_3$ and amorphous species formed, illustrating that the CHA structure of Cu/SSZ-13 was destroyed by high contents of Na. This is consistent with our previous study on Na impregnated Cu/SAPO-34 [11]. As a result, a greater exchange process, as well as the CHA structure damage with high Na intake, was owed to the inferior acidity and low intensity of acid sites compared with other samples (Figures 2 and 3).

To comprehend the effect of Na ions on the CHA structure, the chemical environmental of ^{29}Si and ^{27}Al on Na-Cu-1.0 and Na-Cu-1.5 were compared to those of F-Cu in Figure 10. F-Cu, the fresh catalyst, presented two framework Si features, assigned to Si(3Si, 1Al) and Si(4Si, 0Al), at −105 and −111 ppm, respectively, as well as one framework Al feature at 57.5 ppm [3]. However, the spectroscopy pattern changed when Na loading increased up to 1.5 mmol/g (Na-Cu-1.5). The peak of ^{29}Si NMR spectra became broad due to the breakage of Si–O–Al bonds and formation of amorphous silicon complexes, which resulted in the non-horizontal baseline XRD pattern of Na-Cu-1.5 [32]. Meanwhile, the ^{27}Al NMR spectrum (Figure 10b) shows non-framework Al centered at −36.5 ppm [13], which illustrates the dealumination process and broken Si–O–Al bonds.

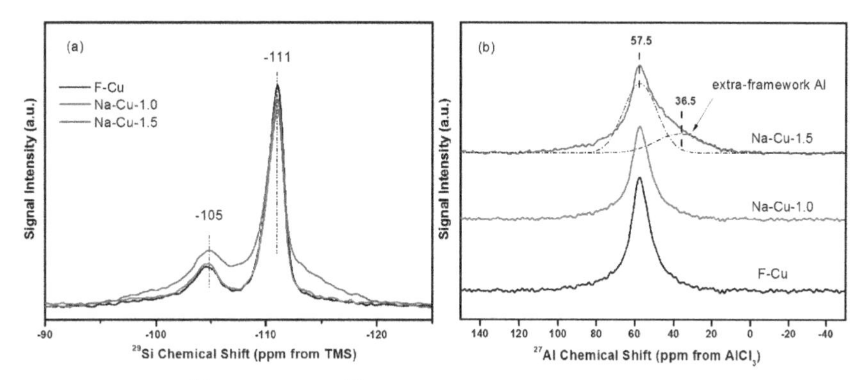

Figure 10. (**a**) ^{29}Si NMR and (**b**) ^{27}Al NMR spectra of F-Cu, Na-Cu-1.0, and Na-Cu-1.5.

Hydrothermal stability is an essential issue for Cu/SSZ-13 catalysts. It is interesting to note that the chemical environmental of Al and Si did not change over Cu/SSZ-13 with hydrothermal treatment at 600 °C. However, the framework of Cu/SSZ-13 was damaged when Na loading was higher than 1.0 mmol/g, as shown in our XRD and NMR results (Figures 1 and 10). The results suggest high Na loading is not beneficial for Cu/SSZ-13 stability against hydrothermal aging. Furthermore, the impact of low Na loading on hydrothermal stability of Cu/SSZ-13 was studied by hydrothermally aging Na-Cu-0.3 at 750 °C for 16 h in air containing 10% water vapor. According to the XRD results in Figure S2, low Na contents also damaged the CHA structure and formed $SiO_2/NaSiO_3$, revealing the stability reduced when the hydrothermal temperature increased. In conclusion, the impregnated Na reduced the hydrothermal stability of Cu/SSZ-13, especially at high temperatures when DPF was applied upstream of the SCR applications.

3.2. *Effect of Na Ions on Acid Sites over Cu/SSZ-13*

H_2-TPR results show that two reduction peaks (≈220 °C and ≈350 °C) were assigned to two different types of Cu^{2+} ions present on F-Cu, which suggests the state of copper species on fresh catalysts was an ionic state. The similar content of copper loading obtained using EPR and ICP results (2.01% vs 1.91%) also confirmed this suggestion. When Na was introduced to Cu/SSZ-13 catalysts, Na competitively took over the exchange sites and influenced the distribution and amounts of acid sites, including Brönsted acid sites and Cu^{2+} ions. Our NH_3-TPD and ex situ DRIFTs results (Figures 2 and 3) showed that Na intakes reduced the both amounts of Brönsted acid sites and Cu^{2+} ions. Moreover, Brönsted acid sites reduced more than that of Cu^{2+} based on a greater decrease of desorption peak B and the intensity of bands (3604 and 3580 cm^{-1}) as shown in Figures 2 and 3. The results show that Na could easily exchange with H ions (Si–O(H)–Al) rather than Cu^{2+} ions on Cu/SSZ-13 due to the stronger acidity of H ions [11].

The change processes of Cu^{2+} ions contents and distributions could be monitored using H_2-TPR results (Figure 6). The amounts of both types of Cu^{2+} ions, $[Cu(OH)]^+$ in 8 MR and Cu^{2+} in 6 MR, decreased when Na loading increased to 1.0 mmol/g, and Cu^{2+} in 6 MR completely disappeared when Na contents reached up to 1.5 mmol/g. Based on the results of ex situ DRIFTs, the regulation of Cu^{2+} ions changed: the intensity of both $[Cu(OH)]^+$ (3655 cm^{-1}) and Cu^{2+} (900 cm^{-1}) declined at the initial stage when Na impregnated and only the weak intensity of $[Cu(OH)]^+$ was left on Na-Cu-1.5 (Figure 3). In addition, the property of Cu^{2+} ions on catalysts was also confirmed using EPR and H_2-TPR results (Figures 5 and 6). Since all catalysts had the same hyperfine splitting with $g_{//}$ = 2.39 and $A_{//}$ = 136 G, as well as a similar reduction peak, the property of the left Cu^{2+} ions was not affected by Na. All results showed that Na impregnation influenced the Cu^{2+} ions amount, but had no effect on the remaining Cu^{2+} ions. Moreover, Cu_xO and CuO present on all Na-impregnated catalysts and their contents increased with increasing Na loading as shown in the H_2-TPR and UV-Vis DRS results (Figures 6 and 7). Note that on Na-Cu-1.5, besides Cu_xO and CuO, $CuAlO_x$ species formed a new hyperfine splitting (g// = 2.32, A// = 147 G), and showed a reduction peak at ≈410 °C in the EPR and H_2-TPR results. In combination with the non-framework Al centered at −36.5 ppm (Figure 10), $CuAlO_x$ species came from the reactions with Cu_xO, CuO, and non-framework Al.

The results above reveal that the reduced Cu^{2+} ions caused by Na introducing transformed to other copper species (Cu_xO, CuO, and $CuAlO_x$ species). To further prove this variation regulation, NH_3 oxidation over all catalysts are shown in Figure 11. Olsson and Lerstner [33,34] have pointed out that NH_3 oxidation at low temperature (≤300 °C) occurs on Cu^{2+} ions in 6 MR, while $[Cu(OH)]^+$ in 8 MR and copper oxide species were responsible for NH_3 oxidation at a higher temperature (>400 °C). At 300 °C, the NH_3 conversion of Na impregnated catalysts was lower than that of F-Cu and NH_3 conversion showed a decreasing trend with increasing Na contents, which confirmed the decreasing amounts of Cu^{2+} ions in 6 MR. When the temperature was higher than 400 °C, Cu_xO and CuO improved the NH_3 conversion on all impregnated catalysts, because the NH_3 oxidation reaction could easily occur on copper oxide species. Based on all the conclusions above, the deactivation mechanism of Na on Cu/SSZ-13 is proposed in Scheme 1.

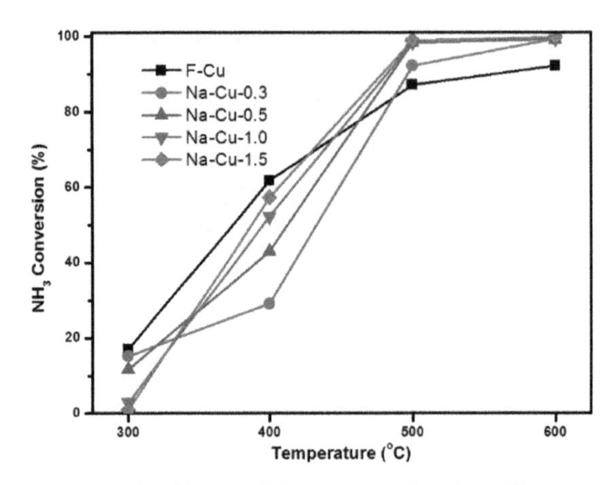

Figure 11. NH_3 oxidation results of fresh and Na-impregnated catalysts. The reaction was carried out with a feed containing 500 ppm NH_3, 10% O_2, 3% H_2O, balance N_2, and with GHSV = 80,000 h^{-1}.

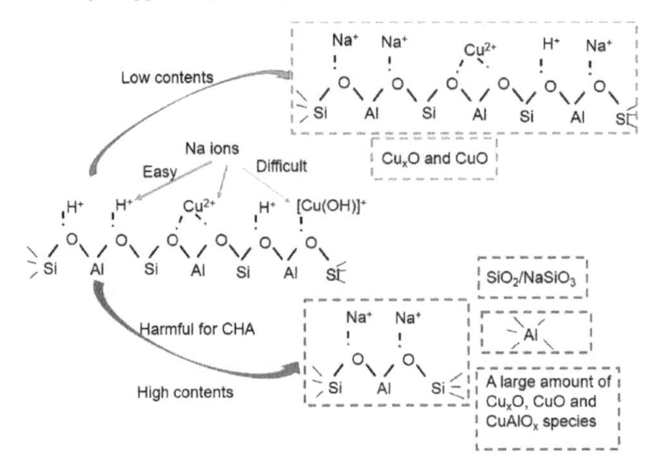

Scheme 1. Nature of the Na ions on Cu/SSZ-13 catalysts.

3.3. *Effect of Na on the NH_3-SCR Reaction*

Based on the conclusions above, the impregnated Na's impact on NH_3-SCR reaction (at low and high temperature) can be addressed. According to Ea measurement results, four catalysts—F-Cu, Na-Cu-0.3, Na-Cu-0.5, and Na-Cu-1.0—had the same Ea value, suggesting the same NH_3-SCR reaction mechanism. Therefore, as shown in Figure 12, the turnover frequencies (TOFs) of these four catalysts were calculated based on their reaction rate and the number of Cu^{2+} ions from the EPR results (Figure 5). The identical TOFs for F-Cu, Na-Cu-0.3, Na-Cu-0.5, and Na-Cu-1.0 illustrate that the Cu^{2+} ions were the active sites for NH_3-SCR reaction at kinetic temperature region and the existing Na ions could not influence the nature of the remaining Cu^{2+} ions that were not occupied by Na^+.

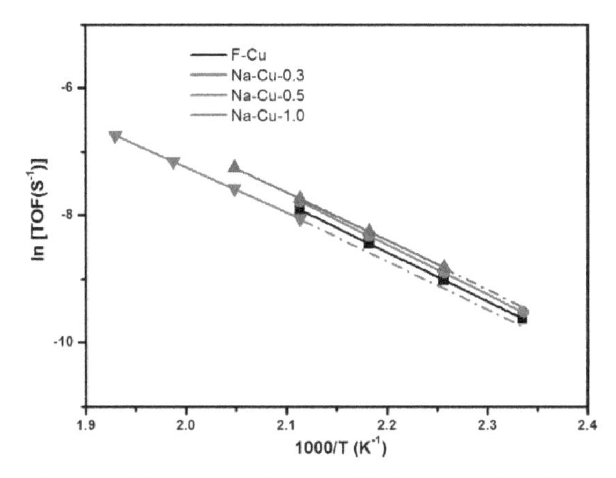

Figure 12. TOFs results of fresh and Na-impregnated catalysts.

It is also worth noting that Na-Cu-1.5 showed a different Ea value compared with others, demonstrating a different rate-determining step occurred for Na-Cu-1.5. Recently, Paolucci [35] found that NH_3-SCR reactions become second-order in low Cu contents when Cu/Al is lower than 0.1. They also pointed out that NH_3-solvated Cu could migrate in the CHA structure to form a Cu dimer species and the formation of Cu-dimer was the rate-determining step. According to our H_2-TPR and ex situ DRIFTs results of Na-Cu-1.5, only Cu^{2+} in 8 MR remained, and the amounts of Cu^{2+} ions should be less than half of the content of Cu^{2+} ions on Na-Cu-1.0. Based on the stoichiometry of SSZ-13 ($H_{2.77}Al_{2.77}Si_{33.23}O_{72}$), the ion-exchange level (Cu/Al) of Na-Cu-1.5 was lower than 0.03. Therefore, it is reasonable to believe that the formation of the Cu-dimer species was the rate-determining step on this catalysts. Moreover, Na occupied all Brönsted acid sites, as shown in our NH_3-TPR results (Figure 2), and the migration of solvated Cu to Cu-dimer species became difficult due to the larger steric hindrance of Na ions. Besides that, the pore diffusion limitations caused by the damage of CHA structure and formation of another phase also contributed to the lower Ea value, which agreed well with the report by Chen [23], who found the high content of phosphorus impregnated Cu/SSZ-13 showed lower Ea compared with the fresh one. In their study, the damage of zeolitic structure, the pore diffusion limitations, the formation of new Cu copper species, and the reduction in the number of acid sites and Cu^{2+} ions could explain the lower Ea value.

When the temperature was higher than that of the kinetic region, it clearly showed that the NO_x conversion became lower when Na metal loadings increased. Since isolated Cu^{2+} monomers were the active sites for NH_3-SCR activity at high temperatures [1] and a higher amount of Brönsted acid sites is beneficial to SCR activity [10], Na-impregnated catalysts with a gradual decrease of Cu^{2+} contents and Brönsted acid sites had an inferior SCR performance. Moreover, the continuous increase of Cu_xO and CuO amounts on Na-impregnated catalysts had a higher NH_3 oxidation ability, as shown in Figure 11, and this competitive reaction also helped to explain the low SCR activity at high temperatures.

4. Materials and Methods

4.1. Materials

In this study, Cu/SSZ-13 samples with Si/Al = 12 were prepared using a two-step solution ion exchange method. First, Na/SSZ-13 with Si/Al = 12 were synthesized using a hydrothermal method and the detailed synthesized processes can be found in Reference [27]. Briefly, the synthesis gel composition was listed as follows: 1 SiO_2: 0.04 Al_2O_3: 0.025 Na_2O: 0.1 SDA: 20 H_2O. Before the ion exchange process, the obtained powder was dried in an oven at 100 °C for 12 h and calcined in a muffle furnace with air at 650 °C for 8 h to remove the template agent. A proportion of 1.91% Cu/SSZ-13 was prepared via exchanging Na/SSZ-13 in a 1 M $(NH_4)_2SO_4$ solution at 80 °C for 4 h and 0.5 M $Cu(NO_3)_2$ at 80 °C for 1 h, respectively. Finally, the samples were calcined at 550 °C for 6 h in a muffle furnace to obtain Cu/SSZ-13.

In order to probe the nature of Na ions on Cu/SSZ-13 catalysts, four different Na loading Cu/SSZ-13 catalysts of 0.3, 0.5, 1.0, and 1.5 mmol/g were prepared using an incipient wetness impregnation method. Briefly, 4 g of fresh Cu/SSZ-13 was put into the required amount solution of $NaNO_3$ and the obtained slurry was dried at 100 °C for 24 h. Then, the catalysts were calcined in a muffle furnace with air at 550 °C for 5 h. Finally, all Na impregnated and reference (Cu/SSZ-13) catalysts were treated at 600 °C for 16 h in 10% H_2O/air. Their corresponding catalyst nomenclatures is shown in Table 1. The reference catalyst was named "F-Cu" and the impregnated catalysts were denoted as "Na-Cu-X," where X stands for metal loading.

4.2. Methods

The compositions of Cu/SSZ-13 catalysts were determined using ICP-AES.

The BET surface areas of Cu/SSZ-13s were measured using Brunauer–Emmett–Teller measurements (Norcross, GA, USA) after dehydration of the catalysts at 300 °C for 24 h under vacuum.

The XRD spectra were collected using an X' Pert Pro diffractometer with nickel-filtered Cu Kα radiation (λ = 1.5418 A), operating at 40 kV and 40 mA in the range of 5–50° with a step size of 0.01°.

The structural damage mechanism was investigated using ^{29}Si and ^{27}Al NMR measurements. ^{29}Si and ^{27}Al NMR measurements were tested on a Varian Infinity plus 300 WB spectrometer at resonance frequencies of 59.57 and 78.13 MHz, respectively, with sample-spinning rates of 4 kHz for ^{29}Si and 8 kHz for ^{27}Al. Tetramethylsilane (TMS) and $Al(NO_3)_3$ aqueous solutions were used as a chemical reference for ^{29}Si and ^{27}Al NMR spectroscopy, respectively.

DRIFTs tests were performed using Nicolet 6700 FTIR (Waltham, MA, USA). KBr was chosen as the standard substance to make the background for ex situ tests. For CO_2-DRIFTs, the samples were initially treated in 5% O_2/N_2 at 500 °C for 30 min and then cooled to 50 °C in N_2. The background spectrum was collected first when the temperature became stable, then the sample was purged in 1% CO_2/N_2 for 15 min and N_2 for only 30 min, and the spectra were collected.

NH_3-TPD experiments were performed to reveal the acidity of catalysts. The catalysts/supports were dehydrated at 250 °C for 30 min in 5% O_2/N_2, and then cooled to 100 °C in N_2. The samples were purged using 500 ppm NH_3/N_2 at 100 °C until NH_3 concentration was stable. Then, the samples were purged with N_2 to remove any weakly absorbed NH_3 at 100 °C. When the NH_3 concentration was lower than 10 ppm, the samples were heated from 100 °C to 550 °C at a ramping rate of 10 °C/min.

The overall changing tendency of active sites was measured by EPR at −150 °C when hydrated Cu/SSZ-13s were used and quantitative results were obtained using $CuSO_4{\cdot}5H_2O$ as a reference.

H_2-TPR were further used to probe the copper species variation. Prior to reduction, the samples (100 mg) were first treated at 250 °C under 5% O_2/N_2. Then, the samples were cooled down to room temperature, followed by purging in N_2. Finally, the samples were measured in a flow of 5% H_2/N_2 (10 mL/min^{-1}) from 30 °C to 900 °C at a ramping rate of 10 °C/min.

Copper oxide species were particularly tested using Shimadzu UV-3600 spectrometer (Kyoto, Japan) equipped with $BaSO_4$ as the reference.

NH_3-SCR and kinetic tests were performed on all catalysts. The detailed description of the experimental process and calculation about NO_x conversion, NH_3-SCR reaction rates, and turnover frequency (TOF) can be found in our previous studies [11,36]. NH_3 oxidation measurements were also performed to clearly explain the NH_3-SCR activity.

5. Conclusions

The Na effect on a Cu/SSZ-13 catalyst after mild hydrothermal aging has been investigated with respect to different Na loadings. The main conclusions of this study are listed below:

1. The impregnated Na exchanged with Brönsted acid sites and Cu^{2+} ions as Si-O(Na^+)-Al. The Na intakes declined with the number of acid sites, including Brönsted acid sites and Cu^{2+} ions. The reduction degree of the number of Brönsted acid sites were greater than that of Cu^{2+} ions contents.
2. Except for sample with highest Na loading (1.5 mmol/g), all Na-impregnated catalysts remained as two types of Cu^{2+} ions and the overall amounts of Cu^{2+} ions declined with Na and increase in contents. Na-Cu-1.5 only contained Cu^{2+} ions in 8 MR and it had the lowest amounts of Cu^{2+}.
3. The Na introduced damage to the framework of catalysts and formed $SiO_2/NaSiO_3$ and amorphous species when the contents of Na $\geq$ 1.0 mmol/g, as confirmed using XRD, BET, and NMR results.
4. The reduced Cu^{2+} ions changed into other copper species, Cu_xO, CuO, and $CuAlO_x$ species. Furthermore, $CuAlO_x$ species were only generated on Na-Cu-1.5 catalyst.
5. For Na-impregnated catalysts, the loss of Cu^{2+} contents contributed to the inferior SCR activity at low temperatures, while at high temperatures, the reduction in the number of acid sites and Cu^{2+} ions as well as CuO_x species formation lead to low SCR performance.

Supplementary Materials: The following are available online at http://www.mdpi.com/2073-4344/8/12/593/s1, Figure S1: DRIFTS spectra of NH_3-TPD over Cu/SSZ-13 catalysts, Figure S2: XRD results of Na-Cu-0.3 after hydrothermal aging at 750 °C for 16 h.

Author Contributions: Conceptualization, C.W.; Formal analysis, J.W. (Jianqiang Wang); Funding acquisition, J.W. (Jun Wang), M.S. and C.W.; Investigation, Z.C., Z.W., X.L., and X.K.; Methodology, J.W. (Jun Wang) and M.S.; Project administration, J.W. (Jun Wang); Resources, M.S.; Supervision, C.W. and M.S.; Validation, J.W. (Jianqiang Wang), W.Y. and X.K.; Visualization, C.W.; Writing—review and editing, C.W.

Funding: This research was funded by National Key Research and Development program, grant number 2017YFC0211302; National Natural Science Foundation of China, grant number 21676195; Science Fund of State Key Laboratory of Engine Reliability, grant number skler-201714; and the State Key Laboratory of Advanced Technology for Comprehensive Utilization of Platinum Metals, grant number SKL-SPM-2018017.

Conflicts of Interest: The authors declare no conflict of interest.

References

1. Gao, F.; Walter, E.D.; Kollar, M.; Wang, Y.; Szanyi, J.; Peden, C.H.F. Understanding ammonia selective catalytic reduction kinetics over Cu/SSZ-13 from motion of the Cu ions. *J. Catal.* **2014**, *319*, 1–14. [CrossRef]
2. Han, S.; Cheng, J.; Zheng, C.; Ye, Q.; Cheng, S.; Kang, T.; Dai, H. Effect of Si/Al ratio on catalytic performance of hydrothermally aged Cu-SSZ-13 for the NH_3-SCR of NO in simulated diesel exhaust. *Appl. Surf. Sci.* **2017**, *419*, 382–392. [CrossRef]
3. Song, J.; Wang, Y.; Walter, E.D.; Washton, N.M.; Mei, D.; Kovarik, L.; Engelhard, M.H.; Prodinger, S.; Wang, Y.; Peden, C.H.F.; et al. Toward Rational Design of Cu/SSZ-13 Selective Catalytic Reduction Catalysts: Implications from Atomic-Level Understanding of Hydrothermal Stability. *ACS Catal.* **2017**, *7*, 8214–8227. [CrossRef]
4. Leistner, K.; Kumar, A.; Kamasamudram, K.; Olsson, L. Mechanistic study of hydrothermally aged Cu/SSZ-13 catalysts for ammonia-SCR. *Catal. Today* **2018**, *307*, 55–64. [CrossRef]
5. Luo, J.; Gao, F.; Kamasamudram, K.; Currier, N.; Peden, C.H.F.; Yezerets, A. New insights into Cu/SSZ-13 SCR catalyst acidity. Part I: Nature of acidic sites probed by NH_3 titration. *J. Catal.* **2017**, *348*, 291–299. [CrossRef]

6. Fan, C.; Chen, Z.; Pang, L.; Ming, S.; Zhang, X.; Albert, K.B.; Liu, P.; Chen, H.; Li, T. The influence of Si/Al ratio on the catalytic property and hydrothermal stability of Cu-SSZ-13 catalysts for NH_3-SCR. *Appl. Catal. A Gen.* **2018**, *550*, 256–265. [CrossRef]
7. Gargiulo, N.; Caputo, D.; Totarella, G.; Lisi, L.; Cimino, S. Me-ZSM-5 monolith foams for the NH_3-SCR of NO. *Catal. Today* **2018**, *304*, 112–118. [CrossRef]
8. Lin, Q.; Liu, J.; Liu, S.; Xu, S.; Lin, C.; Feng, X.; Wang, Y.; Xu, H.; Chen, Y. Barium-promoted hydrothermal stability of monolithic Cu/BEA catalyst for NH_3-SCR. *Dalton Trans.* **2018**, *47*, 15038–15048. [CrossRef] [PubMed]
9. Ma, J.; Si, Z.; Weng, D.; Wu, X.; Ma, Y. Potassium poisoning on Cu-SAPO-34 catalyst for selective catalytic reduction of NO_x with ammonia. *Chem. Eng. J.* **2015**, *267*, 191–200. [CrossRef]
10. Wang, L.; Li, W.; Schmieg, S.J.; Weng, D. Role of Brønsted acidity in NH_3 selective catalytic reduction reaction on Cu/SAPO-34 catalysts. *J. Catal.* **2015**, *324*, 98–106. [CrossRef]
11. Wang, C.; Wang, C.; Wang, J.; Wang, J.; Shen, M.; Li, W. Effects of Na(+) on Cu/SAPO-34 for ammonia selective catalytic reduction. *J. Environ. Sci.* **2018**, *70*, 20–28. [CrossRef] [PubMed]
12. Gao, F.; Wang, Y.; Washton, N.M.; Kollár, M.; Szanyi, J.; Peden, C.H.F. Effects of Alkali and Alkaline Earth Cocations on the Activity and Hydrothermal Stability of Cu/SSZ-13 NH_3–SCR Catalysts. *ACS Catal.* **2015**, *5*, 6780–6791. [CrossRef]
13. Zhao, Z.; Yu, R.; Zhao, R.; Shi, C.; Gies, H.; Xiao, F.-S.; De Vos, D.; Yokoi, T.; Bao, X.; Kolb, U.; et al. Cu-exchanged Al-rich SSZ-13 zeolite from organotemplate-free synthesis as NH_3-SCR catalyst: Effects of Na + ions on the activity and hydrothermal stability. *Appl. Catal. B Environ.* **2017**, *217*, 421–428. [CrossRef]
14. Xie, L.; Liu, F.; Shi, X.; Xiao, F.-S.; He, H. Effects of post-treatment method and Na co-cation on the hydrothermal stability of Cu–SSZ-13 catalyst for the selective catalytic reduction of NO_x with NH_3. *Appl. Catal. B Environ.* **2015**, *179*, 206–212. [CrossRef]
15. Fan, C.; Chen, Z.; Pang, L.; Ming, S.; Dong, C.; Brou Albert, K.; Liu, P.; Wang, J.; Zhu, D.; Chen, H.; et al. Steam and alkali resistant Cu-SSZ-13 catalyst for the selective catalytic reduction of NO_x in diesel exhaust. *Chem. Eng. J.* **2018**, *334*, 344–354. [CrossRef]
16. Liu, Q.; Fu, Z.; Ma, L.; Niu, H.; Liu, C.; Li, J.; Zhang, Z. MnO_x-CeO_2 supported on Cu-SSZ-13: A novel SCR catalyst in a wide temperature range. *Appl. Catal. A Gen.* **2017**, *547*, 146–154. [CrossRef]
17. Wang, J.; Peng, Z.; Qiao, H.; Yu, H.; Hu, Y.; Chang, L.; Bao, W. Cerium-Stabilized Cu-SSZ-13 Catalyst for the Catalytic Removal of NO_x by NH_3. *Ind. Eng. Chem. Res.* **2016**, *55*, 1174–1182. [CrossRef]
18. Tang, J.; Xu, M.; Yu, T.; Ma, H.; Shen, M.; Wang, J. Catalytic deactivation mechanism research over Cu/SAPO-34 catalysts for NH_3-SCR (II): The impact of copper loading. *Chem. Eng. Sci.* **2017**, *168*, 414–422. [CrossRef]
19. Wang, D.; Jangjou, Y.; Liu, Y.; Sharma, M.K.; Luo, J.; Li, J.; Kamasamudram, K.; Epling, W.S. A comparison of hydrothermal aging effects on NH_3-SCR of NO_x over Cu-SSZ-13 and Cu-SAPO-34 catalysts. *Appl. Catal. B Environ.* **2015**, *165*, 438–445. [CrossRef]
20. Luo, J.; Wang, D.; Kumar, A.; Li, J.; Kamasamudram, K.; Currier, N.; Yezerets, A. Identification of two types of Cu sites in Cu/SSZ-13 and their unique responses to hydrothermal aging and sulfur poisoning. *Catal. Today* **2016**, *267*, 3–9. [CrossRef]
21. Kwak, J.H.; Varga, T.; Peden, C.H.F.; Gao, F.; Hanson, J.C.; Szanyi, J. Following the movement of Cu ions in a SSZ-13 zeolite during dehydration, reduction and adsorption: A combined in situ TP-XRD, XANES/DRIFTS study. *J. Catal.* **2014**, *314*, 83–93. [CrossRef]
22. Godiksen, A.; Stappen, F.N.; Vennestrøm, P.N.R.; Giordanino, F.; Rasmussen, S.B.; Lundegaard, L.F.; Mossin, S. Coordination Environment of Copper Sites in Cu-CHA Zeolite Investigated by Electron Paramagnetic Resonance. *J. Phys. Chem. C* **2014**, *118*, 23126–23138. [CrossRef]
23. Chen, Z.; Fan, C.; Pang, L.; Ming, S.; Liu, P.; Li, T. The influence of phosphorus on the catalytic properties, durability, sulfur resistance and kinetics of Cu-SSZ-13 for NO_x reduction by NH_3-SCR. *Appl. Catal. B Environ.* **2018**, *237*, 116–127. [CrossRef]
24. Kim, Y.J.; Lee, J.K.; Min, K.M.; Hong, S.B.; Nam, I.-S.; Cho, B.K. Hydrothermal stability of CuSSZ13 for reducing NO_x by NH_3. *J. Catal.* **2014**, *311*, 447–457. [CrossRef]
25. Gao, F.; Walter, E.D.; Washton, N.M.; Szanyi, J.; Peden, C.H.F. Synthesis and evaluation of Cu/SAPO-34 catalysts for NH_3-SCR 2: Solid-state ion exchange and one-pot synthesis. *Appl. Catal. B Environ.* **2015**, *162*, 501–514. [CrossRef]

26. Zhang, T.; Qiu, F.; Chang, H.; Li, X.; Li, J. Identification of active sites and reaction mechanism on low-temperature SCR activity over Cu-SSZ-13 catalysts prepared by different methods. *Catal. Sci. Technol.* **2016**, *6*, 6294–6304. [CrossRef]
27. Gao, F.; Washton, N.M.; Wang, Y.; Kollár, M.; Szanyi, J.; Peden, C.H.F. Effects of Si/Al ratio on Cu/SSZ-13 NH_3-SCR catalysts: Implications for the active Cu species and the roles of Brønsted acidity. *J. Catal.* **2015**, *331*, 25–38. [CrossRef]
28. Zhang, T.; Qiu, F.; Li, J. Design and synthesis of core-shell structured meso-Cu-SSZ-13@mesoporous aluminosilicate catalyst for SCR of NO_x with NH_3: Enhancement of activity, hydrothermal stability and propene poisoning resistance. *Appl. Catal. B Environ.* **2016**, *195*, 48–58. [CrossRef]
29. Wang, J.; Peng, Z.; Qiao, H.; Han, L.; Bao, W.; Chang, L.; Feng, G.; Liu, W. Influence of aging on in situ hydrothermally synthesized Cu-SSZ-13 catalyst for NH_3-SCR reaction. *RSC Adv.* **2014**, *4*, 42403–42411. [CrossRef]
30. Li, J.; Wilken, N.; Kamasamudram, K.; Currier, N.W.; Olsson, L.; Yezerets, A. Characterization of Active Species in Cu-Beta Zeolite by Temperature-Programmed Reduction Mass Spectrometry (TPR-MS). *Top. Catal.* **2013**, *56*, 201–204. [CrossRef]
31. Leistner, K.; Xie, K.; Kumar, A.; Kamasamudram, K.; Olsson, L. Ammonia Desorption Peaks Can Be Assigned to Different Copper Sites in Cu/SSZ-13. *Catal. Lett.* **2017**, *147*, 1882–1890. [CrossRef]
32. Wang, J.; Fan, D.; Yu, T.; Wang, J.; Hao, T.; Hu, X.; Shen, M.; Li, W. Improvement of low-temperature hydrothermal stability of Cu/SAPO-34 catalysts by Cu^{2+} species. *J. Catal.* **2015**, *322*, 84–90. [CrossRef]
33. Xie, K.; Leistner, K.; Wijayanti, K.; Kumar, A.; Kamasamudram, K.; Olsson, L. Influence of phosphorus on Cu-SSZ-13 for selective catalytic reduction of NO_x by ammonia. *Catal. Today* **2017**, *297*, 46–52. [CrossRef]
34. Leistner, K.; Mihai, O.; Wijayanti, K.; Kumar, A.; Kamasamudram, K.; Currier, N.W.; Yezerets, A.; Olsson, L. Comparison of Cu/BEA, Cu/SSZ-13 and Cu/SAPO-34 for ammonia-SCR reactions. *Catal. Today* **2015**, *258*, 49–55. [CrossRef]
35. Paolucci, C.; Khurana, I.; Parekh, A.A.; Li, S.; Shih, A.J.; Li, H.; di Iorio, J.R.; Albarracin-Caballero, J.D.; Yezerets, A.; Miller, J.T.; et al. Dynamic multinuclear sites formed by mobilized copper ions in NO_x selective catalytic reduction. *Science* **2017**, *357*, 898–903. [CrossRef] [PubMed]
36. Wang, C.; Wang, J.; Wang, J.; Yu, T.; Shen, M.; Wang, W.; Li, W. The effect of sulfate species on the activity of NH_3-SCR over Cu/SAPO-34. *Appl. Catal. B Environ.* **2017**, *204*, 239–249. [CrossRef]

Article

Influence of Sulfur-Containing Sodium Salt Poisoned V_2O_5–WO_3/TiO_2 Catalysts on SO_2–SO_3 Conversion and NO Removal

Haiping Xiao [1], Chaozong Dou [1,*], Hao Shi [1], Jinlin Ge [1] and Li Cai [2]

[1] School of Energy, Power and Mechanical Engineering, North China Electric Power University, Beijing 102206, China; dr_xiaohaiping@126.com (H.X.); yzsrsh@126.com (H.S.); gejinlin945@163.com (J.G.)

[2] Sichuan Electric Power Consulting Design Co., Ltd., Chengdu 610041, Sichuan, China; caili@ncepu.edu.cn

* Correspondence: dczhebut@163.com; Tel.: +86-188-131-618-10; Fax: +86-10-617-728-11

Received: 5 November 2018; Accepted: 12 November 2018; Published: 13 November 2018

Abstract: A series of poisoned catalysts with various forms and contents of sodium salts (Na_2SO_4 and $Na_2S_2O_7$) were prepared using the wet impregnation method. The influence of sodium salts poisoned catalysts on SO_2 oxidation and NO reduction was investigated. The chemical and physical features of the catalysts were characterized via NH_3-temperature programmed desorption (NH_3-TPD), H_2-temperature programmed reduction (H_2-TPR), X-ray photoelectron spectroscopy (XPS), Brunauer–Emmett–Teller (BET), X-ray diffraction (XRD), and Fourier Transform Infrared Spectroscopy (FT-IR). The results showed that sodium salts poisoned catalysts led to a decrease in the denitration efficiency. The 3.6% Na_2SO_4 poisoned catalyst was the most severely deactivated with denitration efficiency of only 50.97% at 350 °C. The introduction of $SO_4{}^{2-}$ and $S_2O_7{}^{2-}$ created new Brønsted acid sites, which facilitated the adsorption of NH_3 and NO reduction. The sodium salts poisoned catalysts significantly increased the conversion of SO_2–SO_3. 3.6%$Na_2S_2O_7$ poisoned catalyst had the strongest effect on SO_2 oxidation and the catalyst achieved a maximum SO_2–SO_3-conversion of 1.44% at 410 °C. Characterization results showed sodium salts poisoned catalysts consumed the active ingredient and lowered the V^{4+}/V^{5+} ratio, which suppressed catalytic performance. However, they increased the content of chemically adsorbed oxygen and the strength of V^{5+}=O bonds, which promoted SO_2 oxidation.

Keywords: V_2O_5–WO_3/TiO_2 catalysts; poisoning; sulfur-containing sodium salts; SO_3; NO removal

1. Introduction

Nitrogen oxides (NO_x) are recognized as a major air pollutant. They destroy the ozone layer, form acid rain, affect the ecological environment, and endanger human health. The main source of NO_x in China is thermal power plants [1,2]. NO_x are listed as a binding assessment indicator for total air pollution control. Consequently, selective catalytic reduction (SCR) flue gas denitration equipment is being used on a large scale in China's thermal power plants. Catalysts are the heart of SCR flue gas denitration technology. The most extensively used commercial catalyst is V_2O_5–WO_3/TiO_2 [3]. Zhundong coal enriches a large amount of sodium (the total content is higher than 2%) because of the special coal-forming environment and the effect of groundwater [4]. The sodium in the coal is not completely stable in the furnace after burning [5]. The presence of large amounts of fly ash and alkali metals in the flue gas can cause catalyst clogging and poisoning. The former is generally reversible and belongs to physical function. The latter belongs to chemical action. In recent years, the toxic effect of alkali metals on the catalyst has been extensively investigated [6,7]. The mechanism of toxicity can be summarized as follows: (1) The presence of alkali metal causes V–O–H to be replaced by V–O–M and decreases the strength and number of Brønsted acid sites. This leads to the reduction of denitration efficiency [8]. (2) Alkali metal can weaken the intensity of V^{5+}=O bonds, decreasing the oxidation ability of the catalysts. Moreover, alkali metals interact with the active ingredient on the catalyst surface. This causes the chemical valence of the elements and the concentration of the active ingredient to change [9,10].

Peng et al. [11,12] studied the mechanism of alkali metals poisoning catalysts. They concluded that after the alkali metals were added they would interact with the V species, causing a reduction in the surface acidity and inhibition of the adsorption of NH_3. This is thought to have resulted in the decreased activity of the catalysts. According to a series of alkali metal bromide poisoning results obtained by Chang et al. [13], the addition of alkali metal compounds decreased the intensity of the V=O bonds and the content of chemically adsorbed oxygen on the catalyst surface. Consequently, the redox ability of the vanadium-based catalysts was weakened. When considering catalysts poisoning, most researchers have focused on the toxic effects of different forms of alkali metals on the catalysts. However, the flue gas contains a large amount of SO_2. Therefore, it will react with gas phase NaCl to form substances such as Na_2SO_4 and $Na_2S_2O_7$ [14–16]. The reaction formulas are as follows:

$$2NaCl + 2SO_2 + O_2 + H_2O \rightarrow Na_2S_2O_7 + 2HCl \quad (1)$$

$$4NaCl + 2SO_2 + O_2 + 2H_2O \rightarrow 2Na_2SO_4 + 4HCl \quad (2)$$

V_2O_5–WO_3/TiO_2 catalysts also have a catalytic effect on the conversion of SO_2 to SO_3. SO_2 conversion is also the main index in evaluating the denitration performance of an SCR catalyst. SO_3 will cause the corrosion of gas pipes. NH_3 will react with SO_3 to produce $(NH_4)_2SO_4$ and NH_4HSO_4. These aforementioned compounds can plug the air preheater and cause great harm. Li et al. [17] found that the presence of SO_2 decreased the catalytic activity of K poisoned catalysts because of the generation of $K_2S_2O_7$. They suggested that $K_2S_2O_7$ inhibited the adsorption of NH_3 and weakened the oxidation ability of the catalysts. However, some researchers concluded that the presence of pyrosulfates would maintain the high oxidizability of V species to a certain degree [18]. Tian et al. [19] studied the influence of different Na salts on the deactivation of SCR catalysts. They concluded that the V–OH bonds were replaced by V–O–Na, reducing the amount of Brønsted acid sites. However, the presence of SO_4^{2-} produced new Brønsted acid sites and promoted the adsorption of NH_3. Hu et al. [20] and Chen et al. [21] specifically studied the role of SO_4^{2-} in catalyst deactivation. Their research found the addition of SO_4^{2-} can create more acid sites on the catalyst surface. These acted as Brønsted acid sites and adsorbed more NH_3. Consequently, the performance of the catalysts was enhanced. Dahlin et al. [22] studied the toxic effects of K, Na, P, S, and other poisons on the catalyst. They concluded that Na and K had the greatest toxic effect on the catalyst. However, sulfates were formed to prevent the alkali metal from interacting with the active sites of the catalyst when Na and S existed simultaneously. Thus, the poisoning effect of the alkali metal decreased.

Most researchers focus on the denitration performance and deactivation of the catalysts [23,24]. The catalytic effect of sulfur-containing sodium salts poisoned catalysts on SO_2 oxidation and the effect of SO_2 on denitration efficiency have rarely been studied. To address this, a series of different concentrations of Na_2SO_4 and $Na_2S_2O_7$ poisoned catalysts were prepared via the wet impregnation method. The influence of sulfur-containing sodium salts poisoned catalysts on SO_2 oxidation and NO removal was investigated experimentally.

2. Results and Discussion

2.1. *Effect of Different Catalysts on SO_3 Generation*

2.1.1. Effect of Temperature on SO_3 Generation

SO_2–SO_3-conversion for different catalysts at different reaction temperatures are shown in Figure 1. The results indicate that the SO_2–SO_3-conversion of different catalysts increases gradually with increasing temperature. The SO_2–SO_3-conversion of the pure catalyst increases from 0.52% at 290 °C to 0.83% at 410 °C. The concentration of SO_3 increases from 9.33 ppm to 14.97 ppm. The most significant increase is for the 3.6% $Na_2S_2O_7$ poisoned catalyst. The SO_2–SO_3-conversion increases from 0.83% to 1.44% and the concentration of SO_3 increases from 15 ppm to 26 ppm. The overall variation of 3.6% Na_2SO_4 poisoned catalyst is lower than that of 1.2% $Na_2S_2O_7$.

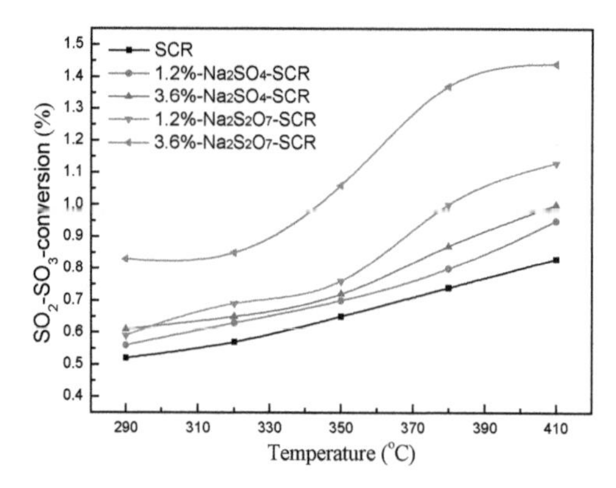

Figure 1. Effect of temperature on SO_3 generation. Reaction conditions: 1800 ppm SO_2, 5% O_2, 500 ppm NO, 500 ppm NH_3, 2% H_2O, total flow gas 1.5 L/min, and gas hourly space velocity (GHSV) = 45,000 h^{-1}.

The sulfur-containing sodium salts poisoned catalysts leads to an increase in the amount of V–O–S bonds. The presence of V–O–S bonds promotes SO_3 generation [25]. Zhang et al. [26] and Ma et al. [27] suggested that the addition of sodium salts (in the presence of SO_2 and O_2) caused the generation of $VOSO_4$. With an increase in temperature, $VOSO_4$ was reoxidized to SO_3 and V_2O_5. Hence, SO_2–SO_3-conversion increased. For $Na_2S_2O_7$ poisoned catalysts, Alvarez et al. [18] and Wang et al. [28] considered that the addition of $Na_2S_2O_7$ inhibited the formation of $VOSO_4$. This would keep the V species in 5+ oxidation state below 400 °C, and ensure the catalytic effect. Meanwhile, the acid strength of the catalyst surface can be enhanced by the induction of the S=O group. The pyrosulfate substance provided stronger acidic sites than the sulfate substance. This resulted in an obvious increase in the SO_2–SO_3-conversion of the pyrosulfates poisoned catalysts.

2.1.2. Effect of SO_2 on SO_3 Generation

Figure 2 shows the effect of SO_2 concentration on both SO_2–SO_3-conversion and SO_3 concentration for different catalysts. It shows that the SO_3 concentration generated with all catalysts used increases gradually. However, it does not increase linearly and a turning point can be identified. SO_2–SO_3-conversion decreases with increasing SO_2 concentration. When the SO_2 concentration is 3000 ppm, the SO_3 generation concentration for 3.6%-$Na_2S_2O_7$-SCR increases to 23 ppm. However, the SO_2–SO_3-conversion is only 0.77%. Therefore, the concentration of SO_3 should also be considered. In high concentration SO_2 flue gas, the diffusion rate of SO_2 is much larger than the reaction rate. Consequently, the main factors in determining the SO_2–SO_3-conversion are the SO_2 oxidation process and the active sites of the catalysts. According to the Le Chatelier's principle, the concentration of the reactant increases. This results in a shift of the equilibrium to the positive reaction direction in the reversible reaction. Hence, the increased reactant further reduces. But the reactant cannot be completely converted. Consequently, the increase of SO_2 is greater than the reacted amount of SO_2. This results in the decrease of SO_2–SO_3-conversion. However, increasing SO_2 can also inhibit the decomposition of SO_3 which can result in the growth of SO_3 concentration [29]. The addition of $Na_2S_2O_7$ causes less reduction in the catalyst activity than addition of Na_2SO_4. Hence, the SO_3 concentration of the $Na_2S_2O_7$ poisoned catalyst is higher than Na_2SO_4 poisoned catalyst at the same temperature. As far as the catalyst is concerned, the limited active sites on the catalyst surface restrict the adsorption of SO_2. This results in low SO_2–SO_3-conversion at high SO_2 concentrations [30].

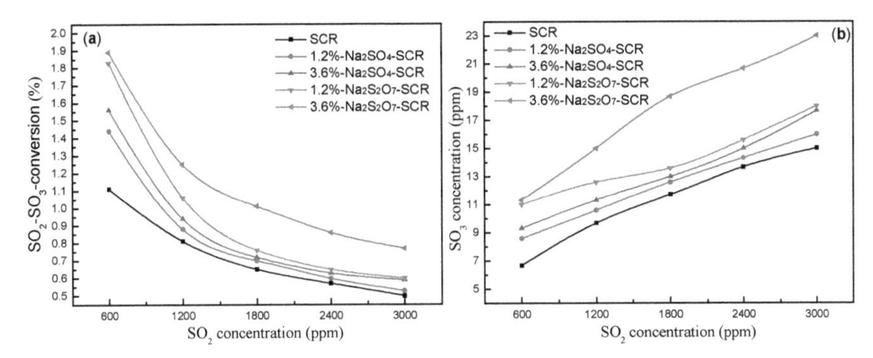

Figure 2. Effect of SO_2 concentration on (**a**) SO_2–SO_3-conversion and (**b**) SO_3 concentration. Reaction conditions: T = 350 °C, 5% O_2, 500 ppm NO, 500 ppm NH_3, 2% H_2O, total flow gas 1.5 L/min, and gas hourly space velocity (GHSV) = 45,000 h^{-1}.

2.2. *Performance of Different Catalysts*

2.2.1. Effect of Temperature on NO Conversion

The denitration efficiency curves for different catalysts at different temperatures are shown in Figure 3. The pure catalyst has a wide range of activity temperature and exhibits favorable denitration performance. The efficiency remains above 95% over the entire temperature range considered and reaches 99.33% at 350 °C. The addition of sodium salts results in the deactivation of catalysts. The activity sequence is as follows: SCR > 1.2%-$Na_2S_2O_7$-SCR > 1.2%-Na_2SO_4-SCR > 3.6%-$Na_2S_2O_7$-SCR > 3.6%-Na_2SO_4-SCR. The denitration efficiency decreases with the increase of sodium salt loading. It initially increases but then decreases with increasing temperature. For 1.2%-Na_2SO_4-SCR and 1.2%-$Na_2S_2O_7$-SCR, the denitration efficiency decreases slightly but remains above 80% (reaching 87.86% and 91.49% at 350 °C, respectively). When the loading increases to 3.6%, the denitration efficiencies of $Na_2S_2O_7$ and Na_2SO_4 poisoned catalysts drop to 75.57% and 50.97% at 350 °C, respectively. Overall, the degree of poisoning of Na_2SO_4 is higher than $Na_2S_2O_7$ under equal loading. Therefore, the degree of poisoning of catalysts is also related to the form of sodium salts used. For instance, $S_2O_7^{2-}$ can provide more Brønsted acid sites and has stronger oxidizability than SO_4^{2-}. This results in more NH_3 being adsorbed which promotes NO reduction [31].

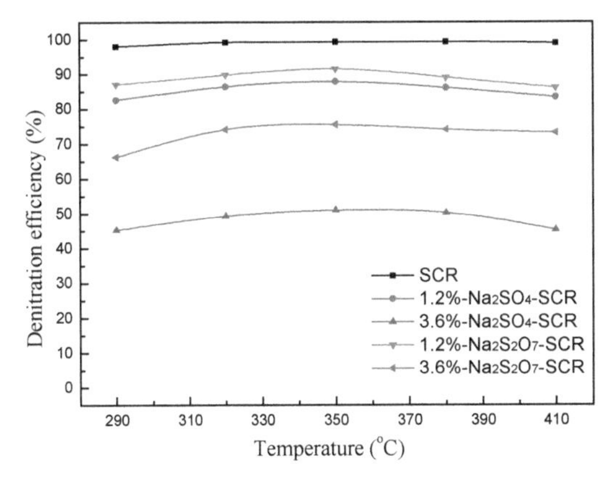

Figure 3. Effect of temperature on NO conversion. Reaction conditions: 500 ppm NO, 500 ppm NH_3, 5% O_2, 2% H_2O, total flow gas 1.5 L/min, and gas hourly space velocity (GHSV) = 45,000 h^{-1}.

2.2.2. Effect of SO_2 on NO Conversion

The influence of SO_2 concentration on the denitration efficiency of different catalysts is shown in Figure 4. The denitration efficiency of pure catalysts remains approximately constant after initially decreasing from 99.33% to 88%. The 3.6% Na_2SO_4 and $Na_2S_2O_7$ poisoned catalysts initially increase before decreasing and then plateauing. The denitration efficiencies (when SO_2 concentration is 1800 ppm) of catalysts poisoned with 3.6% Na_2SO_4 and 3.6% $Na_2S_2O_7$ reach maximum values of 57.03% and 82.95%, respectively. Wu et al. [32] examined pure catalysts after the introduction of SO_2 and found that a large amount of Lewis acid sites on the catalyst surface were covered by ammonium sulfate. This weakened the adsorption of NH_3 and NO by the catalyst. Anstrom et al. [33] and Zhu et al. [34] suggested that due to the addition of SO_2, NO, and SO_2 competed for adsorption on the catalyst surface. This would explain why the adsorption amount of NO was reduced and the denitration efficiency was lowered. For catalysts poisoned with Na salts, Hu et al. [20] concluded that the anions from the sodium salt provided more acidic sites and V–O–S bonds. This was found to promote the catalytic oxidation of SO_2 and enhance the adsorption capacity of NH_3. Consequently, the denitration efficiency improved. The amount of acid sites on the catalyst surface began to decrease when the concentration of SO_2 in the flue gas was increased to a certain extent. The adsorption of SO_2 on V_2O_5 resulted in the formation of an intermediate structure, namely $VOSO_4$. This then reduced the V^{5+} concentration for the SCR reaction, which caused the denitration efficiency to decrease.

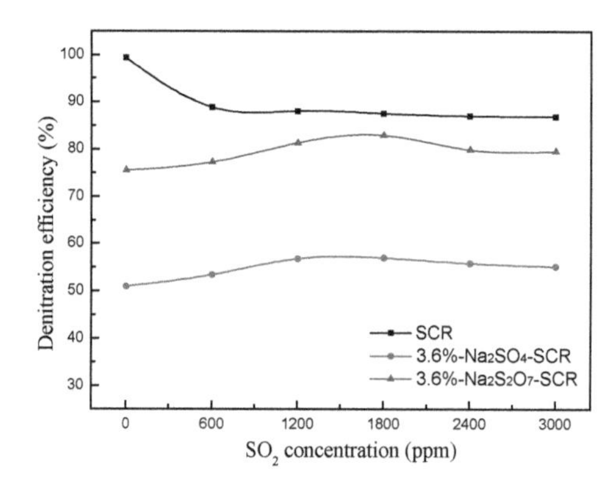

Figure 4. Effect of SO_2 concentration on NO conversion. Reaction conditions: 500 ppm NO, 500 ppm NH_3, 5% O_2, 2% H_2O, T = 350 °C, total flow gas 1.5 L/min, and gas hourly space velocity (GHSV) = 45,000 h^{-1}.

2.3. *Catalyst Characterization*

2.3.1. NH_3-TPD and H_2-TPR Analysis

The adsorption capacity of NH_3 on the acid sites indicates the activity of the catalyst. Therefore, the NH_3-TPD experiment was conducted. The spectrum obtained can indicate the strength and acidity of the acid center. The larger the area of the desorption peak, the higher the corresponding acid concentration. The higher the peak temperature is, the greater the corresponding acid strength will be. Figure 5 shows the NH_3-TPD patterns of different catalysts. It is generally believed that the desorption peak below 200 °C corresponds to the desorption of physisorbed NH_3, the desorption peak in the range of 200 to 350 °C corresponds to the weakly chemisorbed NH_3, and the desorption peak between 350 and 500 °C corresponds to the strongly chemisorbed NH_3 [35,36]. For 1.2% and 3.6% Na_2SO_4 poisoned catalysts, NH_3 adsorption decreased most noticeably. For 1.2% and 3.6% $Na_2S_2O_7$ poisoned catalysts, the amount of physisorbed NH_3 reduces between 100 and 200 °C. In contrast, the amount of strongly chemisorbed NH_3 significantly increased between 350 to 500 °C. According to Zheng et al. [37], this increase can be attributed to the catalyst surface being sulfated through the addition of $Na_2S_2O_7$. The traces of these compounds remaining on the surface of catalysts can provide strong acid sites, which is beneficial to the adsorption of NH_3. Chang et al. [13] have shown that Na preferentially coordinated on V–OH bonds and V–O–H was replaced by V–O–Na after addition of alkali metals. This resulted in a reduced number of acid sites and lower catalytic activity. The reaction formulas are as follows:

$$-V{-}OH + Na_2SO_4 \rightarrow -V{-}O{-}Na + NaHSO_4 \tag{3}$$

$$-V{-}OH + NaHSO_4 \rightarrow -V{-}O{-}Na + H_2SO_4 \tag{4}$$

$$-V{-}OH + Na_2S_2O_7 \rightarrow -V{-}O{-}Na + NaHS_2O_7 \tag{5}$$

$$-V{-}OH + NaHS_2O_7 \rightarrow -V{-}O{-}Na + H_2S_2O_7 \tag{6}$$

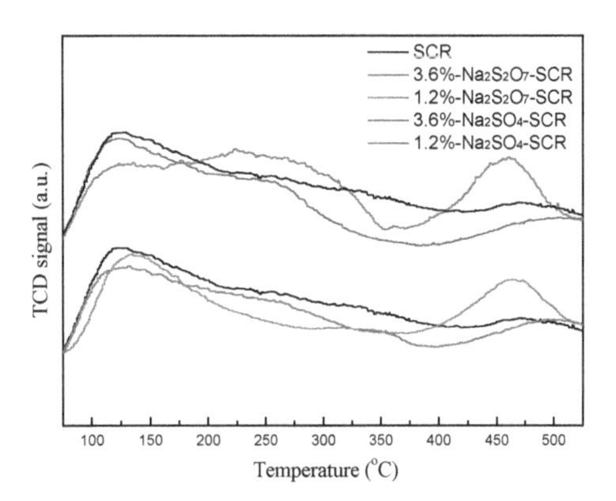

Figure 5. NH_3-TPD profiles of the catalyst samples.

Figure 6 shows the NH_3-TPD curves of different catalysts when the SO_2 concentration is 0 and 1800 ppm. The desorption of the physisorbed and weakly chemisorbed NH_3 on the poisoned catalyst surfaces increased significantly after the introduction of SO_2. This improved the denitration performance of the catalyst to some extent. Giakoumelou et al. [38] noted that the introduction of SO_2 led to the formation of more SO_4^{2-} on the poisoned catalyst surface. SO_4^{2-} can adsorb more NH_3 in the form of NH_4^+ on the catalyst surface to react with NO in the flue gas. For the pure catalyst, the desorption of physisorbed NH_3 increased slightly and the peak area (ranging from 250 °C to 450 °C) clearly decreased. This was unfavorable to denitration performance. After the introduction of SO_2, the denitration efficiency of the pure catalyst is mainly attributed to the competitive adsorption between SO_2 and NO [39]. This agrees with the experimental results.

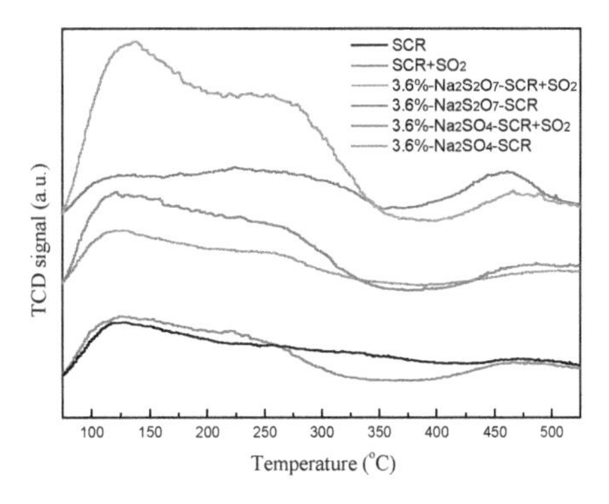

Figure 6. NH_3-TPD profiles of the catalysts in different SO_2 concentration (0 and 1800 ppm).

The redox performance of the catalyst accounts for another important factor in catalytic reduction. The lower the temperature corresponding to the reduction peak is, the easier the catalytic redox reaction of the catalyst will be. Peak area represents H_2 consumption. To test the impact of sulfur-containing sodium salts poisoned catalysts on the redox performance, the H_2-TPR was carried out and the experimental results are shown in Figure 7. The reduction peak temperature of the pure catalysts appeared at 556 °C, which can be attributed to the reduction of V^{5+} to V^{3+}. After the addition of sodium salts, the reduction peak temperature of the catalysts increased. This indicates that sodium salts poisoned catalysts decreased the oxidizability. Chen et al. [40] suggested that the interaction of sodium salts with V species hindered the release of lattice oxygen in the catalyst, making V species more difficult to be reduced. The temperature shifted towards higher values with the increased loading. However, the reduction peak temperatures of the Na_2SO_4 and $Na_2S_2O_7$ poisoned catalysts with the same loading were very similar. The H_2 consumptions were 0.251, 0.897, 1.614, 2.690, and 4.698 mmol/g for pure catalyst, 1.2% Na_2SO_4, 1.2% $Na_2S_2O_7$, 3.6% Na_2SO_4, and 3.6% $Na_2S_2O_7$ poisoned catalyst, respectively. The H_2 consumption significantly increases for sodium salts poisoned catalysts. In particular, the $Na_2S_2O_7$ poisoned catalyst is 1.8 times higher than Na_2SO_4 poisoned catalyst with the same loading. This is because $S_2O_7^{2-}$ has stronger oxidizability than SO_4^{2-}.

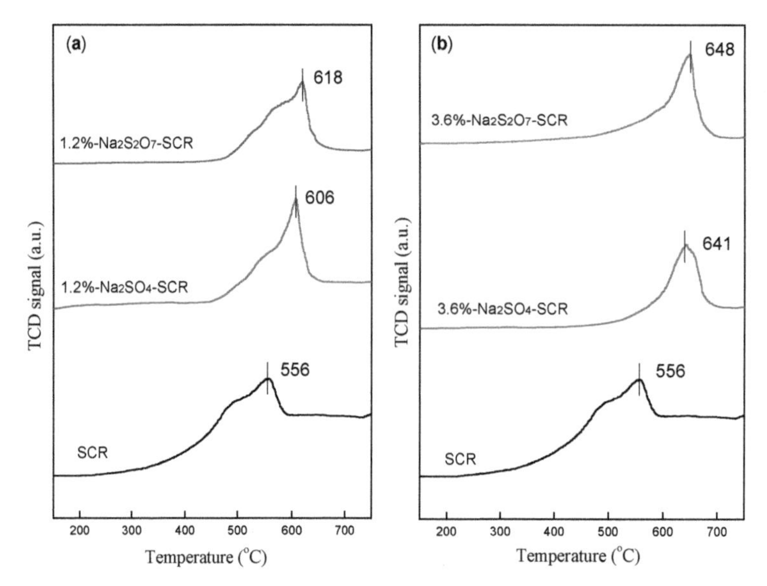

Figure 7. H_2-TPR profiles of the catalyst samples for Na content (**a**) 1.2% and (**b**) 3.6%.

Figure 8 shows the H_2-TPR curves of 3.6% Na_2SO_4 and $Na_2S_2O_7$ poisoned catalysts when the SO_2 concentration is 0 and 1800 ppm. The reduction peak temperature shifted towards the higher end after addition of SO_2, which decreased the denitration performance of the catalyst. However, the H_2 consumption also increased to 3.120 and 5.129 mmol/g, respectively. Yu et al. [41] considered that the introduction of SO_2 made the surface of the poisoned catalyst sulfated, increasing the catalytic performance of the poisoned catalyst. However, this is not the only influence factor for catalytic performance based on catalytic performance results.

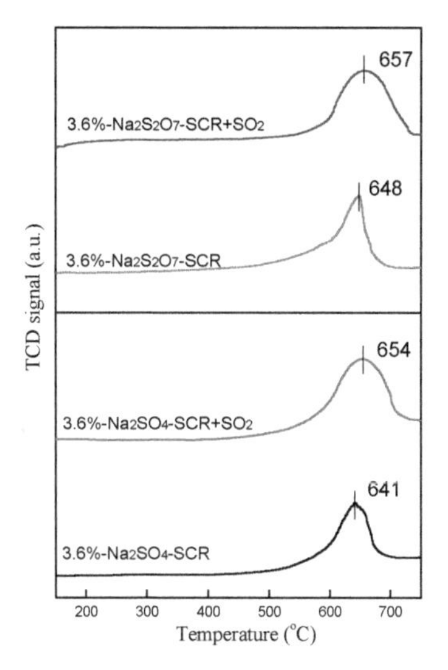

Figure 8. H_2-TPR profiles of the catalysts in different SO_2 concentration (0 and 1800 ppm).

2.3.2. XPS Analysis

The chemical species and surface atomic concentration of several major elements were analyzed using X-ray photoelectron spectroscopy (XPS). Table 1 shows the atomic concentrations on the catalyst surface. Figure 9 shows the spectra of O_{1S} and V_{2P}, respectively. Table 2 shows the states of O and V on these catalyst surfaces. Table 1 shows that with increasing loading of sodium salts, the atomic concentration of the active component on the catalyst surface significantly decreased. This result is consistent with the XRD measurement, indicating that sodium salts interacted with the active component of the catalyst.

Table 1. The surface atomic concentrations for these samples.

Samples	Surface Atomic Concentration (%)					
	O	Na	Ti	V	W	S
Pure catalyst	65.04	0.07	16.27	0.60	3.50	0.05
1.2%-Na_2SO_4-SCR	60.24	4.42	14.32	0.40	2.61	1.13
3.6%-Na_2SO_4-SCR	59.88	5.58	13.01	0.35	2.14	3.54
1.2%-$Na_2S_2O_7$-SCR	61.93	4.17	15.19	0.44	2.77	2.32
3.6%-$Na_2S_2O_7$-SCR	61.08	5.25	14.77	0.38	2.69	5.12

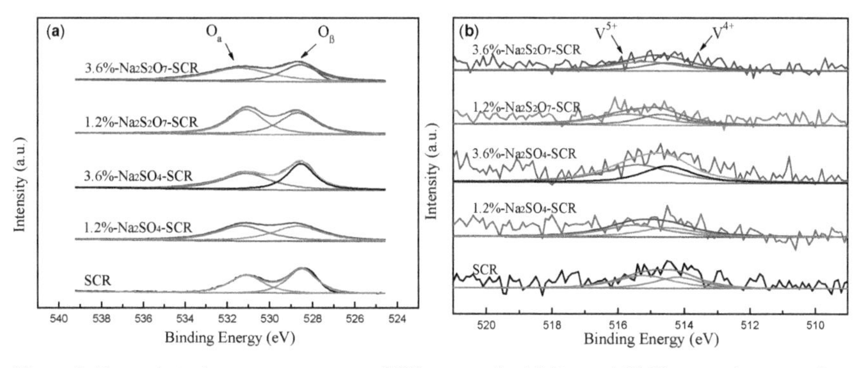

Figure 9. X-ray photoelectron spectroscopy (XPS) spectra for (**a**) O_{1S} and (**b**) V_{2P} over these samples.

Table 2. The states of O and V on the catalyst surfaces.

Samples	Surface Atomic Concentration (%)				Surface Atomic Ratio
	$O_\alpha/(O_\alpha+O_\beta)$	$O_\beta/(O_\alpha+O_\beta)$	V^{4+}	V^{5+}	V^{4+}/V^{5+}
Pure catalyst	44.3	55.7	43.6	56.4	0.77
1.2%-Na_2SO_4-SCR	51.0	49.0	41.4	58.6	0.71
3.6%-Na_2SO_4-SCR	53.1	46.9	40.2	59.8	0.67
1.2%-$Na_2S_2O_7$-SCR	51.8	48.2	40.6	59.4	0.69
3.6%-$Na_2S_2O_7$-SCR	55.3	44.7	39.8	60.2	0.66

As shown in Figure 9a, the spectra of O_{1S} were divided into two peaks. The peak corresponding to 531.0–531.6 eV was attributed to surface chemical adsorption oxygen (defined as O_α). The peak corresponding to 528.4–528.7 eV was attributed to lattice oxygen (defined as O_β) [42,43]. The concentration ratio of chemical adsorption oxygen ($O_\alpha/(O_\alpha+O_\beta)$) was computed using peaking software, which is shown in Table 2. The sodium salts poisoned catalysts increased chemical adsorption oxygen content on the catalyst surface. However, for 1.2% and 3.6% Na_2SO_4 poisoned catalyst the surface oxygen concentration was lower than for other catalysts (see Table 1). Therefore, the chemical adsorption oxygen content of the Na_2SO_4 poisoned catalyst was low. The chemical adsorption oxygen is very active and is indispensable for oxidation reactions. There is a positive correlation between the oxidative properties of the catalyst and the surface chemically adsorbed oxygen content [44]. Therefore, the relatively high chemically adsorbed oxygen content in the sodium salt poisoned catalysts are beneficial for SO_2 oxidation.

Figure 9b shows the spectra of V_{2P}. The spectra were also divided into two peaks. The peak corresponding to 515.3–516 eV belonged to V^{5+}. The peak corresponding to 514.0–514.6 eV belonged to V^{4+} [45]. Table 2 shows that the V^{4+} ratio was 43.6% for pure catalyst. It also shows the V^{4+} content of other sodium salt poisoned catalysts were reduced and the corresponding V^{4+}/V^{5+} ratios decreased. Economidis et al. [46] confirmed experimentally that in the process of synthesizing the vanadium-based catalyst, V^{5+} partially transformed into V^{4+}. Equivalently, the reduction of $VO_2{}^+$ to VO^{2+} occurred. The ratio of V^{4+}/V^{5+} plays a pivotal role in the redox reaction of the catalyst. The denitration efficiency could be improved by appropriately increasing the ratio. However, the addition of sodium salts decreased the ratio (causing the decrease in denitration performance). This is consistent with the denitration performance test.

2.3.3. BET and XRD Analysis

The structural characteristics of different catalysts were determined by N_2 adsorption-desorption experiments. The specific surface area and pore structure also affected the denitration performance of the catalyst to an extent. Table 3 describes the specific surface area and pore structure characteristics of different catalysts. For example, the specific surface area, pore volume and pore diameter of the pure catalyst were found to 95 $m^2 \cdot g^{-1}$, 0.21 $cm^3 \cdot g^{-1}$, and 9.3 nm, respectively. After loading with sodium salts, the specific surface area and pore volume of the catalyst significantly reduced. With increased loading, the degree of reduction increased. $Na_2S_2O_7$ poisoned catalyst has a greater influence on the specific surface area and pore volume. Xiao et al. [47] suggested that sodium pyrosulfate had larger molecular size, which led to the catalyst blockage. Conversely, the pore size increased. This is because the addition of sodium salts caused the microporous blockage and the proportion of macropores consequently increased.

Table 3. Structural characteristics of different Na salts loadings.

Samples	S_{BET} ($m^2 \cdot g^{-1}$)	V_{total} ($cm^3 \cdot g^{-1}$)	D_p (nm)
Pure catalyst	95	0.21	9.3
1.2%-Na_2SO_4-SCR	75	0.18	10.5
1.2%-$Na_2S_2O_7$-SCR	69	0.17	10.5
3.6%-Na_2SO_4-SCR	63	0.16	11.3
3.6%-$Na_2S_2O_7$-SCR	44	0.12	11.1

Figure 10 shows the Barret–Joyner–Halenda (BJH) pore size curves of different catalysts. The catalyst surface mainly contains mesopores (2–50 nm), as shown in Figure 10. The sodium salts poisoned catalyst increased the pore size compared with the pure catalyst.

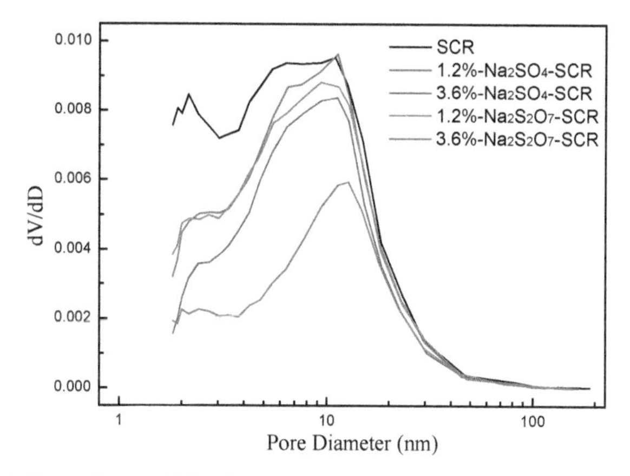

Figure 10. Barret–Joyner–Halenda (BJH) pore size distribution curves of the catalysts.

Figure 11 exhibits the XRD diffractograms of different catalysts. The XRD patterns of all catalyst samples are mostly similar, which demonstrates that the addition of sodium salts did not change the basic structure of the support. However, for the 1.2% and 3.6% Na_2SO_4 poisoned catalyst, the diffraction peak of Na_2SO_4 could be detected. For the pure catalyst and 1.2% $Na_2S_2O_7$ poisoned catalyst, there was no peak of V_2O_5 and no appearance of $Na_2S_2O_7$. But for the 3.6% $Na_2S_2O_7$ poisoned catalyst, the diffraction peak of $Na_4V_3O_9$ could be detected. This was because some molten sodium pyrosulfate interacted with V_2O_5 during the preparation process. The appearance of the new peaks indicates that the accumulation of sodium salts on the catalyst surface blocked the catalyst pores and reduced the specific surface area. This was not conducive to the catalytic reaction. However, Shpanchenko et al. [48] considered that the new vanadium oxide complex $Na_4V_3O_9$ had special structural and magnetic properties. It was made from isolated chains of square $V^{4+}O_5$ pyramids linked by two bridging $V^{5+}O_4$ tetrahedra. This structure had strong magnetic exchange between the V^{4+} along the chain, which was beneficial to catalytic performance.

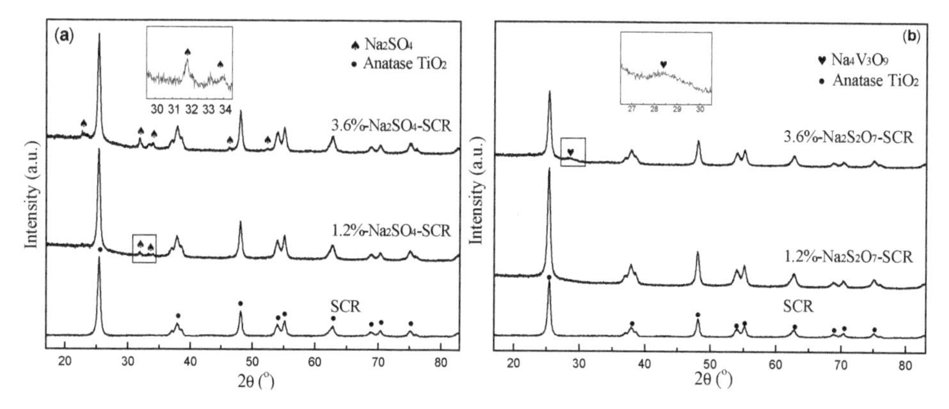

Figure 11. XRD results of the (**a**) Na_2SO_4 and (**b**) $Na_2S_2O_7$ poisoned catalysts.

2.3.4. FT-IR Analysis

The characteristic peaks of different functional groups on different catalyst surfaces were obtained via FT-IR measurements. The results are shown in Figure 12. These catalysts exhibit two key peaks located at 1015 cm^{-1} and 1627 cm^{-1}. The peak at 1015 cm^{-1} is attributed to the terminal vanadium oxy group (V^{5+}=O) [49]. The peak at 1627 cm^{-1} is attributed to the Lewis acid sites [50]. According to the experimental results, the sodium salts poisoned catalyst was found to have reduced the Lewis acid sites. With increased loading, the degree of reduction increased. This indicates that the weak acidic strength of the catalyst surface was lowered. This is consistent with the results of NH_3-TPD. The peak intensity at 1015 cm^{-1} increased and $Na_2S_2O_7$ poisoned catalyst had the most significant effect. The increase of V^{5+}=O bond strength is beneficial for promoting the conversion of SO_2 to SO_3 [51]. Alvarez et al. [18] suggested that the presence of pyrosulfates would maintain the high oxidizability of V species to a certain degree. This result is consistent with the results of SO_2 oxidation experiments.

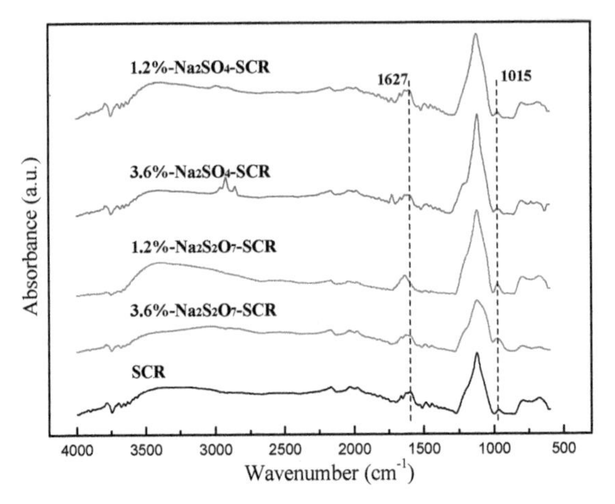

Figure 12. FT-IR spectra of the catalysts.

3. Experimental

3.1. Sample Preparation

The commercial SCR catalysts were obtained from Beijing Nation Power Group Co., Ltd., Beijing, China. The poisoned catalysts were prepared using the wet impregnation method. A certain amount of sodium sulfate or sodium pyrosulfate was weighed according to the mass percentage of Na in the active ingredient of catalysts. After being formulated into solution, they were mixed with catalysts and ultrasonically shaken for 4 h. They were then dried in the blast drying oven at 110 °C for 12 h. Finally, the catalysts were calcined in the muffle furnace at 350 °C for 5 h and ground to 40–60 mesh to obtain x-Na_2SO_4/$Na_2S_2O_7$-SCR poisoned catalysts. Here, x means the mass percentage of sodium element (1.2% or 3.6%).

3.2. Catalyst Characterization

The experiment was performed using the model tp-5080 temperature programmed adsorption instrument (manufactured by Tianjin Xianquan Company, Tianjin, China) for NH_3-TPD and H_2-TPR of different catalysts. For a NH_3-TPD test, 0.1 g samples were prepared and then pretreated for 1 h at 250 °C. Next, they were cooled to ambient temperature in a pure N_2 atmosphere (30 mL/min). Then 10%NH_3 (N_2 as balance gas) was passed over the samples for 1 h. After NH_3 was cut off, the catalysts were warmed to 60 °C and purged in pure N_2 for 20 min. Finally, they were heated to 800 °C (at a rate of 10 °C/min) and maintained at 800 °C for 5 min. The consumption of NH_3 was recorded. For a H_2-TPR test, 0.05 g samples were prepared and then pretreated for 1 h at 250 °C. Next, they were cooled to ambient temperature in pure N_2 atmosphere (30 mL/min). Then 5%H_2 (N_2 as balance gas) was passed over the samples as a reducing agent. Finally, the catalysts were heated to 800 °C (at a rate of 10 °C/min) and maintained at 800 °C for 5 min. The consumption of H_2 was recorded.

The pore structural parameters of the samples were determined via the specific surface area and pore size analyzer (ASAP 2010, Micromeritics Instrument Corporation, Norcross, GA, USA). The specific surface area of the catalyst was obtained by linear regression via the Brunauer–Emmett–Teller (BET) equation. The pore size was calculated using the BJH model.

The XRD patterns of samples were obtained using the Bruker D8 (Bruker AXS Company, Karlsruhe, Germany) Advance to determine the crystallinity and dispersion of the surface material of the samples. The measurement was performed using a Cu Kα irradiation source with a scan range of 10–80°. XPS was performed using the AXIS ULTRADLD (Kratos Company, Shimadzu, Kyoto, Japan). An X-ray source was used as a monochromatic Al target and C1s (284.8 eV) was used for correction when fitting the peak. FT-IR spectra were recorded in a Nicolet Nexus 670 FT-IR (Nicolet Company, Madison, WI, USA). The catalysts were ground and then blended with KBr powder at the mass ratio of 1:100 with a resolution of 4 cm^{-1}. The recorded spectral range was 600–4000 cm^{-1}, and the number of scans was 32.

3.3. Test. Setup

Catalytic performance and SO_2–SO_3 conversion test apparatus is presented in Figure 13. Each catalyst (2 mL, 40–60 mesh) was laid in a quartz tube reactor. The simulated flue gas used in the experiment included 500 ppm NH_3, 500 ppm NO, 5% O_2, 2% H_2O, and the SO_2 concentration ranged from 0 to 3000 ppm (N_2 acted as a balance gas). The total gas flow rate was 1.5 L/min, resulting in a GHSV of 45,000 h^{-1}. The experimental test temperature was 250–410 °C. The inlet and outlet flue gas (O_2, SO_2, NO) concentrations were monitored using the Testo 350 flue gas analyzer (Testo AG, Lenzkirch, Germany). SO_3 was gathered by the Graham Condenser, which was placed in a constant temperature 80 °C water bath. The gathered SO_3 was converted to SO_4^{2-} using an 80% isopropanol solution. Then the SO_4^{2-} content in the solution was measured using an ion chromatograph to determine the SO_3 content, which was averaged over multiple measurements.

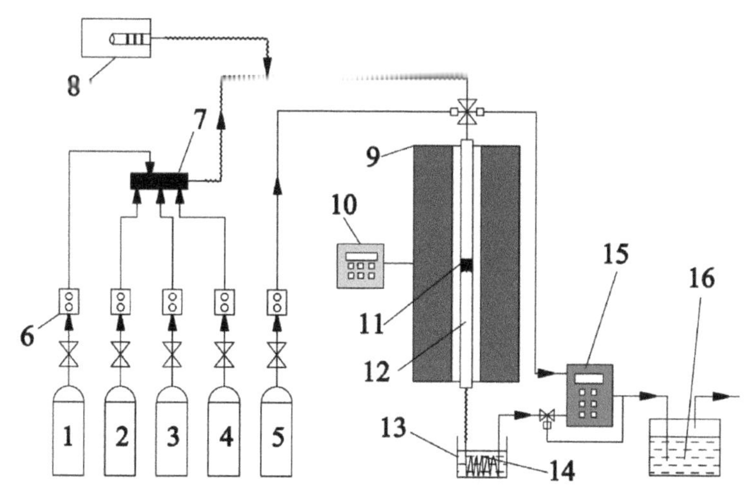

Figure 13. Catalytic performance and SO_2 oxidation tests device. 1. NO/N_2; 2. O_2; 3. N_2; 4. SO_2/CO_2; 5. NH_3/N_2; 6. Mass flowmeter; 7. Gas mixer; 8. Peristaltic pump; 9. Tubular resistance furnace; 10. Temperature controller; 11. Catalyst; 12. Quartz tube; 13. Water heater; 14. Graham condenser; 15. Gas analyzer; 16. Absorption liquid.

The catalytic performance of the catalyst is represented by the conversion of NO, which is defined as:

$$\eta_{NO}(\%) = \left(\frac{[NO]_{inlet} - [NO]_{outlet}}{[NO]_{inlet}} \right) \times 100\% \quad (7)$$

The SO_2–SO_3 conversion is indirectly expressed by the SO_2 oxidation, which is defined as:

$$SO_2\text{–}SO_3\text{-conversation}(\%) = \left(\frac{[SO_3]_{outlet}}{[SO_2]_{inlet}} \right) \times 100\% \quad (8)$$

4. Conclusions

The influence of different sulfur-containing sodium salts (Na_2SO_4 and $Na_2S_2O_7$) poisoned catalysts on SO_2 oxidation and NO reduction was investigated. Sodium salts poisoned catalysts led to a decrease in the denitration efficiency, while significantly improving the SO_2–SO_3-conversion. The degree of change is related to the loading and form of the sodium salts poisoned catalysts. The degree of poisoning of $Na_2S_2O_7$ poisoned catalyst was weaker because $S_2O_7{}^{2-}$ can create more Brønsted acid sites and has stronger oxidizability than $SO_4{}^{2-}$. The introduction of SO_2 clearly increased the number of surface acid sites. However, it had little effect on the redox capacity of the catalyst. Hence, SO_2 slightly enhanced the denitration efficiency of the sodium salts poisoned catalysts. According to analysis of NH_3-TPD, FT-IR, and H_2-TPR results, sodium salts poisoned catalysts reduced the Lewis acid sites and redox capacity. They also increased Brønsted acid sites and V^{5+}=O bonds strength. The presence of the V^{5+}=O bonds facilitates SO_2–SO_3 conversion. XPS results showed that sodium salts poisoned catalysts increased the chemically adsorbed oxygen content and promoted SO_2 oxidation. However, sodium salts poisoned catalysts reduced the V^{4+}/V^{5+} ratio and inhibited denitration performance.

Author Contributions: H.X. conceived and designed the experiments; C.D. performed the experiments and wrote the paper; H.S., J.G. and L.C. contributed reagents/materials/analysis tools.

Acknowledgments: This work was supported by National Natural Science Foundation of China (no. 51206047).

Conflicts of Interest: The authors declare no conflicts of interest.

References

1. Wang, J.; Qiu, Y.; He, S.; Liu, N.; Xiao, C.; Liu, L. Investigating the driving forces of NO_x generation from energy consumption in China. *J. Clean. Prod.* **2018**, *184*, 836–846. [CrossRef]
2. Yang, J.; Sun, R.; Sun, S.; Zhao, N.; Hao, N.; Chen, H.; Wang, Y.; Guo, H.; Meng, J. Experimental study on NO_x reduction from staging combustion of high volatile pulverized coals. Part 1. Air staging. *Fuel Process. Technol.* **2014**, *126*, 266–275. [CrossRef]
3. Kim, M.H.; Yang, K.H. The role of Fe_2O_3 species in depressing the formation of N_2O in the selective reduction of NO by NH_3 over V_2O_5/TiO_2-based catalysts. *Catalysts* **2018**, *8*, 134. [CrossRef]
4. Qi, X.; Song, G.; Song, W.; Lu, Q. Influence of sodium-based materials on the slagging characteristics of Zhundong coal. *J. Energy Inst.* **2017**, *90*, 914–922. [CrossRef]
5. Xu, L.; Liu, H.; Zhao, D.; Cao, Q.; Gao, J.; Wu, S. Transformation mechanism of sodium during pyrolysis of Zhundong coal. *Fuel* **2018**, *233*, 29–36. [CrossRef]
6. Zhang, S.; Liu, S.; Hu, W.; Zhu, X.; Qu, W.; Wu, W.; Zheng, C.; Gao, X. New insight into alkali resistance and low temperature activation on vanadia-titania catalysts for selective catalytic reduction of NO. *Appl. Surf. Sci.* **2019**, *466*, 99–109. [CrossRef]
7. Lisi, L.; Lasorella, G.; Malloggi, S.; Russo, G. Single and combined deactivating effect of alkali metals and HCl on commercial SCR catalysts. *Appl. Catal. B Environ.* **2004**, *50*, 251–258. [CrossRef]
8. Lei, T.; Li, Q.; Chen, S.; Liu, Z.; Liu, Q. KCl-induced deactivation of V_2O_5–WO_3/TiO_2 catalyst during selective catalytic reduction of NO by NH_3: Comparison of poisoning methods. *Chem. Eng. J.* **2016**, *296*, 1–10. [CrossRef]

9. Klimczak, M.; Kern, P.; Heinzelmann, T.; Lucas, M.; Claus, P. High-throughput study of the effects of inorganic additives and poisons on NH_3-SCR catalysts—Part I: V_2O_5–WO_3/TiO_2 catalysts. *Appl. Catal. B Environ.* **2010**, *95*, 39–47. [CrossRef]
10. Liu, Y.; Liu, Z.; Mnichowicz, B.; Harinath, A.V.; Li, H.; Bahrami, B. Chemical deactivation of commercial vanadium SCR catalysts in diesel emission control application. *Chem. Eng. J.* **2016**, *287*, 680–690. [CrossRef]
11. Peng, Y.; Li, J.; Huang, X.; Li, X.; Su, W.; Sun, X.; Wang, D.; Hao, J. Deactivation Mechanism of Potassium on the V_2O_5/CeO_2 Catalysts for SCR Reaction: Acidity, Reducibility and Adsorbed-NO_x. *Environ. Sci. Technol.* **2014**, *48*, 4515–4520. [CrossRef] [PubMed]
12. Peng, Y.; Li, J.; Shi, W.; Xu, J.; Hao, J. Design strategies for development of SCR catalyst: Improvement of alkali poisoning resistance and novel regeneration method. *Environ. Sci. Technol.* **2012**, *46*, 12623–12629. [CrossRef] [PubMed]
13. Chang, H.; Shi, C.; Li, M.; Zhang, T.; Wang, C.; Jiang, L.; Wang, X. The effect of cations (NH^{4+}, Na^+, K^+, and Ca^{2+}) on chemical deactivation of commercial SCR catalyst by bromides. *Chin. J. Catal.* **2018**, *39*, 710–717. [CrossRef]
14. Li, G.; Wang, C.A.; Wang, P.; Du, Y.; Liu, X.; Chen, W.; Che, D. Ash deposition and alkali metal migration during Zhundong high-alkali coal gasification. *Energy Procedia* **2017**, *105*, 1350–1355. [CrossRef]
15. Yang, S.; Song, G.; Na, Y.; Yang, Z. Alkali metal transformation and ash deposition performance of high alkali content Zhundong coal and its gasification fly ash under circulating fluidized bed combustion. *Appl. Therm. Eng.* **2018**, *141*, 29–41. [CrossRef]
16. Niu, Y.; Gong, Y.; Zhang, X.; Liang, Y.; Wang, D.; Hui, S.E. Effects of leaching and additives on the ash fusion characteristics of high-Na/Ca Zhundong coal. *J. Energy Inst.* **2018**. [CrossRef]
17. Li, Q.; Chen, S.; Liu, Z.; Liu, Q. Combined effect of KCl and SO_2 on the selective catalytic reduction of NO by NH_3 over V_2O_5/TiO_2 catalyst. *Appl. Catal. B Environ.* **2015**, *164*, 475–482. [CrossRef]
18. Alvarez, E.; Blanco, J.; Avila, P.; Knapp, C. Activation of monolithic catalysts based on diatomaceous earth for sulfur dioxide oxidation. *Catal. Today* **1999**, *53*, 557–563. [CrossRef]
19. Tian, Y.; Yang, J.; Yang, C.; Lin, F.; Guang, H.; Kong, M.; Liu, Q. Comparative study of the poisoning effect of NaCl and Na_2O on selective catalytic reduction of NO with NH_3 over V_2O_5–WO_3/TiO_2 catalyst. *J. Energy Inst.* **2018**, 1–8. [CrossRef]
20. Hu, W.; Gao, X.; Deng, Y.; Qu, R.; Zheng, C.; Zhu, X.; Cen, K. Deactivation mechanism of arsenic and resistance effect of SO_4^{2-} on commercial catalysts for selective catalytic reduction of NO_x with NH_3. *Chem. Eng. J.* **2016**, *293*, 118–128. [CrossRef]
21. Chen, J.P.; Buzanowski, M.A.; Yang, R.T.; Cichanowicz, J.E. Deactivation of the vanadia catalyst in the selective catalytic reduction process. *J. Air Waste Manag. Assoc.* **2012**, *40*, 1403–1409. [CrossRef]
22. Dahlin, S.; Nilsson, M.; Bäckström, D.; Bergman, S.L.; Bengtsson, E.; Bernasek, S.L.; Pettersson, L.J. Multivariate analysis of the effect of biodiesel-derived contaminants on V_2O_5–WO_3/TiO_2 SCR catalysts. *Appl. Catal. B Environ.* **2016**, *183*, 377–385. [CrossRef]
23. Li, X.; Li, X.; Yang, R.; Mo, J.; Li, J.; Hao, J. The poisoning effects of calcium on V_2O_5–WO_3/TiO_2 catalyst for the SCR reaction: Comparison of different forms of calcium. *Mol. Catal.* **2017**, *434*, 16–24. [CrossRef]
24. Nicosia, D.; Czekaj, I.; Kröcher, O. Chemical deactivation of V_2O_5/WO_3–TiO_2 SCR catalysts by additives and impurities from fuels, lubrication oils and urea solution. Part II. Characterization study of the effect of alkali and alkaline earth metals. *Appl. Catal. B Environ* **2008**, *77*, 228–236. [CrossRef]
25. Dunn, J.P.; Koppula, P.R.; Stenger, H.G.; Wachs, I.E. Oxidation of sulfur dioxide to sulfur trioxide over supported vanadia catalysts. *Appl. Catal. B Environ.* **1998**, *19*, 103–117. [CrossRef]
26. Zhang, L.; Li, L.; Cao, Y.; Yao, X.; Ge, C.; Gao, F.; Deng, Y.; Tang, C.; Dong, L. Getting insight into the influence of SO_2 on TiO_2/CeO_2 for the selective catalytic reduction of NO by NH_3. *Appl. Catal. B Environ.* **2015**, *165*, 589–598. [CrossRef]
27. Ma, J.; Liu, Z.; Liu, Q.; Guo, S.; Huang, Z.; Xiao, Y. SO_2 and NO removal from flue gas over V_2O_5/AC at lower temperatures—Role of V_2O_5 on SO_2 removal. *Fuel Process. Technol.* **2008**, *89*, 242–248. [CrossRef]
28. Wang, Y.; Ma, J.; Liang, D.; Zhou, M.; Li, F.; Li, R. Lewis and Brønsted acids in super-acid catalyst SO_4^{2-}/ZrO_2–SiO_2. *J. Mater. Sci.* **2009**, *44*, 6736–6740. [CrossRef]
29. Xiao, H.; Ru, Y.; Cheng, Q.; Zhai, G.; Dou, C.; Qi, C.; Chen, Y. Effect of sodium sulfate in ash on sulfur trioxide formation in the postflame region. *Energy Fuels* **2018**, *32*, 8668–8675. [CrossRef]

30. Schwaemmle, T.; Heidel, B.; Brechtel, K.; Scheffknecht, G. Study of the effect of newly developed mercury oxidation catalysts on the $DeNO_x$-activity and SO_2–SO_3-conversion. *Fuel* **2012**, *101*, 179–186. [CrossRef]
31. Guo, X.; Bartholomew, C.; Hecker, W.; Baxter, L. Effects of sulfate species on V_2O_5/TiO_2 SCR catalysts in coal and biomass-fired systems. *Appl. Catal. B Environ.* **2009**, *92*, 30–40. [CrossRef]
32. Wu, Z.; Jin, R.; Wang, H.; Liu, Y. Effect of ceria doping on SO_2 resistance of Mn/TiO_2 for selective catalytic reduction of NO with NH_3 at low temperature. *Catal. Commun.* **2009**, *10*, 935–939. [CrossRef]
33. Anstrom, M.; Topsøe, N.; Dumesic, J.A. Density functional theory studies of mechanistic aspects of the SCR reaction on vanadium oxide catalysts. *J. Catal.* **2003**, *213*, 115–125. [CrossRef]
34. Huang, Z.; Zhu, Z.; Liu, Z.; Liu, Q. Formation and reaction of ammonium sulfate salts on V_2O_5/AC catalyst during selective catalytic reduction of nitric oxide by ammonia at low temperatures. *J. Catal.* **2003**, *214*, 213–219. [CrossRef]
35. Peng, Y.; Li, J.; Si, W.; Luo, J.; Dai, Q.; Luo, X.; Liu, X.; Hao, J. Insight into deactivation of commercial SCR catalyst by arsenic: An experiment and DFT study. *Environ. Sci. Technol.* **2014**, *48*, 13895–13900. [CrossRef] [PubMed]
36. Du, X.; Gao, X.; Qiu, K.; Luo, Z.; Cen, K. The Reaction of poisonous alkali oxides with vanadia SCR catalyst and the afterward influence: A DFT and experimental study. *J. Phys. Chem. C* **2015**, *119*, 1905–1912. [CrossRef]
37. Zheng, Y.; Jensen, A.D.; Johnsson, J.E.; Thøgersen, J.R. Deactivation of V_2O_5–WO_3/TiO_2 SCR catalyst at biomass fired power plants: Elucidation of mechanisms by lab- and pilot-scale experiments. *Appl. Catal. B Environ.* **2008**, *83*, 186–194. [CrossRef]
38. Giakoumelou, I.; Fountzoula, C.; Kordulis, C.; Boghosian, S. Molecular structure and catalytic activity of V_2O_5/TiO_2 catalysts for the SCR of NO by NH_3: In situ Raman spectra in the presence of O_2, NH_3, NO, H_2, H_2O, and SO_2. *J. Catal.* **2006**, *239*, 1–12. [CrossRef]
39. Huang, Z.; Zhu, Z.; Liu, Z. Combined effect of H_2O and SO_2 on V_2O_5/AC catalysts for NO reduction with ammonia at lower temperatures. *Appl. Catal. B Environ.* **2002**, *39*, 361–368. [CrossRef]
40. Chen, L.; Li, J.; Ge, M. The poisoning effect of alkali metals doping over nano V_2O_5–WO_3/TiO_2 catalysts on selective catalytic reduction of NOx by NH_3. *Chem. Eng. J.* **2011**, *170*, 531–537. [CrossRef]
41. Yu, Y.; Miao, J.; He, C.; Chen, J.; Li, C.; Douthwaite, M. The remarkable promotional effect of SO_2 on Pb-poisoned V_2O_5–WO_3/TiO_2 catalysts: An in-depth experimental and theoretical study. *Chem. Eng. J.* **2018**, *338*, 191–201. [CrossRef]
42. Gan, L.; Guo, F.; Yu, J.; Xu, G. Improved low-temperature activity of V_2O_5–WO_3/TiO_2 for denitration using different vanadium precursors. *Catalysts* **2016**, *6*, 25. [CrossRef]
43. Wang, T.; Zhang, X.; Liu, J.; Liu, H.; Guo, Y.; Sun, B. Plasma-assisted catalytic conversion of NO over Cu-Fe catalysts supported on ZSM-5 and carbon nanotubes at low temperature. *Fuel Process. Technol.* **2018**, *178*, 53–61. [CrossRef]
44. Wan, Q.; Duan, L.; Li, J.; Chen, L.; He, K.; Hao, J. Deactivation performance and mechanism of alkali (earth) metals on V_2O_5–WO_3/TiO_2 catalyst for oxidation of gaseous elemental mercury in simulated coal-fired flue gas. *Catal. Today* **2011**, *175*, 189–195. [CrossRef]
45. Qi, C.; Bao, W.; Wang, L.; Li, H.; Wu, W. Study of the V_2O_5-WO_3/TiO_2 catalyst synthesized from waste catalyst on selective catalytic reduction of NO*x* by NH_3. *Catalysts* **2017**, *7*, 110. [CrossRef]
46. Economidis, N.V.; Peña, D.A.; Smirniotis, P.G. Comparison of TiO_2-based oxide catalysts for the selective catalytic reduction of NO: Effect of aging the vanadium precursor solution. *Appl. Catal. B Environ.* **1999**, *23*, 123–134. [CrossRef]
47. Xiao, H.; Chen, Y.; Qi, C.; Ru, Y. Effect of Na poisoning catalyst (V_2O_5–WO_3/TiO_2) on denitration process and SO_3 formation. *Appl. Surf. Sci.* **2018**, *433*, 341–348. [CrossRef]
48. Shpanchenko, R.V.; Chernaya, V.V.; Antipov, E.V.; Hadermann, J.; Kaul, E.E.; Geibel, C. Synthesis, structure and magnetic properties of the new mixed-valence vanadate $Na_2SrV_3O_9$. *J. Solid State Chem.* **2003**, *173*, 244–250. [CrossRef]
49. Lewandowska, A.E.; Calatayud, M.; Lozano-Diz, E.; Minot, C.; Bañares, M.A. Combining theoretical description with experimental in situ studies on the effect of alkali additives on the structure and reactivity of vanadium oxide supported catalysts. *Catal. Today* **2008**, *139*, 209–213. [CrossRef]

50. Busca, G.; Saussey, H.; Saur, O.; Lavalley, J.C.; Lorenzelli, V. FT-IR characterization of the surface acidity of different titanium dioxide anatase preparations. *Appl. Catal.* **1985**, *14*, 245–260. [CrossRef]
51. Liu, Y.; Shu, H.; Xu, Q.; Zhang, Y.; Yang, L. FT-IR study of the SO_2 oxidation behavior in the selective catalytic reduction of NO with NH_3 over commercial catalysts. *J. Fuel Chem. Technol.* **2015**, *43*, 1018–1024. [CrossRef]

Article

Water: Friend or Foe in Catalytic Hydrogenation? A Case Study Using Copper Catalysts

Alisa Govender [1,2], Abdul S. Mahomed [1] and Holger B. Friedrich [1,*]

[1] School of Chemistry and Physics, University of KwaZulu-Natal, Durban 4041, South Africa; Alisa.Govender@sasol.com (A.G.); mahomeda1@ukzn.ac.za (A.S.M.)
[2] Group Technology, Research & Technology, Sasol South Africa (Pty) Ltd., 1 Klasie Havenga Road, Sasolburg 1947, South Africa
* Correspondence: Friedric@ukzn.ac.za; Tel.: +27-31-260-3107

Received: 14 September 2018; Accepted: 4 October 2018; Published: 19 October 2018

Abstract: Copper oxide supported on alumina and copper chromite were synthesized, characterized, and subsequently tested for their catalytic activity toward the hydrogenation of octanal. Thereafter, the impact of water addition on the conversion and selectivity of the catalysts were investigated. The fresh catalysts were characterized using X-ray diffraction (XRD), BET surface area and pore volume, SEM, TEM, TGA-DSC, ICP, TPR, and TPD. An initial catalytic testing study was carried out using the catalysts to optimize the temperature and the hydrogen-to-aldehyde ratio—which were found to be 160 °C and 2, respectively—to obtain the best conversion and selectivity to octanol prior to water addition. Water impact studies were carried out under the same conditions. The copper chromite catalyst showed no deactivation or change in octanol selectivity when water was added to the feed. The alumina-supported catalyst showed no change in conversion, but the octanol selectivity improved marginally when water was added.

Keywords: hydrogenation; copper; catalyst; water; deactivation; octanal; octanol

---

1. Introduction

Catalytic hydrogenation reactions are often used in the chemical industry to convert productswith little commercial importance, obtained from other processes, to products with an increased demand and need in the chemical industry [1]. Such reactions find applications in the preparation of pharmaceuticals and fine chemicals [2]. Oxygenated compounds, such as aldehydes, are an example of the starting material utilized in catalytic hydrogenation in the fine chemicals sector to produce alcohols. Typically, Ni systems are utilized for aldehyde hydrogenation, however, they sometimes, due to their high activity, lead to abnormal levels of side products, such as esters, acetals, and other aldol condensation type products. Consequently, Cu has been used in certain instances for the hydrogenation of aldehydes or similar compounds to limit such side reaction products [3,4].

However, due to their lower activity, Cu catalysts need to operate at higher temperatures. In industry this may require higher steam pressures in heat exchangers, which are responsible for heating up the feed to the reactors [5]. It is possible for these heat exchangers to develop leaks, resulting in water ingress into the reactor via the feed, thus affecting the activity of the catalyst [6]. Sometimes, the effect can be detrimental, depending on how much water has ingressed into the reactor. Another source of water could recycle to the hydrogenation reactor when distillation processes upfield of the reactor experience process upsets, for example, reboiler tube leaks [7]. In any case, water is certainly a potential problem in any catalytic process, and its effect warrants some investigation [8–10].

In regards to copper catalysts, much of the deactivation studies have been focused on catalysts for CO hydrogenation for methanol synthesis, low-temperature water-gas shift, and selective hydrogenation catalysts, amongst others [11–14]. A fundamental question that arises is whether water can actually be identified as a poison or not. In fact, in some hydrogenation reactions, the addition of water enhances the activity [15]. Water as a poison is usually associated with the oxidation of the active metal, acceleration of sintering, and even leaching of the active metals [16].

Typically, copper chromite catalysts have been used for hydrogenation applications, however, Cr has been identified as being environmentally unfriendly [17]. Nonetheless, copper chromite catalysts are still being used in various catalytic processes globally [4]. Thus, various other relatively safer supports have been studied to replace chromium, e.g., alumina [18]. However, in so doing, its impact against poisons will also need consideration when compared to copper chromite.

Hence, for this study, as part of our interest in catalyst deactivation and regeneration in catalytic hydrogenation [19], octanal hydrogenation to octanol was used as a model reaction to investigate the influence of water during the hydrogenation process using copper catalysts supported on Al_2O_3 and as copper chromite, and to ascertain its effect on the octanal conversion and product selectivity.

2. Results and Discussion

2.1. Catalyst Characterization

The copper loading for each catalyst is listed in Table 1 and was between 24 and 26 wt. %, reasonably close to the nominal loading of 25%. From the BET surface area measurements listed in Table 1, it can be observed that the surface area of CuO/Al_2O_3 is significantly higher than that of $CuCr_2O_4$. Both the variance in surface area and in crystallite size is ascribed to the manner in which the materials were synthesized, with CuO supported for the alumina material, compared to the particles of co-precipitated $CuCr_2O_4$.

Table 1. Physical characterization data.

Catalyst	Cu Loading/wt. %	BET Surface Area/m^2 g^{-1}	Total Pore Volume/cm^3 g^{-1}	Average TEM Particle Size/nm
CuO/Al_2O_3	23.5	128.8	0.4285 (0.6530) [a]	16
$CuCr_2O_4$	26.3	25.3	0.3697	30

[a] Pore volume of alumina in parenthesis.

The diffractograms for CuO/Al_2O_3, Al_2O_3, and $CuCr_2O_4$are shown in Figure S1a–c, respectively, in the Supplementary Information. Individual peaks corresponding to CuO are observed in Figure S1a, however, peaks corresponding to gamma alumina (γ-Al_2O_3) are not seen in the diffractogram [20]. This is most probably due to the overlap of these peaks with those of the CuO, since the JCPDS files for each compound list similar d-spacing values. Upon examining the diffractogram of γ-Al_2O_3 (Figure S1b), broad low-intensity peaks are seen. The position of these peaks coincides with the high-intensity CuO peaks, thus confirming the overlap of these peaks with those of CuO. Peaks corresponding to $CuCr_2O_4$ are observed in Figure S1c,confirming this phase as the dominant phase for this catalyst system [21]. CuO could not be detected, however, overlapping of the peaks with those of $CuCr_2O_4$ does not preclude its presence.

The TPR profiles for CuO/Al_2O_3, $CuCr_2O_4$, and unsupported CuO are presented in Figure 1a–c, respectively. The data obtained from these profiles (temperature at maximum and degree of reducibility) are given in Table 2. The reduction peak seen for CuO/Al_2O_3 (Figure 1a) at 231 °C and the first shoulder peak for $CuCr_2O_4$ (Figure 1b) seen at 240 °C correspond to the reduction of dispersed CuO. The reduction peaks at 298 °C and 308 °C, respectively, correspond to the reduction of bulk CuO that interacts with the support [22–24]. The $CuCr_2O_4$ catalyst shows another reduction peak between 400 and 500 °C, which can be ascribed to the bulk reduction of $CuCr_2O_4$. The TPR profile of unsupported CuO (as shown in Figure 1c) displays a reduction peak at 416 °C that corresponds to the reduction of bulk CuO. The broadness of the peak is an indication of the particle size range and the inherent difficulty associated with the reduction of unsupported CuO. From Table 2, it is observed that both catalysts show very similar values with respect to the degree of reducibility, suggesting that the active site density is very similar for both systems.

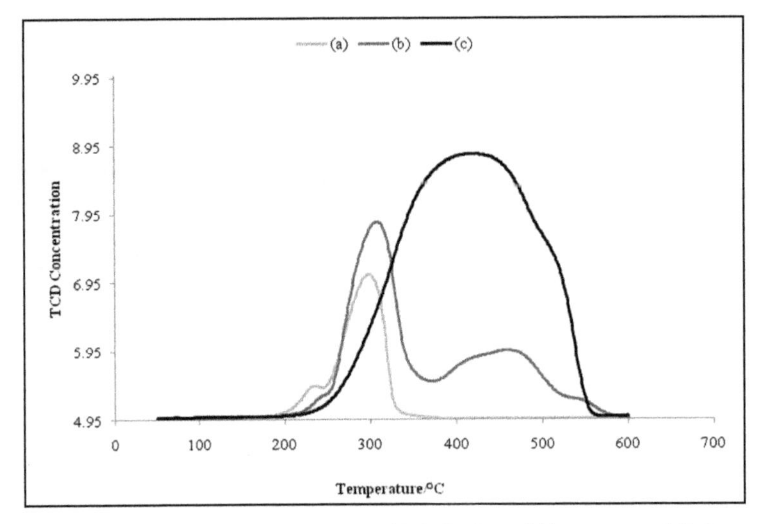

Figure 1. TPR profile of (**a**) CuO/Al_2O_3; (**b**) $CuCr_2O_4$; and (**c**) unsupported CuO.

Table 2. List of temperature at maximum (Tm) and degree of reduction of Cu, as determined by H_2-TPR.

Catalyst	Temperature at Maximum (Tm)/°C	Degree of Reduction [a]/%
Unsupported CuO (Bulk)	416	-
CuO/Al_2O_3	231, 298	84.4
$CuCr_2O_4$	308, 468, shoulder peaks at 240 and 541	85.8

[a] Determined using the equation: moles H_2 consumed/moles reducible Cu*100.

The NH_3 TPD results of the catalysts indicating acid strength and acid site concentration are presented in Table 3. The acid strength of the catalysts is described by three regions, namely, weak, medium, and strong acid sites over the temperature ranges 200–310 °C, 310–500 °C, and 500–1000 °C, respectively. The data listed in Table 3 show that the alumina-supported CuO has both weak and strong acid sites, whilst $CuCr_2O_4$ shows the presence of weak, medium, and strong acid sites. Although the total acidity is higher for $CuCr_2O_4$, it is the acid site density of the different regions that is significant for the reactions occurring during the process.

Table 3. The acid strength and acidity of each catalyst.

	Acid Strength Temperature (°C)			Acid Sites Concentration (μmol g^{-1} of Cat)			
Catalyst	Tm at A	Tm at B	Tm at C	Acidity at A	Acidity at B	Acidity at C	Total Acidity
CuO/Al_2O_3	244	-	636	1288	-	504	1791
$CuCr_2O_4$	261	464	987	847	1014	78	1938

A = Weak Acid Site (200–310 °C); B = Medium Acid Site (310–500 °C); and C = Strong Acid Site (500–1000 °C).

The backscattered SEM images for each of the different catalytic systems are presented in Figure 2. For CuO supported on alumina, the backscattered SEM image shows brighter regions, which correspond to CuO particles, and darker/gray regions corresponding to alumina (Figure 2a). The CuO particles are seen to exist as clusters and appear to be present on the surface of the alumina particles. The SEM image for the copper chromite catalyst does not show any distinct morphology. Owing to the synthesis method to prepare this catalyst, it appears quite chunky and irregular. This corresponds well to the lower surface area obtained for this system. Due to the similarity in atomic number between Cu and Cr, the backscattered SEM image of the catalyst (Figure 2b) provides no unambiguous information.

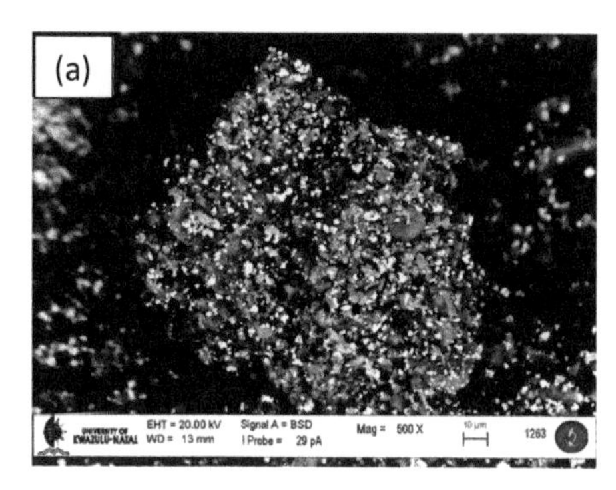

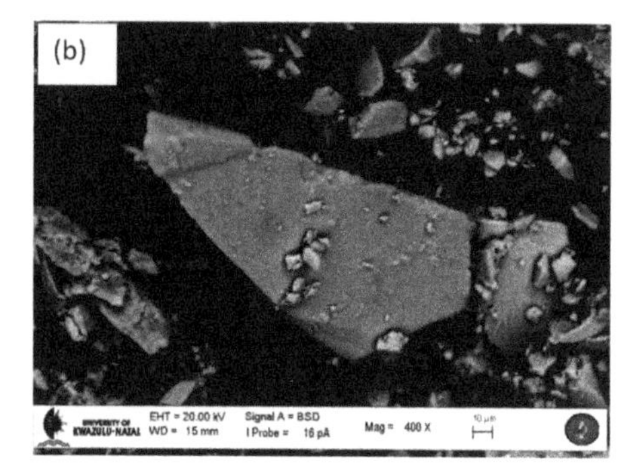

Figure 2. (**a**) Backscattered SEM images of CuO/Al_2O_3; (**b**) backscattered SEM image of $CuCr_2O_4$.

The TEM images for the catalysts are given in Figure 3. The CuO particles appear spherical or somewhat lobed shaped for both systems, the difference being the particle sizes. Table 1 gives the average particle size obtained from TEM for each of the catalysts, with the alumina-supported catalyst having the smaller average size.

The chromia catalyst has a larger average particle size owing to the synthesis method, and this is also reflected in the surface area.

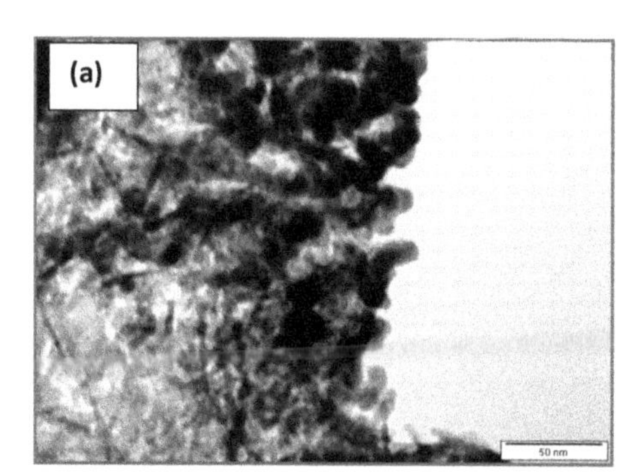

Figure 3. (**a**) TEM images of CuO/Al_2O_3; (**b**) TEM image for $CuCr_2O_4$.

2.2. Catalytic Testing—Water Impact Studies

Water impact studies were carried out at 160 °C and with a hydrogen-to-aldehyde ratio of 2 (H_2:octanal, 2:1). These conditions were established in a previous study as the optimum for octanal conversion and octanol selectivity [25]. Once steady-state conversion was reached using the fresh feed, the water-spiked feed was introduced to the system.

2.2.1. Effect of Water on the Hydrogenation of Octanal Using CuO/Al_2O_3

Figure 4 shows the conversion of octanal and the selectivity to octanol when the reaction was carried out using the fresh feed and the water-spiked feed. Prior to water addition, the conversion of octanal reached steady state after 1 h on stream at a value of 99% and remained steady at this value for 30 h. After this time, the water-spiked feed was introduced into the system and the reaction was allowed to proceed for a further 70 h. During this time, the conversion remained at the high level

(99%), indicating that the presence of water in the feed did not have a negative impact on the catalyst. The selectivity to octanol reached a value of approximately 97% during the first 30 h of the reaction in the absence of water. However, once the water-spiked feed was introduced, the selectivity increased to approximately 98.5% and remained at this value for the duration of the reaction.

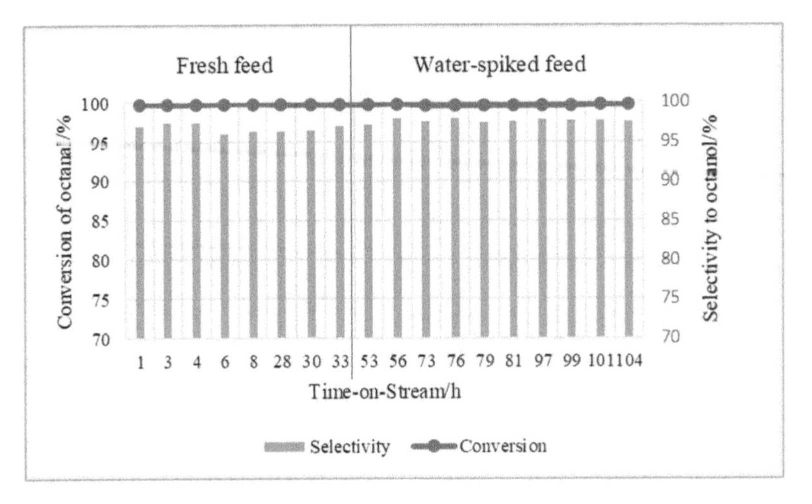

Figure 4. Conversion of octanal and the selectivity to octanol for the hydrogenation of octanal using the fresh feed and the water-spiked feed over CuO/Al_2O_3. (60 bars, 160 °C, H_2:octanal ratio of 2:1).

With the slight increase in the selectivity to octanol, there was a corresponding decrease in the acid/base catalyzed reaction products, such as C16 diol and C24-acetal [26–28]. It is clear from the results that the presence of water did not oxidize the available Cu during the reaction, considering that the conversion remained unchanged. Wang et al. reported in their study on the hydrogenation of hexanal and propanal using sulfided Ni–Mo/Al_2O_3 as catalysts that water proved beneficial in improving the selectivity to the alcohols, however, they noticed a decrease in conversion [26]. It is surprising, however, that the selectivity to octanol does not decrease owing to the possibility of forming more Brønsted sites with the addition of water. This is reasoned by water interacting with the Brønsted site via hydrogen bonding. The interaction occurs either via one molecule of water or clusters that may have formed, depending on the proximity of the acid sites and the concentration of water on the surface. Similar behavior has been reported for zeolite materials [29]. The result of this behavior caused a slight reduction in the total acidity of the catalyst and thus allowed an increase in the octanol selectivity. A more comprehensive list of other byproducts are shown in the Supplementary Information, Figure S2, for one of the data points.

2.2.2. Effect of Water on the Hydrogenation of Octanal Using $CuCr_2O_4$

The conversion of octanal and the selectivity to octanol during the hydrogenation of octanal using the fresh feed and the water-spiked feed are shown in Figure 5. The octanal conversion reached 99% over the $CuCr_2O_4$ catalyst and remained steady at this value for 26 h. Upon introduction of the water-spiked feed, the conversion remained at 99%. The selectivity to octanol reached approximately 96.3% during the first 31 h of the reaction when using the fresh feed, and after introducing the water-spiked feed, an insignificant increase in selectivity was observed.

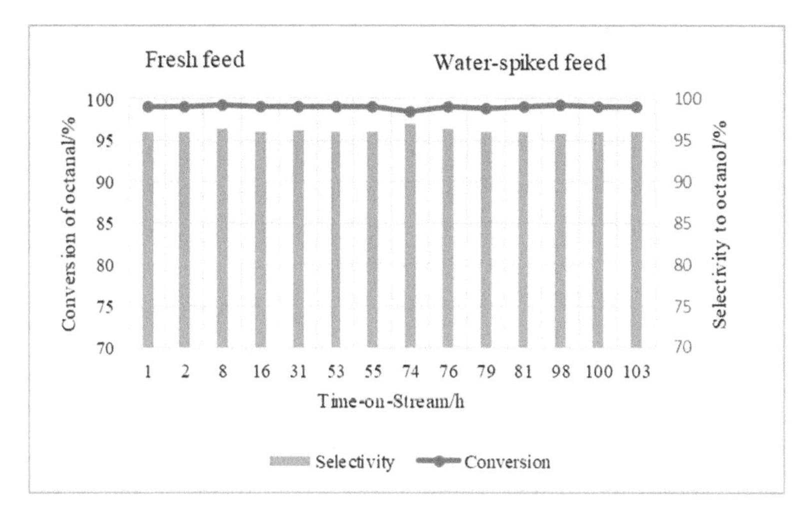

Figure 5. Conversion of octanal and selectivity to octanol for the hydrogenation of octanal using fresh feed and water-spiked feed over $CuCr_2O_4$. (60 bars, 160 °C, H_2:octanal ratio of 2:1).

The major byproduct formed during the reaction with the fresh feed and the water-spiked feed was the C16 diol. The selectivity to the C16 diol when using the fresh feed was about 2%, which remained unchanged when water was introduced. Similarly, all other byproducts showed only minor changes in selectivity after introducing the water-spiked feed into the system. The full selectivity list of byproducts for an arbitrary data point is shown in the Supplementary Information, Figure S3.

For the alumina catalyst, it was observed that the presence of water in the feed stream improved the selectivity to octanol by suppressing the formation of byproducts. However, a key factor favoring this trend is the presence of surface hydroxyls on the catalyst support. A minor increase in the selectivity to octanol was obtained when using $CuCr_2O_4$ as the catalyst, indicating that the effect of water on the hydrogenation of octanal was not as pronounced as when CuO/Al_2O_3 was used as the catalyst.

2.3. *Used Catalyst Characterization*

The diffractograms for Cu/Al_2O_3 and $CuCr_2O_4$after the reaction with the fresh feed and the water-spiked feed are shown in Figures 6 and 7, respectively. These diffractograms show the presence of characteristic copper (Cu^0) peaks (JCPDS 4-0836) and some CuO peaks. In addition, the diffractogram for the $CuCr_2O_4$ catalyst (Figure 7) reveals peaks attributed to the phases $CuCr_2O_4$, Cr_2O_3, and Cu_2O.The full width at half-maximum (FWHM) and the X-ray diffraction (XRD) crystallite size are listed in Table 4. The FWHM for Cu/Al_2O_3 used for the reaction with the water-spiked feed was negligibly different to the value obtained for the catalyst used for the reaction with the fresh feed, however, a significant change was observed for the $CuCr_2O_4$, with an increase of 0.5° for the FWHM of the catalyst used for the reaction with the water-spiked feed.

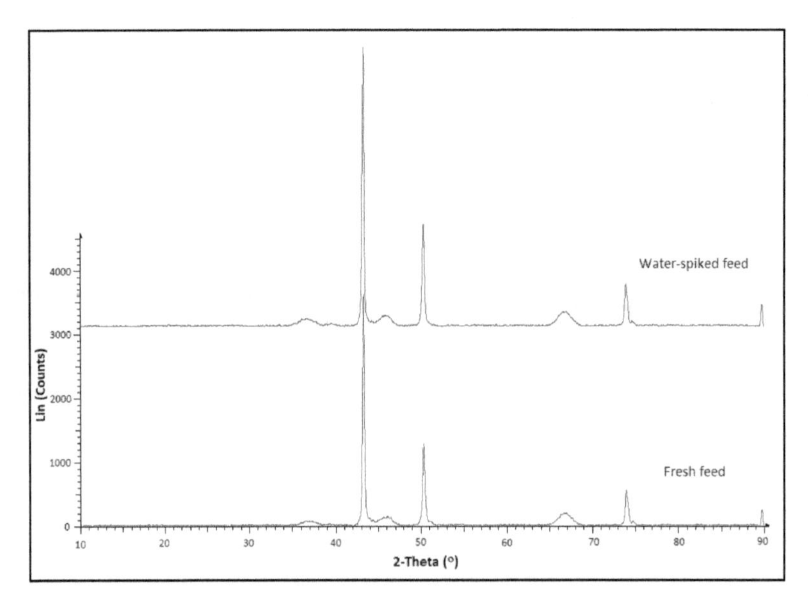

Figure 6. Diffractogram of the used Cu/Al_2O_3 after the reaction with fresh feed only and water-spiked feed.

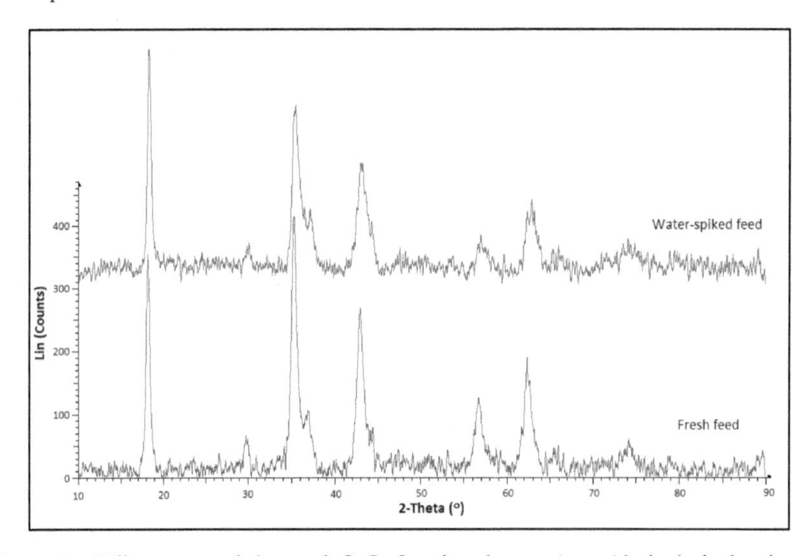

Figure 7. Diffractogram of the used $CuCr_2O_4$ after the reaction with fresh feed only and water-spiked feed.

This indicates that the catalyst suffered a loss of crystallinity during the reaction with the water-spiked feed. The average particle size for both catalysts after the reaction with the water-spiked feed was slightly smaller in comparison to the catalyst used for the reaction with the fresh feed. This smaller particle size translated to a higher BET surface area, shown in Table 4, compared to the catalysts prior to testing. It was noted that the average particle size of the catalyst used with the water-free feed was slightly larger compared to that of the fresh catalyst, yet the BET surface area was higher. This could be attributed to some sintering of the copper allowing the higher-exposed alumina support and, to some extent, the chromite phase, to cause the increase.

Table 4. Full width at half-maximum (FWHM) values and the crystallite size of the highest-intensity Cu peak for each catalyst and BET surface area after the reaction using fresh feed and water-spiked feed.

Catalyst	FWHM/°	Average Particle Size/nm	BET Surface Area/m^2 g^{-1}
Cu/Al_2O_3 fresh feed	0.264	23	141.7
Cu/Al_2O_3 water-spiked feed	0.258	15	143.9
$CuCr_2O_4$ fresh feed	0.799	33	33.5
$CuCr_2O_4$ water-spiked feed	1.335	24	36

The SEM images for Cu/Al_2O_3 and $CuCr_2O_4$ after the reaction with the fresh feed and the water-spiked feed are shown in Figures 8 and 9, respectively. These images show minor differences in the morphology of the used catalyst between Cu/Al_2O_3 and $CuCr_2O_4$ used for the reaction with the fresh feed and the water-spiked feed. This shows that the catalysts were robust to the water, showing no signs of breakage. The EDS composition scanning map data for Cu/Al_2O_3 and $CuCr_2O_4$ after the reaction with the fresh feed and the water-spiked feed are shown in Figures S4–S7 in the Supplementary Information, respectively. These maps show that the distribution of the Cu-rich particles is relatively similar for the catalysts after exposure to both the feeds. However, a change in the morphology of the Cu/Al_2O_3 catalyst could possibly have occurred, thus also influencing the selectivity in that way.

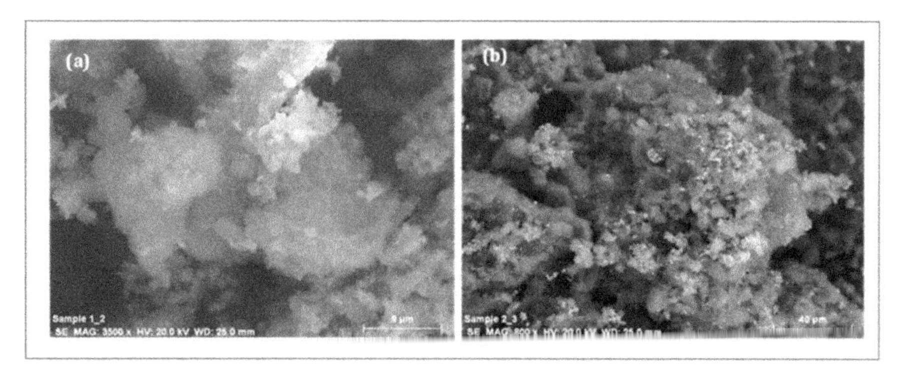

Figure 8. SEM image of Cu/Al_2O_3 after (**a**) the reaction with fresh feed and (**b**) reaction with the water-spiked feed.

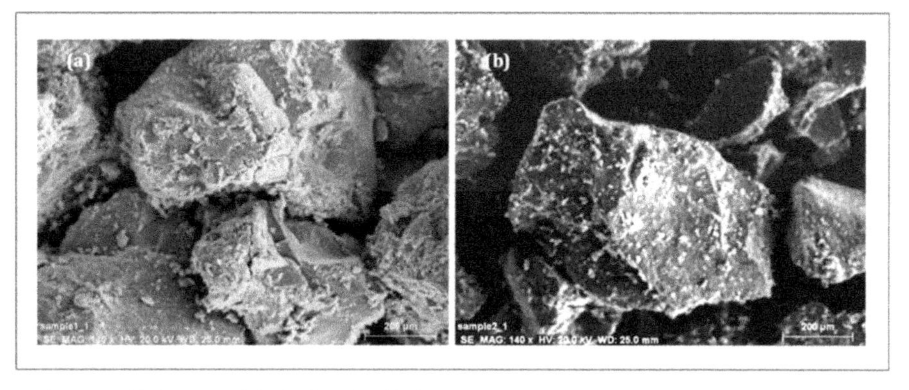

Figure 9. SEM image of $CuCr_2O_4$ after (**a**) the reaction with fresh feed and (**b**) reaction with the water-spiked feed.

3. Materials and Methods

The method of co-precipitation was used to synthesize $CuCr_2O_4$ and the wet impregnation method was used to prepare CuO/Al_2O_3.The syntheses of the catalysts were carried out using modifications of the method of Gredig et al. [30] and Bando et al. [31].

The $CuCr_2O_4$ catalyst was synthesized from metal nitrate solutions with sodium hydroxide (NaOH) at a constant temperature of 298 K and at a pH of approximately 7. The metal nitrate solutions were added dropwise to a round-bottomed flask containing distilled water under vigorous stirring. At the same time, 2 M NaOH was added dropwise to regulate the pH and assist in precipitation. The slurry was aged for half an hour at room temperature and constant pH (7).The resulting precipitate was filtered, washed with distilled water, and dried overnight at 120 °C. The dried catalyst was crushed and thereafter calcined in air for 8 h at 550 °C. For the alumina-supported catalyst, a copper nitrate solution was prepared by dissolving the required amount of copper nitrate in sufficient water to allow dissolution of the metal salt. The solution was then added to the catalyst support under stirring. The slurry obtained was stirred for 24 h. There after, the water was slowly evaporated by gentle heating with vigorous stirring. The resulting solid obtained was oven-dried overnight at 120 °C and thereafter calcined in air for 8 h at 600 °C (alumina catalyst).The catalysts were pelletized and sieved to a particle size of 300–600 microns.

X-ray diffraction (XRD) diffractograms were obtained using a Bruker D8 ADVANCE X-ray diffractometer (Bruker, Karlsruhe, Germany) employing Cu K_α radiation (λ = 1.5418 Å). Prior to the SEM and EDS analyses, the samples were mounted on an aluminium stub using double-sided carbon tape and subsequently gold sputter-coated using a Polaron E5100 coating unit and carbon-coated using a Jeol JEE-4C vacuum evaporator (Joel, Peabody, MA, USA)for the SEM and EDS analyses, respectively. The secondary (normal) and backscattered SEM images were viewed on a Leo 1450 scanning electron microscope (Carl Zeiss AG, Oberkochen, Germany). The EDS composition scanning and certain secondary images were done on a Jeol JSM-6100 scanning microscope (Joel, Peabody, MA, USA) fitted with a Bruker EDX detector. The TEM images were viewed on a Jeol JEM-1010 electron microscope (Joel, Peabody, MA, USA). The elemental composition of the catalysts was determined by ICP-OES on a Perkin Elmer Precisely Optical Emission Spectrometer Optima 5300 DV (PerkinElmer, Shelton, CT, USA). Prior to analysis, the catalyst samples were digested with concentrated nitric acid. The specific surface area and pore volumes of the catalyst were obtained by using the BET nitrogen physisorption analysis using a Micrometrics Gemini instrument (Micrometrics, Atlanta, GA, USA) with nitrogen adsorbate at −196 °C. TPR and TPD were carried out on a Micrometrics Autochem II Chemisorption Analyzer (Micrometrics, Atlanta, GA, USA).Prior to both analyses, the samples were pre-treated with argon flowing at a rate of 30 mL/min, upto 350 °C, at a rate of 20 °C/min. For the H_2-TPR, the pre-treated sample was then subjected to 5% H_2 in argon flowing at a rate of 30 mL/min, whilst the temperature was linearly increased from 90 °C to 600 °C at a rate of 10 °C/min. The hydrogen consumption was monitored by a thermal conductivity detector. For the NH_3-TPD, the pre-treated sample was reduced with 5% H_2 in argon and thereafter treated with helium for 1 h to remove excess hydrogen. A 4% NH_3 in helium gas mixture was then passed through the system and allowed to adsorb onto the surface of the catalyst for half an hour. Helium gas was then passed through at a flow rate of 30 mL/min, whilst the temperature was linearly increased from 100 °C to 1000 °C at a rate of 10 °C/min. The amount of ammonia desorbed was monitored using a thermal conductivity detector.

The hydrogenation reaction was carried out using a fixed bed reactor. A stainless-steel (grade 316) tube with an inner diameter of 15 mm and a length of 325 mm was used as the reactor tube. The amount of gas entering the reactor was controlled by means of Brooks's mass flow meters (0–300 mL/min) (Brooks Instruments, Hatfield, PA, USA), which were calibrated before use. The liquid feed was introduced to the reactor by means of a LabAlliance Series II isocratic HPLC pump (ASI, Richmond, CA, USA). Due to the high exotherm associated with the hydrogenation of octanal, a 10 wt. % octanal in octanol (diluent) feed was used as the fresh feed, whilst for the water impact studies, a water-spiked feed with1.8 wt. % water in the fresh feed was used, which is below the saturation point at 25 °C [32].The product leaving the reactor tube was collected in a 500 mL sampling cylinder. The back-pressure regulator maintained the pressure within the system to the desired value. The gas leaving the back-pressure regulator entered a Ritter drum-type wet gas flow meter (TG1—model 5 with a PVC drum and casing) which allowed for the quantification of off-gas flowing during the run time of the experiment.

The catalyst was nitrogen-treated in situ at 170 °C at a flow to give a GHSV of 1000 h^{-1}. After this time, the reactor was cooled to 130 °C and hydrogen was introduced into the system in increments of 5 vol. %. Once the system stabilized, the nitrogen flow was decreased accordingly, until the hydrogen concentration was 100%, whilst maintaining a constant GHSV. The temperature was then increased to 200 °C and the catalyst was allowed to reduce overnight with pure hydrogen. Once this reduction process was complete, the reactor system was pressurized to 60 bar, the temperature was reduced to 160 °C, and the feed was introduced. The hydrogen flow rate was set to the desired value (GHSV = 464 h^{-1}) and the liquid feed was passed through the reactor (LHSV = 16 h^{-1}) to maintain a H_2:octanal ratio of 2:1. The liquid product formed was collected in the sampling cylinder and was sampled in approximately 1.5 h intervals. The product was analyzed by GC Perkin Elmer Clarus 500 with an FID detector (Perkin Elmer, Akron, OH, USA). A Petro-Elite column (length = 50 m and diameter = 200 μm) was used for the analyses. The typical reaction time was 5 days, which included the baseline testing as well as the effect of the water-spiked feed.

The used catalysts were dried under a flow of ultrapure nitrogen (99.999%, Afrox, Durban, South Africa) and removed from the reactor under a nitrogen blanket. The catalysts were analyzed immediately to avoid or minimize the oxidation of the copper.

4. Conclusions

Copper oxide supported on alumina and chromia were synthesized and characterized. XRD showed characteristic peaks for CuO for the alumina-supported catalysts and established the chromite phase for the $CuCr_2O_4$ catalyst. This was also confirmed by microscopy analyses of the two systems. The catalysts contained between 24 and 26 wt. % copper, and the BET surface area was higher for the alumina-supported catalyst, however, both catalysts showed a high degree of reduction under H_2. The exposure of the catalysts to a water-spiked feed with approximately 1.8 wt. % water showed very little variance in the conversion of the catalyst when compared to the feed without water. For the reaction over CuO/Al_2O_3, the conversion remained unchanged (at 99%) after the introduction of the water-spiked feed, however, an increase in the selectivity to octanol (1.5%) was observed and was attributed to the interaction of the water molecules with the surface hydroxyls on alumina. For the reaction over $CuCr_2O_4$, the conversion of octanal, as well as the octanol selectivity, remained essentially unchanged after the introduction of the water-spiked feed. From characterization of the used catalysts by SEM and XRD, it was determined that the presence of water did not negatively impact the physical structure of the catalyst. The particle size was observed to have decreased for the water-spiked feed and increased marginally for the feed without water when compared to the fresh catalyst. These results indicate that Cu/Al_2O_3 maybe a robust, viable alternative to the copper chromite catalyst for these hydrogenation reactions.

Supplementary Materials: The following are available online at http://www.mdpi.com/2073-4344/8/10/474/s1, Figure S1: Diffractograms of (a) CuO/Al_2O_3; (b) the Al_2O_3 support and (c) $CuCr_2O_4$, Figure S2: Selectivity to the various by-products formed during the hydrogenation of octanal using the fresh feed and the water-spiked feed over CuO/Al_2O_3, Figure S3: Selectivity to the various by-products formed during the hydrogenation of octanal using the fresh feed and the water-spiked feed over $CuCr_2O_4$, Figure S4: (a–c) EDS composition map data for Cu/Al_2O_3 used for the reaction with fresh feed, Figure S5: (a–c) EDS composition map data for Cu/Al_2O_3 used for the reaction with water-spiked feed, Figure S6: (a–c) EDS composition map data for $CuCr_2O_4$ used for the reaction with fresh feed, Figure S7: (a–c) EDS composition map data for $CuCr_2O_4$ used for the reaction with water-spiked feed.

Author Contributions: H.B.F. and A.S.M. led the project and coordinated the study. A.G. conceived and executed the experiments, including catalyst characterization and catalytic testing. A.G. and A.S.M. prepared the manuscript, and editing was done by A.S.M. and H.B.F. The manuscript submission was coordinated by H.B.F. All authors read and approved the final manuscript.

Funding: This research was funded by SASOL R&D and THRIP.

Acknowledgments: We thank James Wesley-Smith, Sharon Eggers, and Priscilla Martins (EM Unit at UKZN) for the TEM and SEM analyses. We also thank Francois Human and Enrico Caricato of SASOL R&D for assistance with the reactor setup.

Conflicts of Interest: The authors declare no conflict of interest.

References

1. Rioux, R.M.; Vannice, M.A. Hydrogenation/dehydrogenation reactions: Isopropanol dehydrogenation over copper catalysts. *J. Catal.* **2003**, *216*, 362–376. [CrossRef]
2. Satterfield, C.N. *Heterogeneous Catalysis in Industrial Practice*; McGraw Hill: New York, NY, USA, 1991.
3. Lok, M. Structure and Performance of Selective Hydrogenation Catalysts. Available online: https://www.topsoe.com/sites/default/files/martin_lok_structure_and_performance_of_selective_hydrogenation_catalysts.pdf (accessed on 5 October 2018).
4. Prasad, R.; Singh, P. Applications and Preparation Methods of Copper Chromite Catalysts: A Review. *Bull. Chem. React. Eng. Catal.* **2011**, *6*, 63–113. [CrossRef]
5. Zhang, D.; Liu, G. Heat Exchanger Network Integration of a Hydrogenation Process of Benzene to Cyclohexene Considering the Reactor Conversion. *Chem. Eng. Trans.* **2017**, *61*, 283–288.
6. Woods, D.R. *Successful Trouble Shooting for Process Engineers: A Complete Course in Case Studies*; Wiley: Darmstadt, Germany, 2006.
7. Kister, H.Z. What Caused Tower Malfunctions in the Last 50 Years? *Chem. Eng. Res. Des.* **2003**, *81*, 5–26. [CrossRef]
8. Chang, J.-R.; Lin, T.-B.; Cheng, C.-H. Pd/Δ-Al_2O_3 Catalysts for Isoprene Selective Hydrogenation: Regeneration of Water-Poisoned Catalysts. *Ind. Eng. Chem. Res.* **1997**, *36*, 5096–5102. [CrossRef]
9. Wang, W.-J.; Qiao, M.-H.; Li, H.-X.; Dai, W.-L.; Deng, J.-F. Study on the deactivation of amorphous NiB/SiO_2 catalyst during the selective hydrogenation of cyclopentadiene to cyclopentene. *Appl. Catal. A: Gen.* **1998**, *168*, 151–157. [CrossRef]
10. Ma, Z.; Jia, R.; Liu, C.; Mi, Z. Production of hydrogen peroxide from carbon monoxide, water, and oxygen over alumina supported amorphous ni catalysts. *Chem. Lett.* **2002**, *31*, 884–885. [CrossRef]
11. Twigg, M.V.; Spencer, M.S. Deactivation of supported copper metal catalysts for hydrogenation reactions. *Appl. Catal. A Gen.* **2001**, *212*, 161–174. [CrossRef]
12. Argyle, M.; Bartholomew, C. Heterogeneous Catalyst Deactivation and Regeneration: A Review. *Catalysts* **2015**, *5*, 145–269. [CrossRef]
13. Rao, R.S.; Walters, A.B.; Vannice, M.A. Influence of crystallite size on acetone hydrogenation over copper catalysts. *J. Phys. Chem. B* **2005**, *109*, 2086–2092. [CrossRef] [PubMed]
14. Vasiliadou, E.S.; Lemonidou, A.A. Investigating the performance and deactivation behaviour of silica-supported copper catalysts in glycerol hydrogenolysis. *Appl. Catal. A Gen.* **2011**, *396*, 177–185. [CrossRef]
15. Masson, J.; Cividino, P.; Court, J. Selective hydrogenation of acetophenone on chromium promoted raney nickel catalysts. III. The influence of the nature of the solvent. *Appl. Catal. A Gen.* **1997**, *161*, 191–197. [CrossRef]
16. Bartholomew, C.H. Mechanisms of catalyst deactivation. *Appl. Catal. A Gen.* **2001**, *212*, 17–60. [CrossRef]
17. Rao, R.; Dandekar, A.; Baker, R.T.K.; Vannice, M.A. Properties of Copper Chromite Catalysts in Hydrogenation Reactions. *J. Catal.* **1997**, *171*, 406–419. [CrossRef]
18. Rajkhowa, T.; Marin, G.B.; Thybaut, J.W. A comprehensive kinetic model for Cu catalyzed liquid phase glycerol hydrogenolysis. *Appl. Catal. B Environ.* **2017**, *205*, 469–480. [CrossRef]
19. Mahlaba, S.V.L.; Valand, J.; Mahomed, A.S.; Friedrich, H.B. A study on the deactivation and reactivation of a Ni/Al_2O_3 aldehyde hydrogenation catalyst: Effects of regeneration on the activity and properties of the catalyst. *Appl. Catal. B Environ.* **2018**, *224*, 295–304. [CrossRef]
20. Suresh, S.; Karthikeyan, S.; Jayamoorthy, K. FTIR and multivariate analysis to study the effect of bulk and nano copper oxide on peanut plant leaves. *J. Sci. Adv. Mater. Devices* **2016**, *1*, 343–350. [CrossRef]

21. Ma, Z.; Xiao, Z.; van Bokhoven, J.A.; Liang, C. A non-alkoxide sol-gel route to highly active and selective Cu-Cr catalysts for glycerol conversion. *J. Mater. Chem.* **2010**, *20*, 755–760. [CrossRef]
22. Chen, L.; Horiuchi, T.; Osaki, T.; Mori, T. Catalytic selective reduction of no with propylene over Cu-Al_2O_3 catalysts: Influence of catalyst preparation method. *Appl. Catal. B: Environ.* **1999**, *23*, 259–269. [CrossRef]
23. Luo, M.-F.; Fang, P.; He, M.; Xie, Y.-L. In situ XRD, Raman, and TPR studies of CuO/Al_2O_3 catalysts for CO oxidation. *J. Mol. Catal. A Chem.* **2005**, *239*, 243–248. [CrossRef]
24. Zhou, R.; Yu, T.; Jiang, X.; Chen, F.; Zheng, X. Temperature-programmed reduction and temperature-programmed desorption studies of CuO/ZrO_2 catalysts. *Appl. Surf. Sci.* **1999**, *148*, 263–270. [CrossRef]
25. Chetty, T.; Dasireddy, V.D.B.C.; Callanan, L.H.; Friedrich, H.B. Continuous Flow Preferential Hydrogenation of an Octanal/Octene Mixture Using Cu/Al_2O_3 Catalysts. *ACS Omega* **2018**, *3*, 7911–7924. [CrossRef]
26. Wang, X.; Saleh, R.Y.; Ozkan, U.S. Reaction network of aldehyde hydrogenation over sulfided Ni–Mo/Al_2O_3 catalysts. *J. Catal.* **2005**, *231*, 20–32. [CrossRef]
27. Chang, Y.-C.; Ko, A.-N. Vapor phase reactions of acetaldehyde over type X zeolites. *Appl. Catal. A Gen.* **2000**, *190*, 149–155. [CrossRef]
28. Luo, S.; Falconer, J.L. Acetone and Acetaldehyde Oligomerization on TiO_2 Surfaces. *J. Catal.* **1999**, *185*, 393–407. [CrossRef]
29. Nascimento, M.A. *Theoretical Aspects of Heterogeneous Catalysis*; Kluwer: Dordrecht, The Netherlands, 2001.
30. Gredig, S.V.; Maurer, R.; Koeppel, R.; Baiker, A. Copper-catalyzed synthesis of methylamines from CO_2. H_2 and NH_3. Influence of support. *J. Mol. Catal. A Chem.* **1997**, *127*, 133–142. [CrossRef]
31. Bando, K.K.; Sayama, K.; Kusama, H.; Okabe, K.; Arakawa, H. In-situ FT-IR study on CO_2 hydrogenation over Cu catalysts supported on SiO_2, Al_2O_3, and TiO_2. *Appl. Catal. A Gen.* **1997**, *165*, 391–409. [CrossRef]
32. Lang, B.E. Solubility of Water in Octan-1-ol from (275 to 369) K. *J. Chem. Eng. Data* **2012**, *57*, 2221–2226. [CrossRef]

Review

Activation, Deactivation and Reversibility Phenomena in Homogeneous Catalysis: A Showcase Based on the Chemistry of Rhodium/Phosphine Catalysts †

Elisabetta Alberico [1], Saskia Möller [2], Moritz Horstmann [2], Hans-Joachim Drexler [2] and Detlef Heller [2,*]

[1] Istituto di Chimica Biomolecolare—CNR, Tr. La Crucca 3, 07100 Sassari, Italy
[2] Leibniz-Institut für Katalyse e.V., Albert-Einstein-Strasse 29a, 18059 Rostock, Germany
* Correspondence: detlef.heller@catalysis.de

† Dedicated to Piet W. N. M. van Leeuwen.

Received: 5 June 2019; Accepted: 29 June 2019; Published: 30 June 2019

Abstract: In the present work, the rich chemistry of rhodium/phosphine complexes, which are applied as homogeneous catalysts to promote a wide range of chemical transformations, has been used to showcase how the in situ generation of precatalysts, the conversion of precatalysts into the actually active species, as well as the reaction of the catalyst itself with other components in the reaction medium (substrates, solvents, additives) can lead to a number of deactivation phenomena and thus impact the efficiency of a catalytic process. Such phenomena may go unnoticed or may be overlooked, thus preventing the full understanding of the catalytic process which is a prerequisite for its optimization. Based on recent findings both from others and the authors' laboratory concerning the chemistry of rhodium/diphosphine complexes, some guidelines are provided for the optimal generation of the catalytic active species from a suitable rhodium precursor and the diphosphine of interest; for the choice of the best solvent to prevent aggregation of coordinatively unsaturated metal fragments and sequestration of the active metal through too strong metal–solvent interactions; for preventing catalyst poisoning due to irreversible reaction with the product of the catalytic process or impurities present in the substrate.

Keywords: Rh; homogeneous catalysis; catalyst deactivation

---

1. Introduction

The term "catalysis" was first introduced by the Swedish chemist Jöns Jacob Berzelius in a report published in 1835 by the Swedish Academy of Sciences [1]. The report was a reflection on earlier findings by European scientists on chemical changes both in homogeneous and heterogeneous systems. Although far from being understood, the recorded phenomena were unified, according to Berzelius, by the fact that "*several simple or compound bodies, soluble and insoluble, have the property of exercising on other bodies an action very different from chemical affinity. By means of this action they produce, in these bodies, decompositions of their elements and different recombinations of these same elements to which they remain indifferent.*" [2] Berzelius designated this unifying property as "catalytic force" and coined the term "catalysis" to describe the decomposition of bodies by this force. Catalysis is the macroscopic manifestation of the reduction of the activation energy of an exergonic reaction. It is as such a purely kinetic phenomenon; Berzelius however did not know the concept of reaction rate yet and therefore he could only describe known findings phenomenologically.

It was not until 1894 that Wilhelm Ostwald succeeded in giving an exact description of catalysis [3]. Later, in his Nobel Lecture of 1909, he said about catalysis that " ... the essence of which is not to be sought in the promotion of a reaction, but in its acceleration, ... " [4] Further introduction into the history of catalysis can be found in references [5–8].

An updated definition to be found in the IUPAC Compendium of Chemical Terminology, though, suggests a clear distinction between a catalyst, that is a substance that *increases* the rate of a reaction, and an inhibitor, which instead *reduces* it [9].

Catalysis is widely applied both in industry [10–13] and academia [14,15]: the switch from a stoichiometric process to a catalytic one represents a way to greatly improve efficiency, as less energy input and less feedstock are usually required, less waste is generated, and a greater product selectivity is in general achieved.

Although heterogeneous catalysts were the first to be applied at the industrial level, where they dominated the scene for several decades, homogenous catalysts started to raise increasing attention in the early 1960s when they were considered for the large scale production of bulk chemicals. One major contribution came in those years from Slaugh and coworkers at Shell Research (USA) [16] later on supported by further developments in Wilkinson's laboratory [17], of Co- and Rh-phosphine catalysts which were applied in the rapid, selective liquid-phase hydroformylation and hydrogenation of alkenes under mild conditions. The possibility to tailor the catalyst properties by proper selection of additional ligands triggered further research which led in the following decades to the development of important industrial processes based on the use of homogeneous catalysts such as the SHOP process for the oligomerization of ethylene or the production of enantioenriched herbicide Metholachlor and the aroma ingredient citronellal [11].

The discovery of a novel catalyst for a certain process may be serendipitous or the result of extensive screening. Yet improvement of the catalyst productivity and efficiency, both in terms of activity and selectivity in the desired product, requires knowledge of the chemical transformations the catalyst undergoes during the course of the reaction. The nature and distribution of the resulting species may vary over time due to the interaction with the substrate, the product, the solvent, and additives. Some of these species may be catalytically inactive and hence negatively affect the outcome of the catalytic process. Although the exact nature of these species depends on the specific catalytic process and the experimental conditions under which it is carried out, inactivation of an homogeneous catalyst can be in general ascribed to the following: metal deposition due to reducing reaction conditions or ligand loss; ligand decomposition; reaction of the metal–ligand bond, such as metal–carbon or metal–hydride bond, with polar species such as water, acids, and alcohols; blocking of the catalyst active site by polar impurities, by metal fragments aggregation, and by ligand metalation. However, if studies are carried out that allow identification of the conditions which favor catalyst inactivation, then countermeasures can be adopted. As it was recently written " ... *Since such studies are currently underrepresented in the catalysis literature, our science will advance, and our community will benefit from increased emphasis on the productivity* (activity and selectivity) *metric*" [18].

The topic of catalyst activation, deactivation, and stability for several metals, for homogeneous and heterogeneous catalysts, has been dealt with in excellent review publications [19–22]. Here we would like to report on recent findings in the chemistry of rhodium/diphosphine complexes, both from others and our laboratory [23], which have allowed understanding of the activity profile of these complexes when used as precatalysts. In several cases the acquired knowledge has allowed improvement of the catalyst performance considerably.

2. In Situ Generation of Precatalysts

Neutral dinuclear rhodium complexes of the type [Rh(diphosphine)(μ_2-X)]$_2$ (X = Cl^-, OMe^-, OH^- ...) are very often applied as catalyst precursors in homogeneous catalysis [24–37]. These coordination compounds are usually not commercially available and must therefore be synthesized. This is mostly done in situ by addition of a diphosphine to a suitable precursor complex; previous work from our group has shown that, contrary to common belief, this procedure is far from being fast and selective in the formation of the desired neutral dinuclear rhodium species [Rh(diphosphine)(μ_2-Cl)]$_2$ [23,38–40]. In many cases, depending on the rhodium precursor, the diphosphine ligand, the solvent, the reaction temperature, and reaction time, other unexpected coordination species are formed, either cationic and/or trinuclear, either exclusively or in admixture with other complexes [38–40] (Scheme 1) which can negatively affect the catalytic activity. The range of species displayed in Scheme 1 might be extended as a result of further investigations.

In light of these findings, mechanistic investigations have been carried out on the rhodium-catalyzed propargylic CH activation which relied on the in situ generation of the catalytically active species. This reaction proceeds with good yields (up to 89%) and excellent selectivities (up to 95% ee) and offers great potential as a novel method for the construction of C–C and C–X bonds [41–47]. A broad range of substrates can be employed and, more interestingly, the reaction shows ligand-specific chemoselectivity (Scheme 2).

In the case of the P,N ligand DPPMP (a complete list of abbreviations, including ligand names, used in this article can be found in Table A2) it could be shown that its reaction with the COD precursor [Rh(COD)(μ_2-Cl)]$_2$ does not produce the desired dinuclear neutral complex [Rh(DPPMP)(μ_2-Cl)]$_2$, but an ionic species [Rh(DPPMP)$_2$][Rh(COD)Cl$_2$] in which both the anion and the cation are rhodium coordination compounds (Figure 1, left) [48].

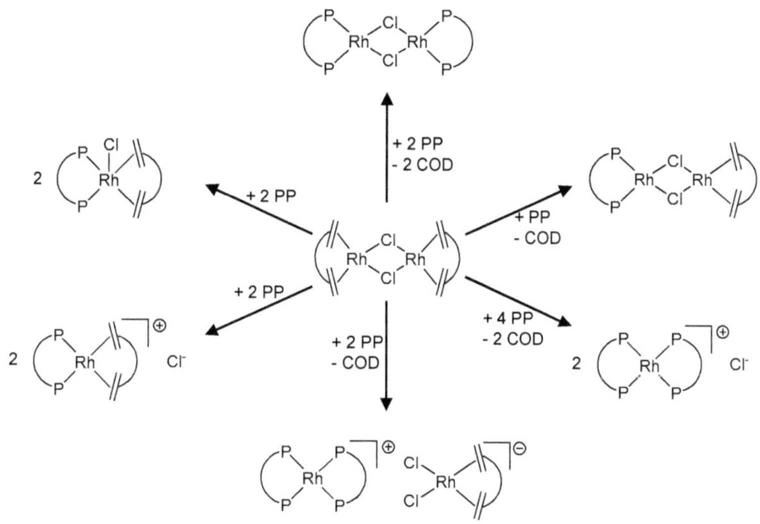

Scheme 1. In situ generation of rhodium/diphosphine complexes—known range of species which can be formed during the in situ reaction of rhodium diolefin precursors with diphosphine ligands.

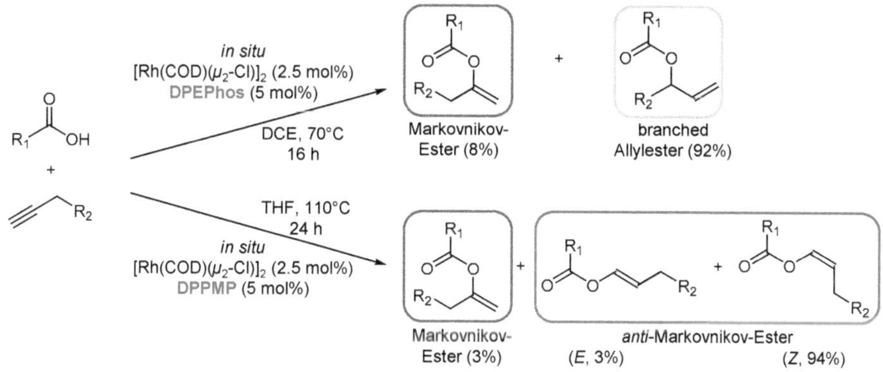

Scheme 2. Schematic representation of the rhodium-catalyzed propargylic CH activation with the ligands DPEPhos, above, and DPPMP, below [41–43].

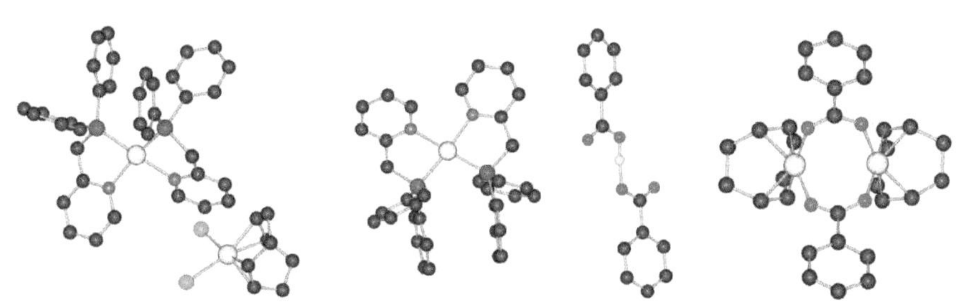

Figure 1. Molecular structure of [Rh(DPPMP)$_2$][Rh(COD)Cl$_2$], left, molecular structure of [Rh(DPPMP)$_2$][H(benzoate)$_2$] and [Rh(COD)(μ_2-benzoate)]$_2$, right. Hydrogen atoms (except *H*(benzoate)$_2$) are omitted for clarity [49].

In addition, it was demonstrated that the chloride ligands in the COD precursor [Rh(COD)(μ_2-Cl)]$_2$ have a deactivating effect on the catalytic reaction, which is why the chloride-free rhodium precursor [Rh(COD)(acac)] has been used instead. The reaction of [Rh(COD)(acac)] with DPPMP in the presence of the substrate benzoic acid leads to a mixture of two complexes [Rh(DPPMP)$_2$][H(benzoate)$_2$] and [Rh(COD)(μ_2-benzoate)]$_2$, which synergistically behave as the active "catalyst" (Figure 1, right) [48].

This example clearly shows that the classical in situ synthesis of rhodium(I) precatalysts can deliver unexpected compounds with poor or no catalytic activity. Only through an in depth investigation of the in situ process can the experimental factors which favor the formation of such species be clarified. Based on this knowledge, conditions can be rationally controlled to achieve better efficiency, activity, and selectivity (Scheme 3). In the following section, the factors which affect the in situ synthesis of rhodium(I) precatalysts will be introduced and their influence will be discussed.

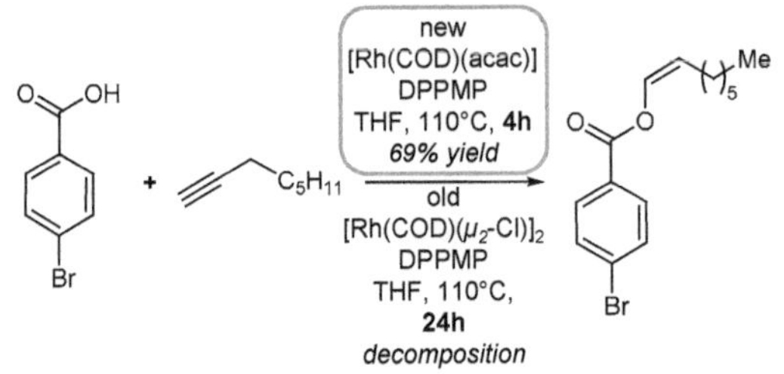

Scheme 3. Optimized reaction conditions for the rhodium catalyzed reaction of 4-bromo-benzoic acid and 1-octyne [48].

2.1. *Influence of Reaction Conditions on Outcome of In Situ Synthesis*

Rhodium(I) precatalysts of the type [Rh(diphosphine)(μ_2-Cl)]$_2$ are usually formed in situ at room temperature by the reaction of a diphosphine ligand with a stable olefin or diolefin-containing rhodium precursor. At first sight it is plausible to assume that a twofold ligand exchange takes place which affords a precatalyst the general structure of which [Rh(diphosphine)(μ_2-Cl)]$_2$ is the same, regardless of the ligand used (Scheme 4). From such a precatalyst the catalytically active species should eventually be formed.

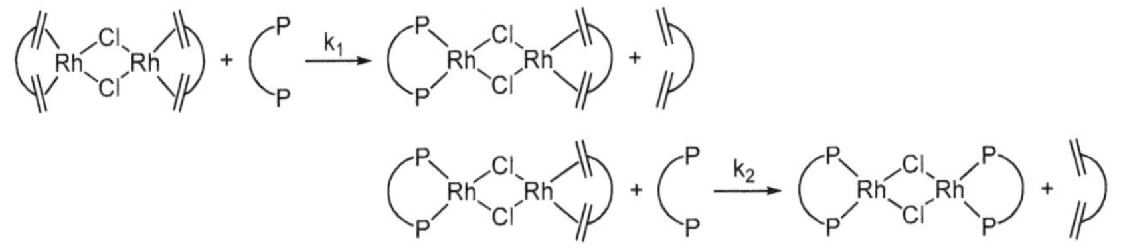

Scheme 4. Sequence of the competing consecutive reactions leading to [Rh(disphosphine)(μ_2-Cl)]$_2$ from [Rh(diolefin)(μ_2-Cl)]$_2$.

Systematic investigations have instead shown that the outcome of the in situ procedure is by no means as straightforward as shown in Scheme 4 and some illustrative cases are discussed below.

When [Rh(NBD)(μ_2-Cl)]$_2$ is reacted with either SEGPhos, DIPAMP, or DPEPhos the product of ligand exchange depends on the diphosphine used. With SEGPhos, when the reaction is carried out in THF at room temperature using a 1:2 metal to ligand ratio, then the desired neutral, dinuclear complex [Rh(SEGPhos)(μ_2-Cl)]$_2$ is obtained at 80% yield. However a small amount of the μ_2-chloro-bridged intermediate [(NBD)Rh(μ_2-Cl)$_2$Rh(SEGPhos)], (Figure 2a) is also formed [38]. When [Rh(NBD)$_2$(μ_2-Cl)]$_2$ is reacted with DIPAMP under the same conditions, a cationic complex results almost quantitatively, in which the rhodium center is coordinated by two DIPAMP molecules and the chloride acts as a counterion (Figure 2b) [38]. With DPEPhos a species is formed in which the diphosphine, the diolefin, and a chloride ligand bind to rhodium to give a penta-coordinated complex (Figure 2c) [50]. Similar coordination compounds of the form [Rh(disphosphine)(diolefin)Cl] have been reported as well and discussed in detail [38,39].

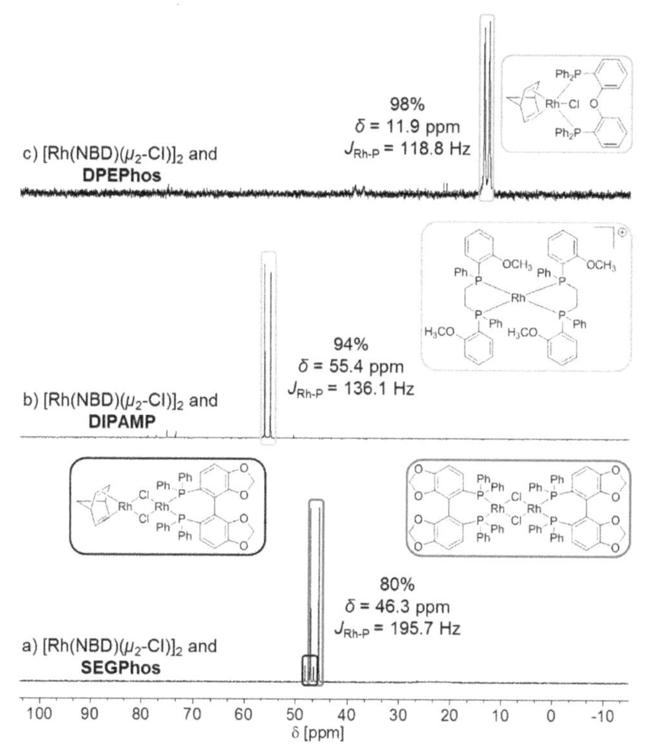

Figure 2. ^{31}P NMR spectra of the reaction solution for the in situ ligand exchange at room temperature in THF-d_8 between [Rh(NBD)(μ_2-Cl)]$_2$ and (**a**) SEGPhos, (**b**) DIPAMP, (**c**) DPEPhos.

The solvent also plays a key role in the outcome of the in situ procedure. Although the course of a catalytic reaction is often solvent-specific, rarely has this correlation been attributed to the formation of different catalytic species [51–53]. For example, when [Rh(COD)(μ_2-Cl)]$_2$ reacts with SEGPhos in a 1:2 ratio at room temperature in THF the desired species, [Rh(SEGPhos)(μ_2-Cl)]$_2$ is produced quantitatively, Figure 3a [38,54]. In DCM an additional monomeric cationic species is formed, in which the rhodium center is coordinated by two SEGPhos molecules and the chloride as possible counterion, Figure 3b [38]. In MeOH the prevailing species is a monomeric cationic complex in which the rhodium center is coordinated by a SEGPhos molecule and COD, Figure 3c [38].

In the literature the used rhodium species is often not discussed and is generalized with the abbreviation "[Rh]" [55–57]. But, as the following example will clearly show, the choice of the rhodium precursor is also decisive for the outcome of the in situ synthesis. When the diphosphine ligand DPPB is reacted with [Rh(COE)(μ_2-Cl)]$_2$ at room temperature in THF, the desired dinuclear rhodium complex [Rh(DPPB)(μ_2-Cl)]$_2$ is formed quantitatively (Figure 4a) [38]. If [Rh(NBD)(μ_2-Cl)]$_2$ is used instead, the reaction affords exclusively a pentacoordinated rhodium species containing chloride and the chelating ligands DPPB and NBD (Figure 4b) [38,39]. With [Rh(COD)(μ_2-Cl)]$_2$ the reaction is unselective and affords several species, as evident from the large number of signals in the corresponding ^{31}P NMR spectrum (Figure 4c) [38].

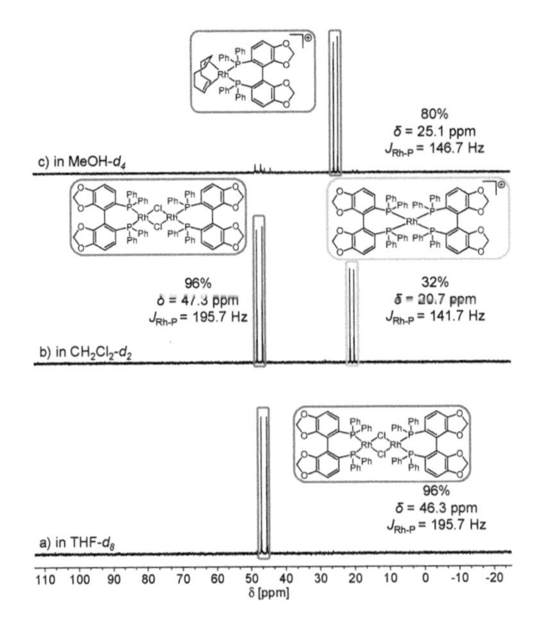

Figure 3. ^{31}P NMR spectrum of the solution resulting from the in situ reaction of [Rh(COD)(μ_2-Cl)]$_2$ and SEGPhos in the ratio 1:2 at room temperature: (**a**) in THF-d_8, (**b**) DCM-d_2, (**c**) MeOH-d_4.

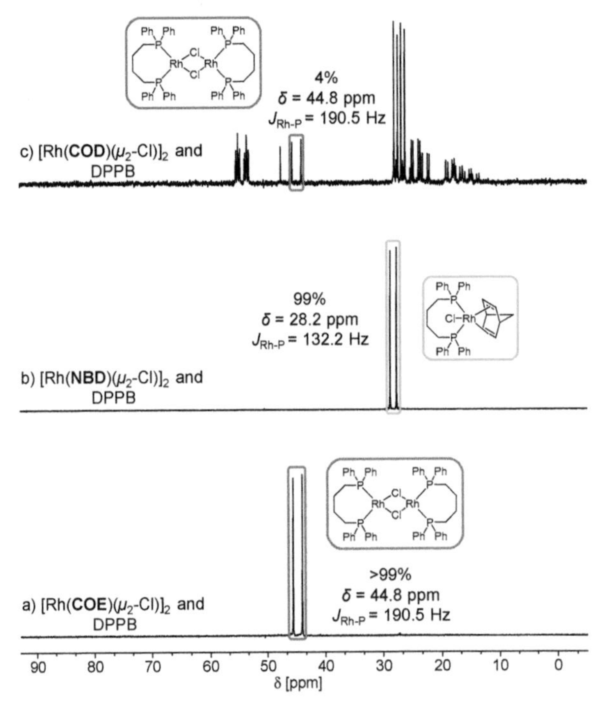

Figure 4. ^{31}P NMR spectrum of the solution resulting from the in situ reaction of DPPB in THF-d_8 with (**a**) [Rh(COE)(μ_2-Cl)]$_2$, (**b**) [Rh(NBD)(μ_2-Cl)]$_2$, (**c**) [Rh(COD)(μ_2-Cl)]$_2$.

The reaction temperature at which the ligand exchange between diolefin and diphosphine takes place may also affect the selectivity of the in situ procedure. Experiments with the chiral ligand DIOP have shown that a mixture of different complexes is formed when the reaction is carried out at 25 °C (Figure 5a). However, if the ligand exchange occurs at −10 °C, then the desired dinuclear precatalyst $[Rh(DIOP)(\mu_2\text{-}Cl)]_2$ is obtained quantitatively (Figure 5b) [38]. With DPPE the in situ procedure does not work at room temperature, instead at 125 °C in toluene it affords the target compound $[Rh(DPPE)(\mu_2\text{-}Cl)]_2$ quantitatively [58].

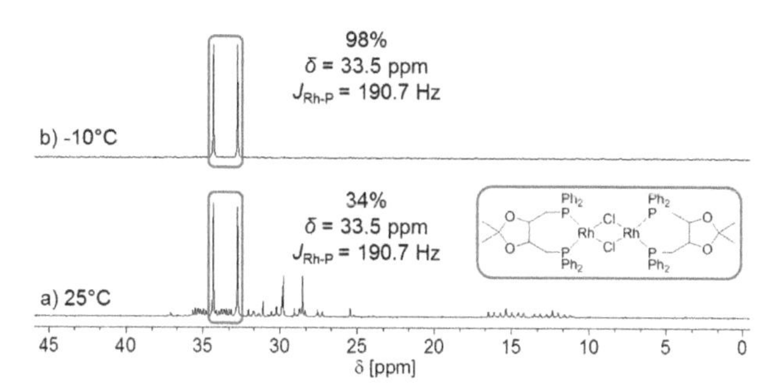

Figure 5. ^{31}P NMR spectrum of the solution resulting from the in situ reaction of $[Rh(COD)(\mu_2\text{-}Cl)]_2$ and DIOP in THF-d_8 at (**a**) 25 °C and (**b**) −10 °C.

2.2. *Mechanistic Investigations into the in situ Generation of Precatalysts*

The examples reported above serve to illustrate how experimental conditions affect the outcome/selectivity of the in situ synthesis of rhodium/diphosphine precatalysts. However, with the ligands BINAP, SEGPhos, DM-SEGPhos, and Difluorphos, the target complex $[Rh(disphosphine)(\mu_2\text{-}Cl)]_2$ is formed quasi quantitatively in THF. Less is known about the rate of formation of such species and the kinetics of the underlying two-stage ligand exchange process (Scheme 4). Quantitative studies of classical ligand exchange reactions can be accomplished using UV-Vis spectroscopy (diode array in combination with a stopped-flow unit) because during the reaction a color change takes place [54,59].

From a kinetic point of view, each stage of the ligand exchange process is a second order reaction, and the two reactions compete with each other. The solid state structure of the intermediate $[(COD)Rh(\mu_2\text{-}Cl)_2Rh(diphosphine)]$ has been determined for various disphosphines [38,60,61]. Only recently has it been shown that indeed an equilibrium exists between the starting and the product complexes on the one hand and the intermediate on the other (Scheme 5) which of course complicates kinetic investigations [54].

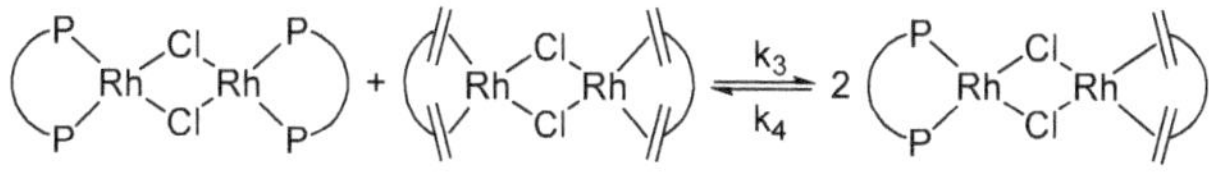

Scheme 5. Equilibrium between the starting $[Rh(diolefin)(\mu_2\text{-}Cl)]_2$ and target complex $[Rh(diphosphine)(\mu_2\text{-}Cl)]_2$ and the intermediate $[(diolefin)Rh(\mu_2\text{-}Cl)_2Rh(diphosphine)]$.

Figure 6 shows the ^{31}P NMR spectrum recorded shortly after crystals of $[(COD)Rh(\mu_2\text{-}Cl)_2Rh(DPEPhos)]$ have been dissolved in benzene-d_6; beside the signals of $[(COD)Rh(\mu_2\text{-}Cl)_2Rh(DPEPhos)]$, black, broad signals due to $[Rh(DPEPhos)(\mu_2\text{-}Cl)]_2$, red, can be observed.

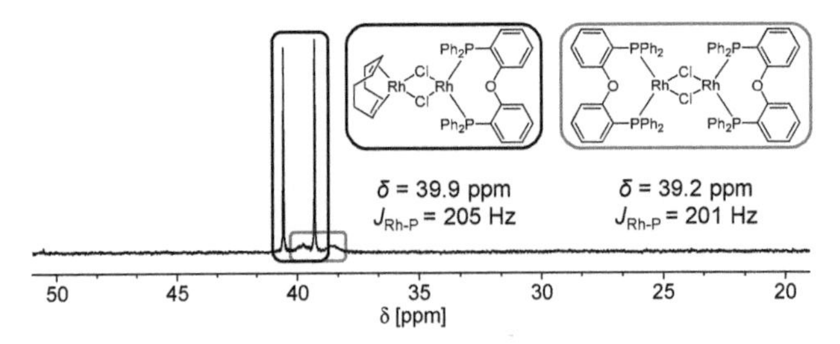

Figure 6. ^{31}P NMR spectrum of the solution obtained by dissolving crystals of [(COD)Rh(μ_2-Cl)$_2$Rh(DPEPhos)], black, in benzene-d_6; following the establishment of an equilibrium, [Rh(DPEPhos)(μ_2-Cl)]$_2$ is formed, red.

The rate constants of the reaction steps described in Schemes 4 and 5 for the ligands BINAP, SEGPhos, DM-SEGPhos, and Difluorphos are collected in Table 1.

Table 1. Rate constants of the ligand exchange reactions described in Schemes 4 and 5 for BINAP, SEGPhos, DM-SEGPhos, and Difluorphos [54].

Ligand	k_1 (L·mol^{-1}·s^{-1})	k_2 (L·mol^{-1}·s^{-1})	k_3 (L·mol^{-1}·s^{-1})	k_4 (L·mol^{-1}·s^{-1})	$t_{98\%\ conv.}$ (min)
BINAP	1790	18	39.3	1.10	22
SEGPhos	10,617	121	16.7	1.43	3
DM-SEGPhos	13,466	14	0.3	0.60	28
Difluorphos	8080	220	-	-	2

These data show that, for all the ligands investigated, the exchange of the first COD ligand to give the intermediate [(COD)Rh(μ_2-Cl)$_2$Rh(diphosphine)] is much faster than the exchange of the second one ($k_1 >> k_2$). The latter is therefore the rate-determining step of the overall reaction. The rapid color change, occasionally described in the literature, is due to the formation of the intermediate and thus cannot be used, in the in situ generation of the precatalyst, as an indication of the formation of the target complex. The experimentally determined rate constants, Table 1, allow then the calculation of the time required for the complete formation of the desired dinuclear precatalysts under normal conditions [62]: 22 min are necessary with BINAP, up to 28 with DM-SEGPhos [54], far more than usually expected.

With the ligands DPEPhos and DIOP the mechanism leading to the formation of the corresponding precatalysts [Rh(disphosphine)(μ_2-Cl)]$_2$, which have been applied in propargylic CH activation, is more complex [41,42,63–66]. Indeed, beside the intermediate [(COD)Rh(μ_2-Cl)$_2$Rh(diphosphine)], a previously unknown diphosphine-bridged species [Rh$_2$(μ_2-diphosphine)(COD)$_2$(Cl)$_2$] could be isolated and characterized [61]. When the reaction of [Rh(COD)(μ_2-Cl)]$_2$ and DPEPhos in benzene is monitored by ^{31}P NMR, 10 min after mixing of the reagents the signals relative to such species (Figure 7, yellow) could be observed, beside those of [Rh(DPEPhos)(μ_2-Cl)]$_2$ (Figure 7, red) and [(COD)Rh(μ_2-Cl)$_2$Rh(DPEPhos)] (Figure 7, black).

Therefore the known twofold ligand exchange sequence described in Scheme 4 must be corrected as shown in Scheme 6, at least for these diphosphines [38,61], to accommodate the newly observed intermediate [Rh$_2$(μ_2-diphosphine)(COD)$_2$(Cl)$_2$].

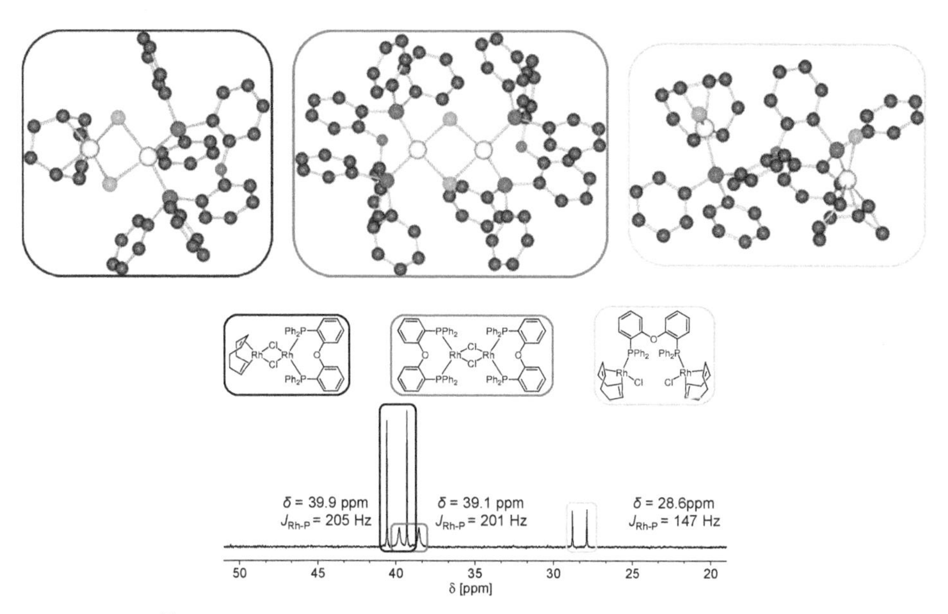

Figure 7. ^{31}P NMR spectrum of the solution obtained from reaction of [Rh(COD)(μ_2-Cl)]$_2$ and DPEPhos, in a 1:2 mixture, 10 min after mixing of the reagents: X-ray structures of the species detected in solution [(COD)Rh(μ_2-Cl)$_2$Rh(DPEPhos)], black, [Rh(DPEPhos)(μ_2-Cl)]$_2$, red, and [Rh$_2$(μ_2-DPEPhos)(COD)$_2$(Cl)$_2$], yellow.

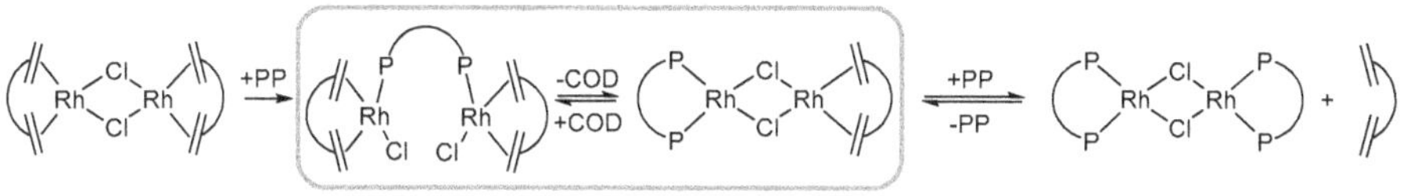

Scheme 6. Schematic representation of the modified ligand exchange mechanism for the formation of the rhodium precatalyst [Rh(diphosphine)(μ_2-Cl)]$_2$.

In the first step, the precursor [Rh(COD)(μ_2-Cl)]$_2$ is cleaved and the first reacting diphosphine acts as a bridging ligand between the two homoleptic rhodium fragments to afford [Rh$_2$(μ_2-diphosphine)(COD)$_2$(Cl)$_2$]. Then, following the loss of one COD ligand and diphosphine rearrangement to become bidentate at one of the two rhodium centers, an equilibrium is established between [Rh$_2$(μ_2-diphosphine)(COD)$_2$(Cl)$_2$] and the newly formed [(COD)Rh(μ_2-Cl)$_2$Rh(diphosphine)]. Such equilibrium can be shifted by altering reaction conditions (COD concentration, solvent, temperature). Exchange of COD for a second molecule of the diphosphine in [(COD)Rh(μ_2-Cl)$_2$Rh(diphosphine)] eventually affords the desired precatalyst [Rh(disphosphine)(μ_2-Cl)]$_2$. While in the case of DPEPhos quantitative formation of the precatalyst requires 30 min, for DIOP the equilibria lie far to the side of the intermediates. This is equivalent to a partial catalyst deactivation.

The examples discussed above clearly show that the in situ synthesis of rhodium precatalysts may not be selective thus affording a complex mixture of species. Even more complex, as shown in Scheme 7, is the network of equilibria which have been documented for the ligand DPPP [39,60,61,67–69]. To the best of our knowledge, the species [Rh(DPPP)(diolefin)]Cl, which is also possible in principle, is not known. It has been described for other diphosphine ligands [70–76]. The nature of the species that are actually present under stationary conditions cannot be predicted a priori and which one, among these, is responsible for the catalytic activity may be difficult to establish. It should also be noted that even the starting diolefin precursor can be catalytically active [77–85]. Because selectivity is mostly determined by the type of ligand coordinated to the metal, if the diolefin precursor is not completely converted, selectivity will be negatively affected.

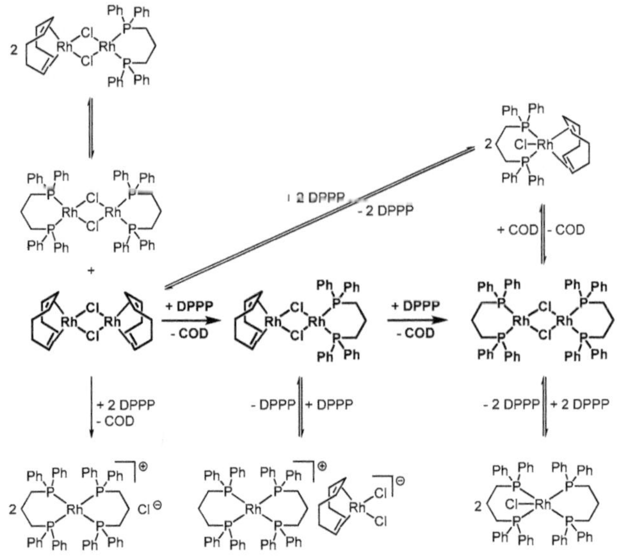

Scheme 7. In situ synthesis of the rhodium catalyst precursor [Rh(disphosphine)(μ_2-Cl)]$_2$ with the ligand DPPP and known by-products arising from side-reactions.

3. Catalyst Activation—Induction Periods

Catalytic processes, either homogeneous or heterogeneous, are often characterized by an initial slow stage after which the reaction accelerates. During the so-called induction period [86], the species which has been added to the reaction medium to function as the catalyst undergoes a series of chemical transformations before it can actually enter the catalytic cycle and promote the catalytic reaction.

For homogeneous catalysts, ligand (reactant) addition and ligand substitution and/or exchange at the catalyst precursor and/or oxidation–reduction of the metal with reactants/co-substrates, ligands, solvents, and other components of the system gradually convert the catalyst precursor into the catalytically active species. The concentration of the latter increases with time in the initial phase of the reaction, hence the induction period [87]. Catalytically active species are characterized by high reactivity/sensitivity, which makes them short-lived and difficult to characterize, hence the need to generate them from suitable catalyst precursors which are more stable, possess a longer shelf life, and are therefore easier to handle.

A classic example is provided by the Wilkinson catalyst [RhCl(PPh_3)$_3$] which, in the solid state, is indefinitely stable in air. It is a suitable precursor both for hydroformylation and hydrogenation of olefins. In solution it dissociates to release one PPh_3 and generate a coordinatively unsaturated species [RhCl(PPh_3)$_2$] [17,88,89]. Under hydroformylation conditions, several equilibria are established which are sensitive functions of experimental conditions: [RhCl(PPh_3)$_2$] reacts with CO to form *trans*-[RhCl(CO)(PPh_3)$_2$]. This species in turn oxidatively adds molecular hydrogen to afford [RhH(CO)(PPh_3)$_2$], and HCl. Hence an induction period is observed, unless an organic base such as triethylamine is used to favor the latter reaction. The dicarbonyl species [RhH(CO)$_2$(PPh_3)$_2$] has also been postulated to be present and active during hydroformylation. Under hydrogenation conditions, [RhCl(PPh_3)$_2$] can oxidatively add H_2 to generate [RhClH$_2$(PPh_3)$_2$], an extremely efficient catalyst for the hydrogenation of non-conjugated olefins and acetylenes at ambient temperature and hydrogen pressures of 1 atm or below. In both cases a rhodium hydride species makes the catalytic cycle turn.

An example from the most recent literature is provided by the following: A rhodium complex [Rh(κ^2-PP-DPEphos){$\eta^2\eta^2$-$H_2B(NMe_3)(CH_2)_2t$-Bu}]$BAr^F{}_4$ is a competent precatalyst for the dehydropolymerization of $H_3B{\cdot}NMeH_2$ to form N-methylpolyaminoborane $(H_2BNMeH)_n$ [90]. Before faster turnover is established, an induction period is observed which becomes longer with increasing total initial catalyst loading. An in-depth kinetic and mechanistic investigation has shown that the induction period is due to the formation, inter alia, of a dimeric rhodium species from which the monomeric catalytically active species is slowly generated. This process can be accelerated by the addition of NMe_2H which breaks down the dimer to form a rhodium amine complex [Rh(κ^2-PP-DPEphos)$H_2(NMeH_2)_2$]$BAr^F{}_4$. The latter is a far better precatalyst than [Rh(κ^2-PP-DPEphos){$\eta^2\eta^2$-$H_2B(NMe_3)(CH_2)_2t$-Bu}]$BAr^F{}_4$. Its isolation, while possible, is tedious but its in situ generation from [Rh(κ^2-PP-DPEphos){$\eta^2\eta^2$-$H_2B(NMe_3)(CH_2)_2t$-Bu}]$BAr^F{}_4$ in the presence of added NMe_2H is equally effective in suppressing the induction period. The structure of the active species though remains unknown.

3.1. *Quantification of Induction Periods*

The asymmetric hydrogenation of prochiral olefins is promoted by rhodium complexes of suitable chiral diphosphines. Such complexes are often commercialized as stable cationic diolefin precatalysts having the general formula [Rh(diphosphine)(diolefin)]$^+$ (dioelfin = COD, NBD); alternatively they can be generated in situ from the corresponding cationic diolefin rhodium complex and the chiral diphosphine. In either case, generation of the active species from the precatalyst requires hydrogenation of the diolefin. In this way, a free coordination site is made available at the metal for the binding of the substrate.

Figure 8, left, shows the difference in hydrogen uptake as a function of time between the asymmetric hydrogenation of mac in MeOH as promoted by the active catalyst [Rh((*R*,*R*)-Et-ButiPhane)$(MeOH)_2$]BF_4, green curve, and the same reaction as promoted by the commercially available precatalyst [Rh((*R*,*R*)-Et-ButiPhane)(COD)]BF_4, blue curve [91]. An induction period is present for the latter, i.e., the hydrogenation accelerates—the moles of hydrogen consumed per unit time as a function of time increase in the initial stages of the reaction (Figure 8, right, blue) as more of the catalytically active species [Rh((*R*,*R*)-Et-ButiPhane)$(MeOH)_2$]BF_4 is generated in the "activation" phase through hydrogenation of COD.

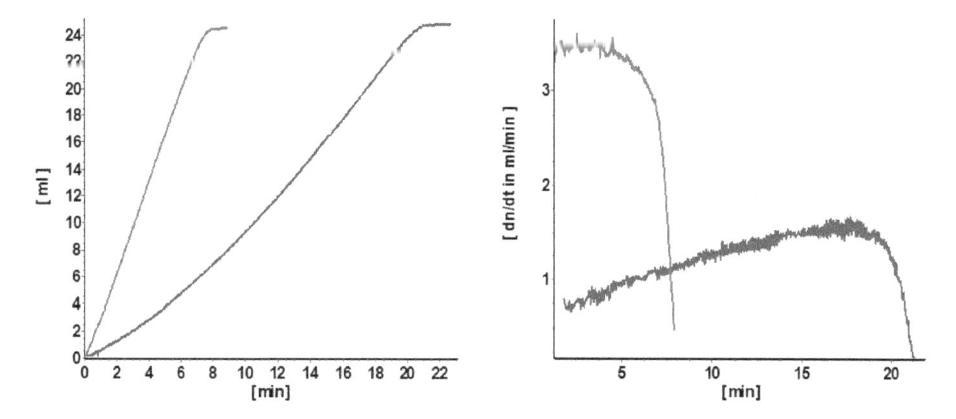

Figure 8. Left: Hydrogen uptake (ml) as a function of time (min) in the hydrogenation of mac with [Rh((*R*,*R*)-Et-ButiPhane)(COD)]BF_4 (99.0% ee, blue) and with [Rh((*R*,*R*)-Et-ButiPhane)$(MeOH)_2$]BF_4 (99.0% ee, green) under standard conditions [92]. Right: moles of hydrogen consumed per unit time (dn/dt) as a function of time (min) in the hydrogenation of mac with [Rh((*R*,*R*)-Et-ButiPhane)(COD)]BF_4 (blue) and [Rh((*R*,*R*)-Et-ButiPhane)$(MeOH)_2$]BF_4 (green).

Induction periods, which have been recognized, at least qualitatively, for a long time [93,94] cause a maximum in the rate profile (Figure 8, right). It should be noted that enantioselectivity is not affected by the different methods used for catalyst preparation (Figure 8, left) [91].

Several factors affect the induction period: the diolefin (stability of the corresponding diolefin rhodium complex), the prochiral olefin (its concentration and the stability of the corresponding rhodium substrate complex), the ligand, the solvent, and the temperature.

An in-depth discussion of the issues concerning diolefin hydrogenation (COD vs. NBD) can be found in references [22,95]. In Table A1 a list of related (pseudo) rate constants for ca. 80 rhodium/diphosphine complexes of the type $[Rh(diphosphine)(diolefin)]^+$ can be found. Rate constants for rhodium complexes of monophosphines, $[Rh(MonoPhos)_2(NBD)]BF_4$ [96], $[Rh(MonoPhos)_2(COD)]BF_4$ [97], and $[Rh(PPh_3)_2(COD)]BF_4$ [98], have also been measured.

Several methods have been reported in the literature which allow, under isobaric conditions, to measure the pseudo 1st order rate constant ($k'_{2diolefin}$) for the diolefin hydrogenation en route to the solvent complexes, according to the following Scheme 8.

$$[Rh(diphosphine)(solvent)_2]^+ + \text{diolefin} \rightleftharpoons [Rh(diphosphine)(diolefin)]^+ + 2\ \text{solvent}$$

$$[Rh(diphosphine)(diolefin)]^+ \xrightarrow{k'_{2diolefin}\,[H_2]} [Rh(diphosphine)(solvent)_2]^+ + \text{alkane}$$

Scheme 8. Generation of the catalytically active species $[Rh(diphosphine)(solvent)_2]^+$ through hydrogenation of the diolefin in $[Rh(diphosphine)(diolefin)]^+$ and related equilibrium.

For most diphosphine ligands, the equilibrium (Michaelis–Menten kinetics) between the solvent complex and the free diolefin lies to the side of the diolefin complex due to the high stability of the latter. Under isobaric conditions, the hydrogen concentration is constant and therefore the stoichiometric reaction is pseudo 1st order in the diolefin concentration. In the presence of a large excess of the diolefin ("catalytic conditions"), the hydrogenation reaction becomes zero order (Michaelis–Menten kinetics in the saturation range).

In principle, it is possible to track either the stoichiometric hydrogenation (by measuring the hydrogen uptake [99] using ^{31}P NMR [100] or UV-Vis spectroscopy [101]) or the catalytic hydrogenation with an automatic recording device [102,103] in the presence of an excess of the diolefin [104] under several pressure regimes [105]. It is likewise possible by means of ^{1}H NMR spectroscopy to assess the H_2 concentration in solution during diolefin hydrogenation under isochoric conditions [106,107]. Through on-line recording of hydrogen consumption and subsequent parameter optimization using a suitable kinetic model, the pseudo rate constant for the diolefin hydrogenation can also be determined [108].

3.2. *Influence of the Diolefin*

Experimental evidence shows that, regardless of the diphosphine ligand, the hydrogenation of COD is always slower than that of NBD, meaning a longer induction period. The reason for such difference is not clear. A perusal of the solid state structures of several rhodium complexes of the general formula $[Rh(diphosphine)(diolefin)]^+$ with diphosphines forming 5-membered chelated rings shows deviations from the expected square-planar structure, that is the centroids of the double bonds, the phosphorus atoms of the diphosphine, and the rhodium central atom are not in the same plane [109]. The rotation of the diolefin away from the ideal square planar coordination serves to partly accommodate the steric demands of the diphosphine ligand while keeping a profitable overlap of the orbitals involved in the binding of the diene to the rhodium atom. Because NBD is smaller than COD, the tetrahedral distortion it brings about is inferior. Therefore it has been suggested that the greater tetrahedral distortion induced by COD might hamper the oxidative addition of dihydrogen from the axial position as the required orbital overlap is not easily achieved.

The very small pseudo rate constants observed for the hydrogenation of COD in $[Rh(diphosphine)(COD)]^+$ in MeOH with DPPE and DIPAMP as diphosphine (Table A1) imply that at room temperature and normal pressure 24 and 30 h are necessary, respectively, for the complete removal of the COD in the two catalyst precursors. This confirms previous findings concerning the challenging quantitative hydrogenation of COD [110,111]: failure to recognize these rather long induction periods may lead to underestimation of the real efficiency of the corresponding rhodium/diphosphine catalyst in the hydrogenation of specific substrates, especially in relation to the application of high throughput methods in catalyst testing [112]. The pseudo rate constants are not affected by the counteranion in the catalyst precursor as shown for the ligand BINAP [113].

Despite their low reactivity, COD containing catalyst precursors are readily available and therefore often preferred in industrial applications. It has been shown for the Rh–DuPhos family that this can still be a sensible choice because the difference between the use of COD and NBD precatalysts becomes increasingly insignificant as the substrate to catalyst ratios increase up to values generally implemented in industry (≥10,000) [114].

An elegant experiment which is very sensitive and allows appreciation of the difference in precatalyst activation rates, that is the difference in COD versus NBD hydrogenation, has been devised. The method requires that equimolar amounts of COD and NBD rhodium precursors $[Rh(diphosphine)(COD)]^+$ and $[Rh(diphosphine)(NBD)]^+$, having the same diphosphine ligand but of opposite chirality, are used in the same hydrogenation reaction [114]. Since the precatalysts generate "enantiomeric" catalytic active species, then the closer the overall productivity given by the NBD and COD precatalysts, the closer the product will be to racemic. For example, in the hydrogenation of dimethyl itaconate with Rh/Me–DuPhos under these circumstances (10,000:1 as substrate/catalyst ratio, 5 bar hydrogen pressure), an almost racemic product was obtained indicating that the active catalyst is formed at almost the same rate from both the COD and NBD precatalysts. However, it must be said that, although very low, the recorded enantiomeric excess (ee < 3%) is not negligible when referred to the absolute product amount (911 mg over 30 g in the example reported above).

During exploratory experiments the substrate/catalyst ratio employed is usually 100. Under these conditions, when using the catalyst precursors $[Rh((S,S)\text{-Me-DuPhos})(COD)]BF_4$ and $[Rh((R,R)\text{-Me-DuPhos})(NBD)]BF_4$ in a 1:1 ratio (0.005 mmol each) in 15 mL MeOH and 1 bar total pressure, the recorded ee is 95.1%. Such value is only slightly lower than the one, 97.4–98% ee, obtained for hydrogenation with only the NBD complex [114]. This can be explained by the fact that the NBD complex is hydrogenated about 320 times faster than the one with COD (Table A1) therefore, when using the equimolar mixture of the two catalyst precursors of opposite chirality, hydrogenation and thus selectivity are determined almost exclusively by the NBD complex.

3.3. *Generation of Solvent Complexes*

In order to avoid undesirable induction periods and benefit from the full activity expected for the amount of "catalyst" added to the reaction vessel, it is advisable to use solvent complexes having the general formula $[Rh(diphosphine)(solvent)_2]^+$. They are prepared through hydrogenation of the corresponding diolefin complexes in the absence of the prochiral olefin. Solvent complexes are also better suited for kinetic investigations. By measuring the rate constant of the precatalyst activation/hydrogenation, it is possible to define the required experimental conditions under which formation of the active species is quantitative before the substrate is added.

While several solvent complexes with different diphosphine ligands have been characterized by ^{31}P NMR spectroscopy, the first X-ray structures of such reactive species have been published only recently [115–117]. In Figure 9 the molecular structures of complex anions $[Rh(DPEPhos)(acetone)_2]^+$ and $[Rh(BINAP)(THF)_2]^+$ are presented.

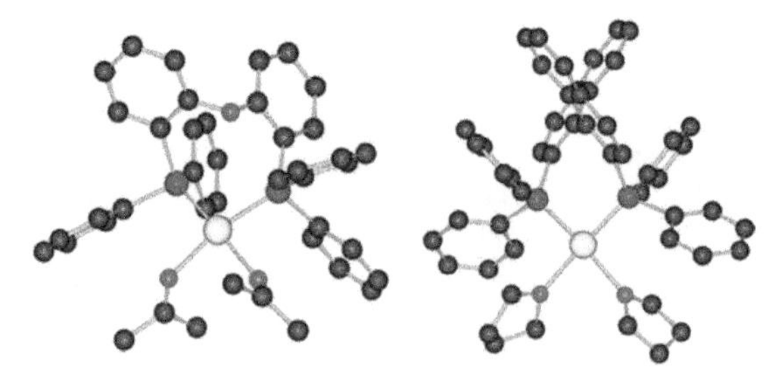

Figure 9. Molecular structure of the cations of the solvent complexes $[Rh(DPEPhos)(acetone)_2]^+$ and $[Rh(BINAP)(THF)_2]^+$. Hydrogen atoms are omitted for clarity.

Besides prochiral olefin hydrogenation, another case in which the use of solvent complexes proved beneficial is the rhodium-promoted asymmetric ring-opening of benzo-7-oxabicyclo-[2.2.1]-heptadiene with nucleophiles. This is an important process for the formation of C–C and C–X bonds which gives access to pharmaceutically relevant hydronaphatelenes. The efficiency and enantioselectivity of the process as originally developed [118] could be improved by using the preformed solvent complex $[Rh((R,S)\text{-PPF-P}(t\text{-Bu})_2)(THF)_2]^+$ instead of the in situ generated dimeric species $[Rh(((R,S)\text{-PPF-P}(t\text{-Bu})_2)(\mu_2\text{-Cl})]_2$ [119].

Data presented in Table A1 allow to calculate for how long the precatalyst $[Rh(diphosphine)(diolefin)]^+$ has to be hydrogenated in order to quantitatively generate the corresponding solvent complex $[Rh(diphosphine)(solvent)_2]^+$ ($t_{1/2} = \ln2/k'$, seven half lives correspond to 99.2% (practically complete) conversion of the diolefin in $[Rh(diphosphine)(diolefin)]^+$).

If hydrogenation lasts longer, then formation of dinuclear or trinuclear hydride complexes may occur (Figure 10) [120–122].

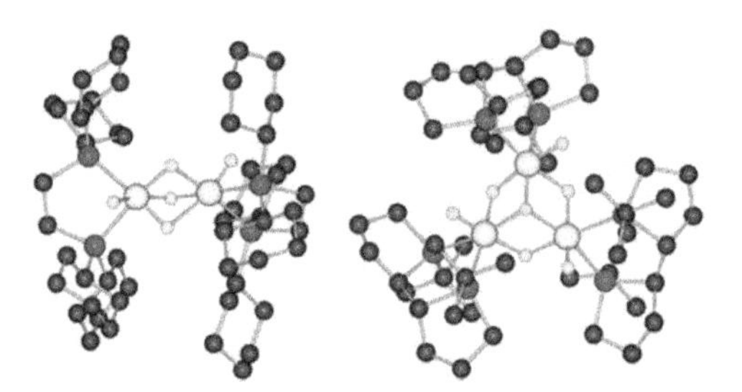

Figure 10. Molecular structure of the hydride complexes $\{[Rh(DCPE)H]_2(\mu_2\text{-H})_3\}^+$, left, and $\{[Rh(Tangphos)H]_3(\mu_2\text{-H})_3\ (\mu_3\text{-H})\}^{2+}$, right. Hydrogen atoms are omitted for clarity.

In some cases the formation of such species starts even before $[Rh(diphosphine)(diolefin)]^+$ is fully hydrogenated. This is the case for the Me–BPE/COD–MeOH system and ^{31}P NMR spectroscopy allows detection of the various species which coexist in solution (Figure 11) [123]. This means that the procedure for the synthesis of the solvent complex is not selective under these conditions. The problem can be solved using the precatalyst $[Rh(Me\text{-BPE})(NBD)]^+$ as hydrogenation of NBD is faster than COD and faster than the formation of polynuclear rhodium hydride.

An alternative entry to the solvent complexes, which circumvents the prehydrogenation step required for diolefin-containing precatalysts and the possible related problems, is represented by ammonia complexes of the general formula $[Rh(diphosphine)(NH_3)_2]^+$ which are very stable species and therefore easy to handle [95]. The solvent complex can be generated in situ by the addition of stoichiometric amounts of acid such as HBF_4 which displace NH_3 as the corresponding ammonium salt (using HBF_4, BF_4^- remains the sole counteranion for any cationic rhodium species in solution). The ammonia precatalyst is in turn easily prepared from $[Rh(diphosphine)(diolefin)_2]^+$ by reaction with a saturated solution of ammonia. Figure 12 shows the solid state structure of the rhodium ammonia catalyst precursor [Rh((*S*c,*R*p)-Duanphos)(NH_3)$_2$]$^+$.

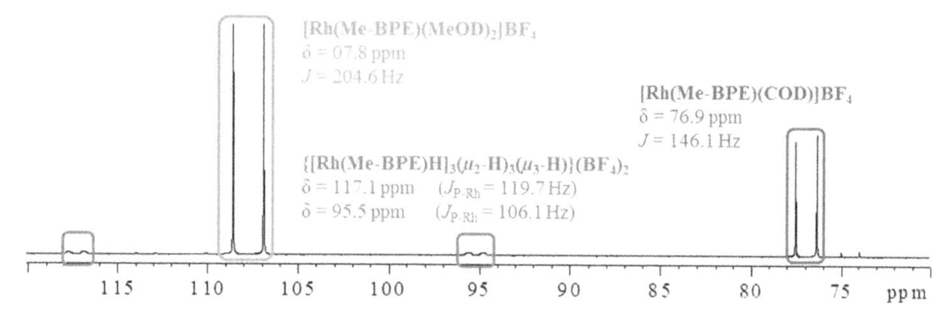

Figure 11. ^{31}P NMR spectrum of the solution obtained following hydrogenation of [Rh(Me-BPE)(COD)]$^+$ in MeOH-d_4 at 25 °C under 1 bar total pressure [123].

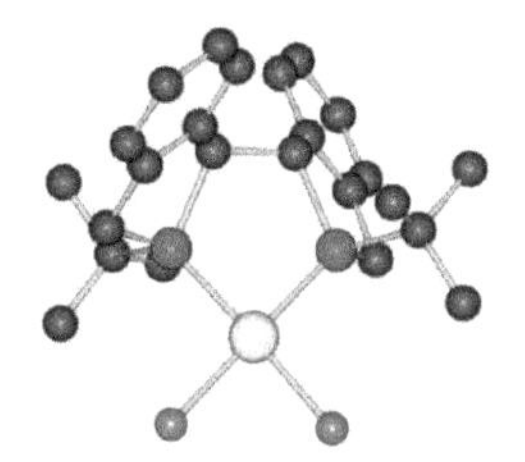

Figure 12. Molecular structure of the rhodium ammonia catalyst precursor [Rh((*S*c,*R*p)-Duanphos)(NH_3)$_2$]$^+$. Hydrogen atoms are omitted for clarity.

This and [Rh((*S*,*S*)-Et-Ferrotane)(NH_3)$_2$]$^+$ proved to be competent catalysts for the [2+2+2] cycloaddition of selected triynes, providing equal and, in some cases, better selectivity than the corresponding independently generated solvent complexes (Scheme 9) [95].

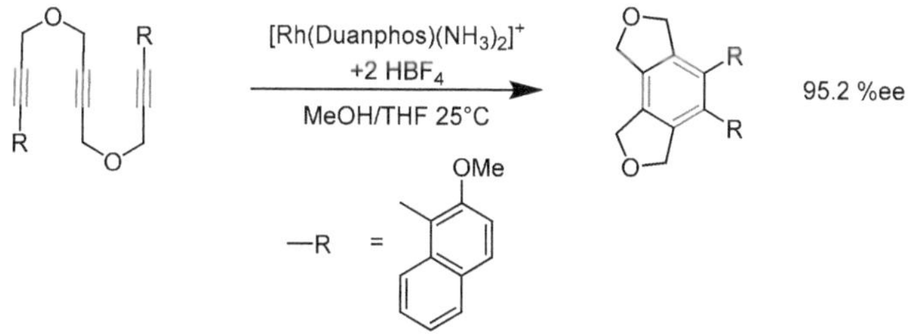

[Rh(Duanphos)(NH_3)$_2$]$^+$
+2 HBF_4
MeOH/THF 25°C

95.2 %ee

—R =

Scheme 9. [2+2+2] Cycloaddition of a triyne catalyzed by the rhodium ammonia catalyst precursor [Rh((*S*c,*R*p)-Duanphos)(NH_3)$_2$]$^+$.

As shown above, hydrogenation of [Rh(diphosphine)(diolefin)]$^+$ in a coordinating solvent such as MeOH or THF generates [Rh(diphosphine)(solvent)$_2$]$^+$. When the diphosphine is electron-rich, that is the substituents at phosphorus are alkyl rather than aryl groups, it contributes to stabilizing the solvent dihydride species [Rh(diphosphine)(H)$_2$(solvent)$_2$]$^+$, now a Rh(III) instead of a Rh(I) species, the product of hydrogen oxidative addition to rhodium. [Rh(diphosphine)(H)$_2$(solvent)$_2$]$^+$ can therefore be detected, although in the majority of cases only at sufficiently low temperatures. [Rh(diphosphine)(H)$_2$(solvent)$_2$]$^+$ is in equilibrium with the parent solvent complex. Examples of such diphosphines are Tangphos, Me–BPE, and *t*-Bu-BisP* [122–124].

As already mentioned, if the solvent complex is left under a hydrogen atmosphere long enough and its diphosphine ligand lacks aryl substituents which would otherwise trigger the formation of stable arene-bridged dimeric species (vide infra), it evolves into a trinuclear polyhydride species $\{[Rh(diphosphine)H]_3(\mu_2\text{-}H)_3(\mu_3\text{-}H)\}^{2+}$ which instead is stable at room temperature [120–122]. The formation of $\{[Rh(diphosphine)H]_3(\mu_2\text{-}H)_3(\mu_3\text{-}H)\}^{2+}$ from the solvent complex is reversible and must proceed through the rhodium dihydride $[Rh(diphosphine)(H)_2(solvent)_2]^+$ mentioned above [125]. In the context of asymmetric olefin hydrogenation, it has been shown that $\{[Rh(diphosphine)H]_3(\mu_2\text{-}H)_3(\mu_3\text{-}H)\}^{2+}$ transfers its hydrogen to the substrate mac quite slowly, over days, as it very likely must first be converted into a species of lower nuclearity to become catalytically active. The formation of these species then withdraws part of the total initial rhodium concentration, negatively affecting activity [126].

4. Catalysis in the Presence of Strongly Coordinating Ligands

Characteristic of each catalytic reaction, either homogeneous, heterogeneous, or enzymatic, is the formation of a catalyst–substrate complex. Through the key steps of the catalytic cycle and the reaction with further reagents, the catalyst–substrate complex evolves into the catalyst–product complex which eventually releases the desired free product molecule. If other species are present in solution which can also coordinate to the catalyst, then such species—inhibitors—compete with the substrate and reduce the amount of catalyst available for the catalytic process of interest. At a macroscopic level, a reduction of activity is therefore observed.

Also in nature, i.e., in the cells of living organisms, biochemical reactions are accelerated by catalysts, the enzymes. Here, too, catalyst deactivation (inhibition) plays an important role, e.g., in the regulation of the enzyme activity. As shown in Scheme 10, different types of inhibition are possible, depending on whether the free enzyme (k_i/k_{-i}) or the enzyme–substrate complex ($k_{i'}/k_{-i'}$) form an adduct with the inhibiting molecule [127].

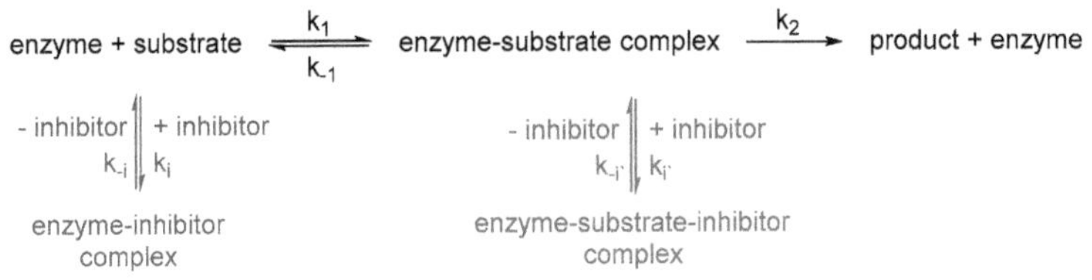

enzyme + substrate $\underset{k_{-1}}{\overset{k_1}{\rightleftharpoons}}$ enzyme-substrate complex $\xrightarrow{k_2}$ product + enzyme

- inhibitor | + inhibitor (k_{-i} / k_i) → enzyme-inhibitor complex

- inhibitor | + inhibitor ($k_{-i'}$ / $k_{i'}$) → enzyme-substrate-inhibitor complex

Scheme 10. Competitive, blue, uncompetitive, red, and mixed, blue and red, inhibition in enzyme catalysis.

4.1. *Formation of Non-Reactive, Monomeric Species*

Well-known inhibitors for cationic Rh-complexes are carbon monoxide CO and diolefins such as COD and NBD, because the corresponding complexes possess very high stability constants. For example, it has been shown that when the hydrogenation of olefins is carried out in primary alcohols with a catalyst prepared in situ from $[Rh(NBD)Cl]_2$ and monophosphines such as PPh_2Et catalyst deactivation takes place due to the formation of $[Rh(CO)(PPh_2Et)_2Cl]$. Under the reaction conditions, 50 °C and 1 bar H_2, the catalyst is also able to promote the reduction of the olefin by transfer of hydrogen from the solvent alcohol. The resulting aldehyde can be decarbonylated by the catalyst with the ensuing formation of $[Rh(CO)(PPh_2Et)_2Cl]$. This species is catalytically inactive. Catalyst deactivation is enhanced at higher temperature and does not take place in the absence of the olefin which acts as hydrogen acceptor, thus promoting the otherwise unfavorable alcohol dehydrogenation [128].

Less known is the fact that molecules which contain aromatic moieties can also act as inhibitors as they form stable η^6-arene rhodium complexes. These are coordinatively saturated 18-electron species which cannot coordinate incoming substrates or undergo oxidative addition and are consequently catalytically inactive. The formation of such species is however reversible; the extent of inactivation is determined by the concentration of the competing complexing agents present in solution which might displace the coordinated arene and the ratio of the stability constants of the corresponding rhodium complexes. Two representative structures of η^6-arene rhodium complexes are shown in Figure 13. Aromatic amines provide an interesting case: coordination of aniline derivatives to rhodium, for example, takes place through the arene, not through the nitrogen lone pair (Figure 13, right).

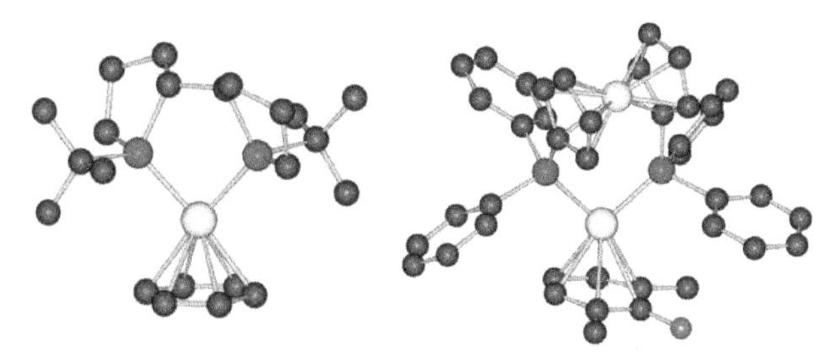

Figure 13. Molecular structure of the cation [Rh(*S*,*S*,*R*,*R*)-Tangphos)(η^6-benzene)]$^+$, left, and of the cation [Rh(DPPF)(2,6-dimethyl-η^6-aniline)]$^+$, right. Hydrogen atoms are omitted for clarity.

The sensitivity of the ^{103}Rh chemical shift to the local organic substituents makes ^{103}Rh NMR spectroscopy a valuable probe to unambiguously assign η^6-arene rhodium complexes [129,130]. Compared to those of the corresponding solvent complexes, the signals of η^6-arene rhodium complexes are shifted to higher fields. ^{103}Rh chemical shifts are also influenced by the size of the chelate ring formed by the coordinated diphosphine [131,132]. The ^{1}H and ^{13}C NMR chemical shifts of the arene provide further evidence of its η^6-coordination to the metal.

The stability of the η^6-arene rhodium complexes as expressed by their stability constants (Scheme 11) can be easily assessed, for example by UV-Vis spectroscopy.

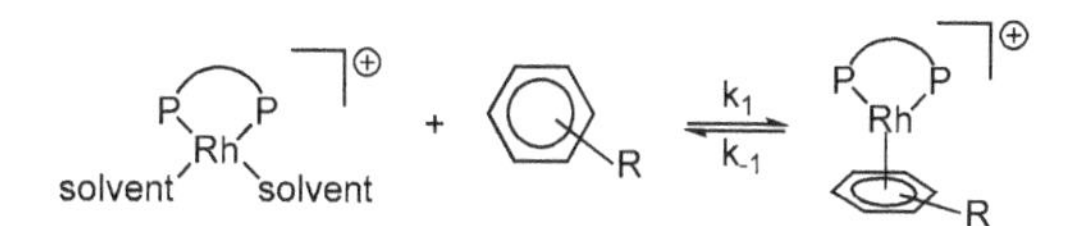

Scheme 11. Equilibrium between solvent complex, free arene, and η^6-arene rhodium complex

The classical method entails the titration of the solvent complex with the aromatics of interest [133–136]. Alternatively it is possible to monitor and quantify the established equilibrium between the solvent complex and the arene-containing molecule as a function of time; this is done using stopped flow techniques under anaerobic conditions which rely on the use of a diode array [59]. It makes no difference from which side the establishment of the equilibrium is examined: both the solvent complex and the isolated η^6-arene rhodium species can be used as starting material. In either case, the method provides the rate constant for the forward and back reaction, the ratio of which corresponds to the stability constant. The data evaluation is carried out either by graphical [137–140] or numerical methods [141–146]. For the η^6-arene complexes shown in Figure 11, the following values were determined: [Rh(*S*,*S*,*R*,*R*)-Tangphos)(MeOH)$_2$]BF$_4$/[Rh(*S*,*S*,*R*,*R*)-Tangphos)(η^6-benzene)]BF$_4$: 70 L/mol; and [Rh(DPPF)(MeOH)$_2$]BF$_4$/[Rh(DPPF)(2,6-dimethyl-η^6-aniline)]BF$_4$: 1500 L/mol [136]. Further values, also for the temperature dependence, can be found in reference [136].

The source of aromatics which can act as inhibitors can be diverse and detailed examples can be found in reference [23]. Aromatic solvents like benzene and toluene are often used in catalytic reactions such as hydrogenations. Many benchmark substrates contain phenyl rings that can lead to coordinatively saturated 18 electron η^6-arene complexes. This is the case, for example, of α-acetamidocinnamic acid: for this substrate, the chelate coordination through the double bond and the oxygen of the amido group is by far preferred over coordination of the phenyl ring. However, once the double bond is reduced, the hydrogenated product coordinates to rhodium through its phenyl substituent and can therefore negatively affect the catalytic activity, especially towards the end of the reaction when its concentration is higher (Figure 14).

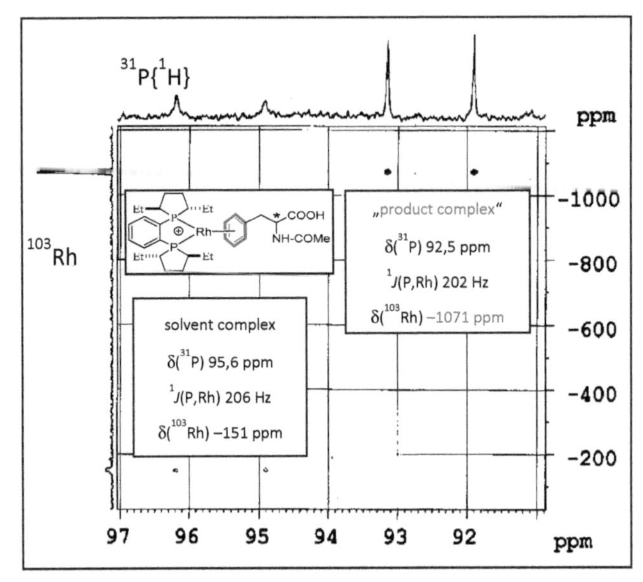

Figure 14. $^{31}P\{^1H\}$-^{103}Rh HMQC spectrum of the rhodium-α-acetamidocinnamic acid $(H)_2$ adduct $[Rh(Et\text{-}DuPhos)(\eta^6\text{-}\alpha\text{-acetamidocinnamic acid }(H)_2)]^+$, in MeOH-$d_4$, its signals are clearly distinguishable from those of the solvent complex $[Rh(Et\text{-}DuPhos)(solvent)_2]^+$.

A similar catalyst–product complex has been observed also in the course of the asymmetric hydrogenation of mac in methanol with the rhodium catalyst precursor $[Rh(t\text{-}Bu\text{-}BisP^*)(NBD)]BF_4$ containing the purely aliphatic ligand *t*-Bu-BisP* (Scheme 12) [147,148].

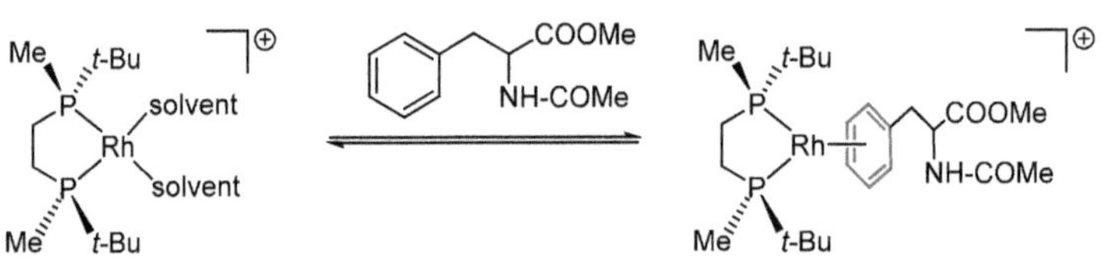

Scheme 12. Equilibrium showing the formation of the rhodium–product complex $[Rh(t\text{-}Bu\text{-}BisP^*)(\eta^6\text{-}mac(H)_2)]BF_4$ following the hydrogenation of mac with $[Rh(t\text{-}Bu\text{-}BisP^*)(NBD)]BF_4$.

The concentration of rhodium in the form of the arene complex $[Rh]_{(arene)}$ as compared to the total or initial rhodium concentration $[Rh]_{(0)}$ can be calculated according to Equation (1) [136].

$$[Rh]_{(arene)} = \frac{[Rh]_0}{\left(\frac{K_{substrate}[substrate]}{K_{arene}[arene]}\right) + 1} \quad (1)$$

Examples of calculated values are reported in reference [23]. The equation clearly shows that deactivation is not only a function of the stability constants of both the substrate- and the η^6-arene complex but also depends on the ratio of substrate- to arene-containing species concentrations. Therefore the extent of deactivation is in principle a dynamic variable which can be controlled. In fact the substrate is consumed during catalysis, which of course alters the ratio of concentrations of substrate- to arene-containing species depending on the degree of conversion. In addition, the effective kinetics, usually of Michaelis–Menten type, influences the macroscopic manifestation of the deactivation process as the following two examples show.

When the stability constant of the catalyst–substrate complex is very low, the reaction rate is 1st order in the substrate concentration. An example is provided by the hydrogenation of dimethyl itaconate with $[Rh(Ph\text{-}\beta\text{-glup-OH})(MeOH)_2]BF_4$ in methanol at room temperature and normal pressure. If a "poisoning" arene is present, then only the latter, but not the substrate, can effectively compete with the solvent, that is largely in excess, for the metal (Scheme 10, blue). The methyl ester of aminocrotonic acid, on the other hand, forms a very stable adduct with the catalyst precursor $[Rh(DIPAMP)(MeOH)_2]BF_4$ which implies that the reaction is zero order in the substrate concentration. When its hydrogenation is carried out in methanol at room temperature and normal pressure, then the substrate can effectively compete with the inhibiting arene for the metal (Scheme 10, red) [149]. In the first case, despite the presence of inhibitors, the reaction, while slower, remains 1st order in the substrate concentration and no apparent change takes place. In the second case, inhibition manifests itself with a continuous decrease in reaction rate instead, because the concentration of the inhibitor remains constant during the reaction while that of the prochiral olefin diminishes with increasing conversion. Therefore an important (tacit) assumption of the formal kinetic treatment of catalysis, that the active catalyst concentration remains constant over the course of the reaction, no longer applies.

Aromatic moieties are also present in polystyrene, commonly used to provide an insoluble support to homogeneous Rh catalysts in order to facilitate their separation from the reaction medium and subsequent reuse. In the case of cationic rhodium complexes, the weakly coordinating counteranion can be replaced by a "polymerizable" one. In this way, the unmodified homogeneous catalyst is attached to the polymer by ionic interactions [150,151]. Alternatively, the diphosphine ligand can be properly derivatized to be covalently attached to a polystyrene support [152–158].

Using soluble, low molecular weight polystyrenic resins, the formation of aromatic complexes with the phenyl rings of the "carrier" could be demonstrated by means of ^{103}Rh NMR spectroscopy. It was also possible to determine the corresponding stability constants which, although not very high, are "compensated" for by the high "concentration" of the phenyl rings. For the complex $[Rh(\mathit{Et}\text{-DuPhos})(\eta^6\text{-PS30000})]BF_4$ the measured equilibrium constant in acetone is 17 L/mol—PS30000 is a polystyrene soluble in THF which contains 288 monomer units. Such a stability constant is comparable with that of the related complex $[Rh(DPPE)(\eta^6\text{-benzene}]BF_4$, 18 L/mol [58,133]. If part of the Rh is bound to the resin as the η^6-coordinated aromatic complex, this is at least the case if Rh is bound to the support by ionic interaction, then the latter is in equilibrium with the solvent complex (Scheme 13) and can be washed out from the resin if the resin is rinsed with the solvent after catalyst loading and recycling [159]. Quantitative estimations of metal leaching can be found in reference [159].

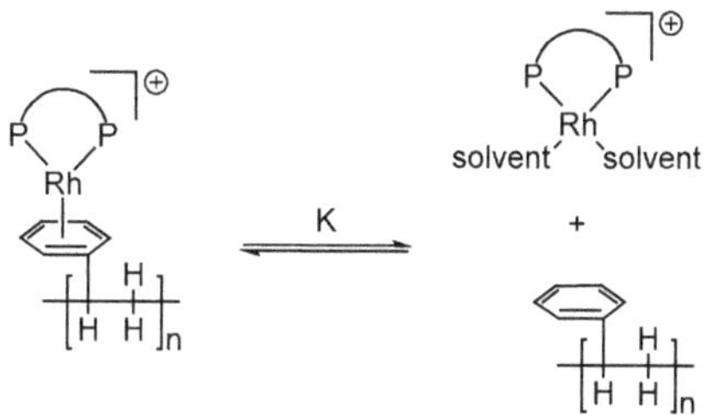

Scheme 13. Equilibrium between the η^6-arene coordinated Rh complex and the solvent complex and the polystyrene polymer matrix.

This gives rise to the well-known problem of "Rh leaching". Each time the resin is rinsed with fresh solvent, a new equilibrium is established that can lead to a further loss of the metal bound to the polystyrene support. This effect is essentially unavoidable.

The hydrogenation of prochiral organic substrates like itaconic acid and dimethyl itaconate can be carried out in aqueous solution in the presence of surfactants like sodium dodecyl sulfate, SDS or, 4-(*1,1,3,3*-Tetramethylbutyl)*phenyl*-polyethylene glycol, Triton X-100. The resulting micelles can be recycled by membrane ultrafiltration [160]. For the rhodium-BPPM/dimethyl itaconate system, the TOF decreases as the Triton X-100 concentration is increased. Selectivity, 60% ee in MeOH, is unaffected. A singlet in the ^{103}Rh NMR spectrum at −1006 ppm suggests the formation of a new η^6-arene complex. Furthermore, in the ^{1}H NMR spectrum the signals for the aromatic hydrogens are shifted to higher fields than expected for the free arene: 7.17, 6.98, 6.37, and 6.05 ppm [160]. The observed catalyst deactivation could then be confidently ascribed to the formation of η^6-arene complexes between rhodium and the phenyl substituents in the additive Triton X-100.

4.2. Formation of Non-Reactive, Multinuclear Species

Dinuclear Species

In 1977 the dimerization of the solvent complex $[Rh(DPPE)(MeOH)_2]^+$ leading to $[Rh(DPPE)]_2^{2+}$ was described for the first time [133]. In the dimer, each DPPE acts as a bridging ligand between the two rhodium centers; it is chelated to one metal through the phosphorus donors and it is coordinated to the second one through one of the phenyl substituents on the phosphorus. Other dimeric rhodium complexes, featuring a similar coordination pattern, which of course is available only when the ligand contains aryl substituents, have been reported, for both bidentate [161–167] and monodentate phosphines [96,168–172]. For the ligands BINAP, Synphos, and DIPAMP the structures of such dimeric species have been thoroughly characterized by means of NMR and X-ray crystallography [116,117,173]. The crystal structure of $[Rh(BINAP)]_2^{2+}$ is presented in Figure 15.

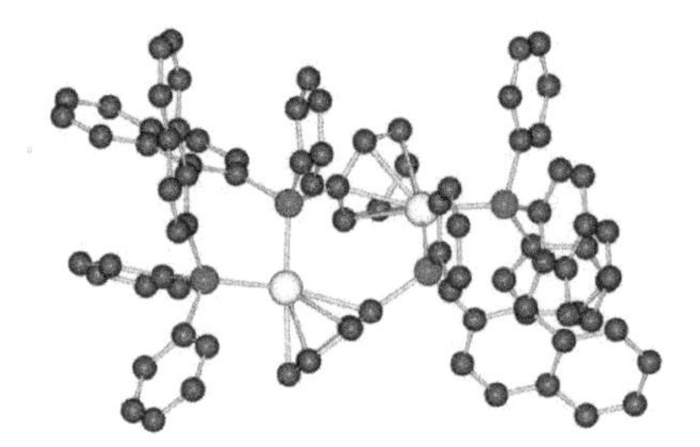

Figure 15. Molecular structure of $[Rh((R)\text{-Binap})]_2^{2+}$. Hydrogen atoms are omitted for clarity.

Such dimers have found application as catalyst precursors, for example in hydroacylation [163,174,175]. They are usually generated in situ in a nonpolar solvent, either DCM or dichloroethane, by hydrogenation of a cationic bis-diolefin rhodium complex, for example, $[Rh(COD)_2]$OTf, in the presence of a diphosphine like BINAP. ([2+2+2]cycloadditions: [176–184], hydrogen-mediated reductive C-C coupling: [185–187], enantioselective reductive cyclization: [188,189], intermolecular cyclotrimerization: [190–193]) It should be noted that the formation of these species and their role in catalysis is often overlooked.

These dimers are saturated 18 electron species which must dissociate into monomers in order to generate a catalytically active species. Such dissociation is hampered by their high stability—the stability constant of $[Rh(DIPAMP)]_2^{2+}$ for example is 52 L/mol [173]—a feature which might negatively affect the catalytic activity. An illustrative example is provided by the rhodium/BINAP promoted reductive cyclization of 1,6 diynes in DCM under a hydrogen atmosphere [194]. Kinetic investigations show that the reaction rate is independent of both the diyne and hydrogen concentration (zero partial order to each): it does not change when, at constant catalyst concentration, the catalyst/enzyme ratio is varied from 1:33 to 1:650 (Figure 16, left) and the hydrogen pressure increases from 1 up to 11 bars (Figure 16, right). More importantly, catalyst productivity increases at lower catalyst loadings [194].

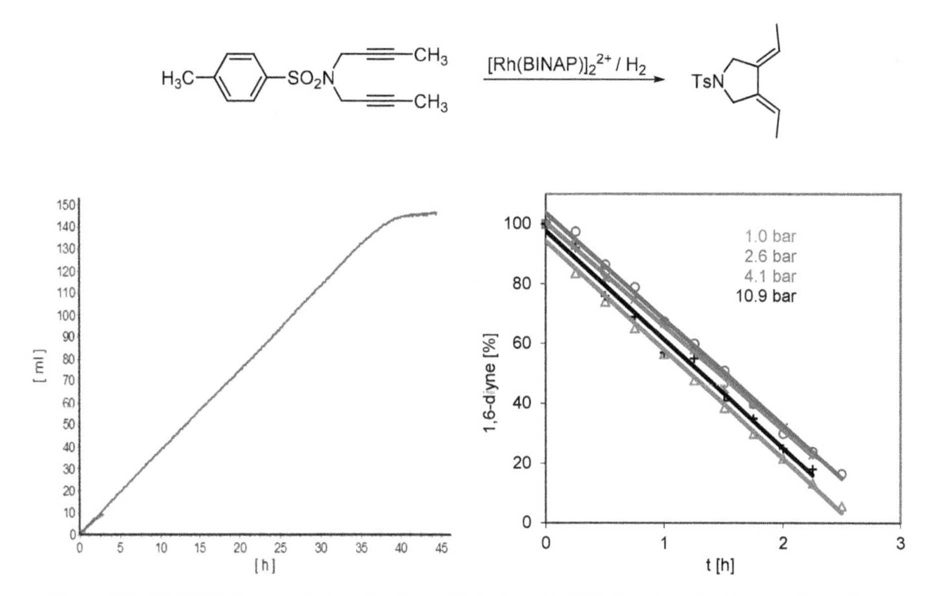

Figure 16. Rh/BINAP promoted cyclization of 1,6-diyne in DCM under a hydrogen atmosphere: hydrogen consumption (ml) as a function of time (h), left (0.01 mmol catalyst ($[Rh(BINAP)]_2^{2+}$), 0.33 or 3.30 mmol diyne, 25 °C in 10 mL DCM at normal hydrogen pressure). 1,6-diyne conversion in % as a function of time, at different hydrogen pressures, right (0.01 mmol catalyst ($[Rh(BINAP)]_2^{2+}$), 0.33 mmol diyne, 25 °C in 20 mL DCM at different hydrogen pressures).

These experimental findings can be easily accounted for if generation of the catalytically active species from the dimeric precursor $[Rh(BINAP)]_2^{2+}$ is the rate-determining step (Scheme 14). Because the steps which make up the catalytic cycle occur after the rate-determining step, disclosure of the actual reaction mechanism is difficult.

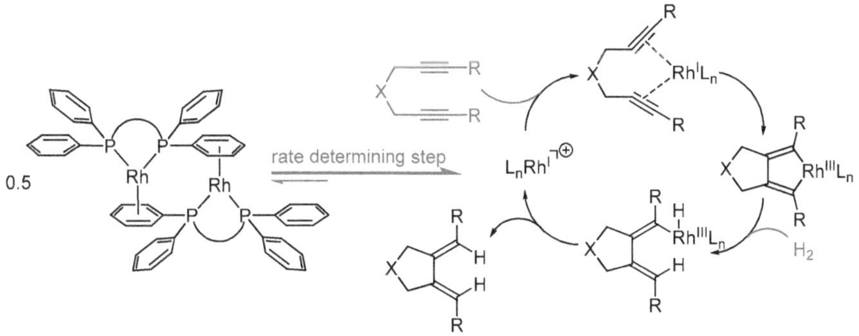

Scheme 14. Mechanism of rhodium/BINAP promoted reductive cyclization of 1,6-diyne in DCM under a hydrogen atmosphere. [195]

Based on this knowledge, the formation of such dimeric species could be prevented by carrying out the reaction in a coordinating solvent, thus preventing the formation of the arene bridged complex in favor of the solvent complex. As shown in Figure 17, by using THF or MeOH, activity could be improved by a factor of ca. 20.

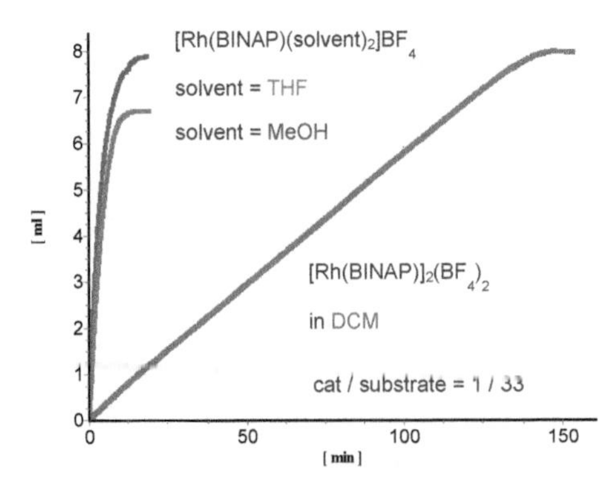

Figure 17. Hydrogen uptake (ml) as a function of time (min) in the reductive cyclization of 1,6 diyne as promoted by $[Rh(BINAP)(THF)_2]BF_4$ in THF, blue curve, $[Rh(BINAP)(MeOH)_2]BF_4$ in MeOH, green curve, and $[Rh(BINAP)]_2(BF_4)_2$ in DCM, red curve. In all cases reactions were run using 0.01 mmol catalyst ($[Rh(BINAP)]_2^{2+}$), 0.33 mmol diyne, at 25 °C in 10 mL solvent under normal hydrogen pressure.

Arene complexes can also play an important role in Rh-catalyzed cross-coupling reactions. Cross-coupling reactions are traditionally promoted by palladium catalysts, yet cationic diphosphine rhodium complexes have been shown to be also competent catalysts for this important transformation [196]. The oxidative addition of haloarenes to rhodium complexes supported by monophosphine [197–200], diphosphine [201], pincer [202–208] and multidentate nitrogen ligands [209,210] has been studied in depth and is assumed to be the first important elementary step in the catalytic cycle of the Suzuki–Miyaura coupling reaction between haloarenes and arylboronic acids. Transmetallation by the nucleophilic partner and subsequent reductive elimination to generate the new C–C bond complete the cycle. When the rhodium catalyst is generated in situ using ligands such as DIPAMP and DPPE, the yields of the biaryl product are strongly affected by the type of aryl halide: it is excellent in the case of iodobenzene (>97%), decreases with bromobenzene (>71%), and is poor with chlorobenzene (4–8%), a trend reflecting previous findings [196]. The stability of the η^6-arene complex $[Rh(DIPAMP)(\eta^6\text{-biphenyl})]BF_4$ with the product biphenyl was determined at room temperature by UV-Vis spectroscopy and is 200 L mol^{-1}. In contrast, the stability constant of the η^6-arene complex $[Rh(DIPAMP)(\eta^6\text{-}C_6H_5Cl)]BF_4$, the formation of which precedes the oxidative addition of the substrate chlorobenzene to rhodium, is only 0.6 L mol^{-1} (Scheme 15). Therefore, in the case under investigation, the poor yield is not only due to the unfavorable activation of the strong C–Cl bond in the substrate, but also to product inhibition. Indeed the active catalyst is completely deactivated after only a few catalytic cycles [196].

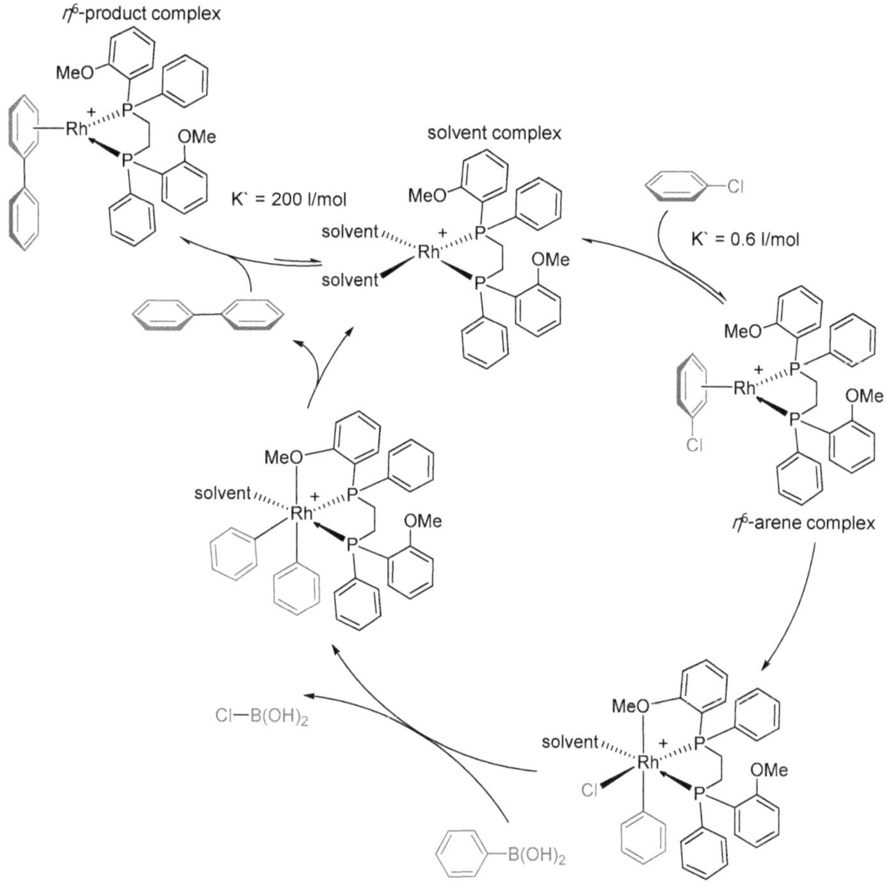

Scheme 15. Proposed catalytic cycle for the rhodium/DIPAMP promoted cross-coupling of chlorobenzene with phenylboronic acid.

4.3. Trinuclear Complexes

Addition of a base to the solvent complex $[Rh(diphosphine)(solvent)_2]^+$ leads to the formation of trinuclear rhodium complexes of the general formula $[Rh_3(diphosphine)_3(\mu_3\text{-}X)_2]^+$ ($X = OMe^-$, OH^- ...). The first example was reported with the ligand DPPE in MeOH [133]. Similar complexes with ligands like *t*-Bu-BisP*, Synphos, BINAP, Me-DuPhos, DIPAMP, and DPPP were published later [40,117,211,212]. In this complex type, an example of which is shown in Figure 18, the three rhodium atoms lie at the vertexes of a regular triangle, each coordinated to a bidentate diphosphine located perpendicular to the Rh_3 plane. Above and below the plane are the two μ_3-bridging anions.

Scheme 16 introduces the general reaction sequence, initially considered as irreversible [211], whereby the trinuclear complexes are formed starting from the solvent complex. In-depth investigations have shown that each step of such sequence is, in principle, reversible [213].

The base is necessary in order to generate the bridging anions either from the solvent (MeO^-) or from adventitious water present in solution (OH^-). ^{103}Rh NMR allows distinguishing among the different species that may be formed with the ligand Me–DuPhos: $[Rh_3(Me\text{-}DuPhos)_3(\mu_3\text{-}OMe)_2]BF_4$, $[Rh_3(Me\text{-}DuPhos)_3(\mu_3\text{-}OMe)(\mu_3\text{-}OH)]BF_4$ and $Rh_3(Me\text{-}DuPhos)_3(\mu_3\text{-}OH)_2]BF_4$ as illustrated in Figure 19.

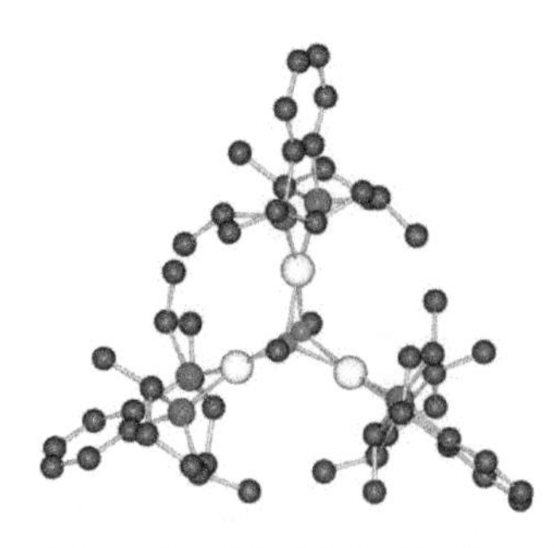

Figure 18. Molecular structure of the cation $[Rh_3(Me\text{-}DuPhos)_3(\mu_3\text{-}OMe)_2]^+$. Hydrogen atoms are omitted for clarity.

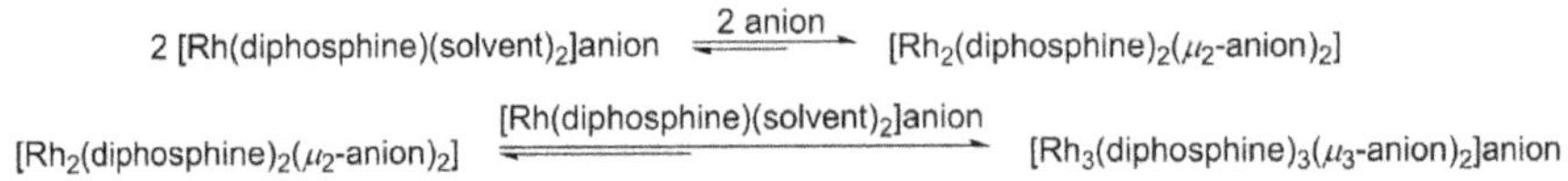

$$2\ [Rh(diphosphine)(solvent)_2]anion \xrightleftharpoons{2\ anion} [Rh_2(diphosphine)_2(\mu_2\text{-}anion)_2]$$

$$[Rh_2(diphosphine)_2(\mu_2\text{-}anion)_2] \xrightleftharpoons{[Rh(diphosphine)(solvent)_2]anion} [Rh_3(diphosphine)_3(\mu_3\text{-}anion)_2]anion$$

Scheme 16. General, reversible reaction sequence for the generation of μ_3-anion bridged trinuclear complexes.

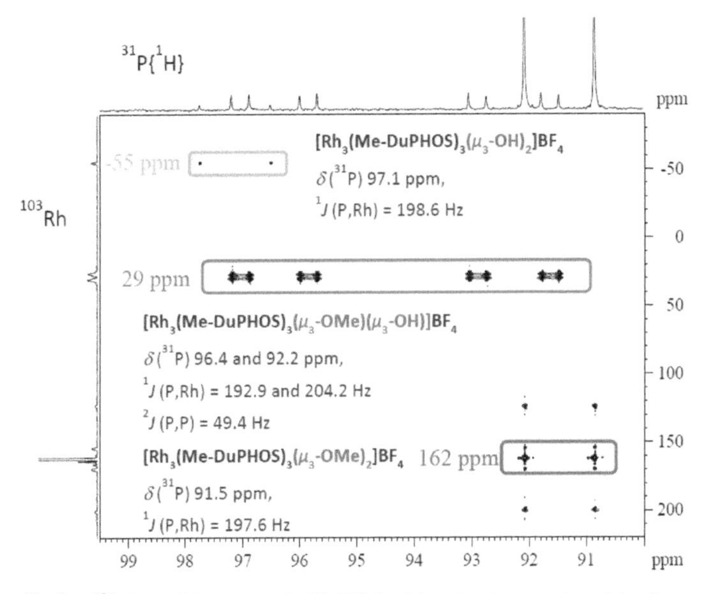

Figure 19. $^{31}P\{^1H\}$-^{103}Rh HMQC spectrum in MeOH-d_4 of the trinuclear species arising from treatment of $[Rh(Me\text{-}DuPhos)(MeOH)_2]BF_4$ with NEt_3/H_2O (1:1) [195].

In the course of Rh promoted catalytic hydrogenation, the formation of such species, which is manifested through a change in the color of the solution from orange to red-brown, leads to a reduction in catalytic activity [212]. Substrates which are sufficiently basic can also trigger the formation of such species, in the absence of any added base! The negative effect on activity though might be difficult to recognize. An example is provided by [2-(3-methoxy-phenyl)-cyclohex-1-enylmethyl]-dimethylamine [214]. When $[Rh(DIPAMP)(MeOH)_2]^+$ is added to a solution of this olefin at room temperature in a ratio 1:20, then NMR spectroscopy shows that up to 75% of the total rhodium content is present as $[Rh_3(DIPAMP)_3(\mu_3\text{-}OMe)_2]^+$ and, to a minor extent, as $[Rh_3(DIPAMP)_3(\mu_3\text{-}OMe)(OH)]^+$ (Figure 20) [40].

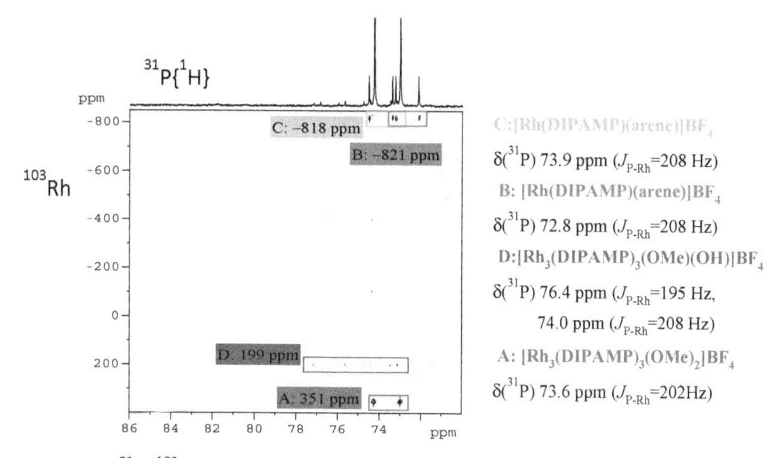

Figure 20. ^{31}P–^{103}Rh correlation spectrum of a mixture of the solvent complex [Rh(DIPAMP)(MeOH)$_2$]BF$_4$ (0.02 mmol) and 100 μL [2-(3-methoxyphenyl)-cyclohex-1-enylmethyl]-dimethylamine (0.4 mmol). The signals highlighted in red and blue correspond to the (μ_3-OMe)$_2$ and (μ_3-OMe)(μ_3-OH) bridged trinuclear complexes (75%), the signals highlighted in green and orange are due to diastereomeric arene complexes (25%). The substrate (purity of >99%) contains traces of water which lead to the formation of the mixed trinuclear complex [40].

Other examples, including the substrate (*E*)-1-(2-methyl-3-phenylallyl)piperidine, are described in reference [212].

Halides have also been recognized as possible sources of deactivation in reactions catalyzed by rhodium complexes. In the asymmetric hydrogenation of 2-methylenesuccinamic acid promoted by [Rh((*S*,*S*)-Et-DuPHOS)(COD)]BF$_4$ [215] activity could be increased by a factor of 26 when the substrate was thoroughly purified from chloride contaminants, leftovers from its synthesis [216]. The reduced activity in the presence of halides may be ascribed to the formation of very stable trinuclear rhodium complexes where halides act as μ_3-bridging anions [40]. The molecular structures of [Rh$_3$((*R*,*R*)*t*-Bu-BisP*)$_3$(μ_3-Cl)$_2$]$^+$ and [Rh$_3$((*R*,*R*)-Me-DuPhos)$_3$(μ_3-Br)$_2$]$^+$ are presented in Figure 21.

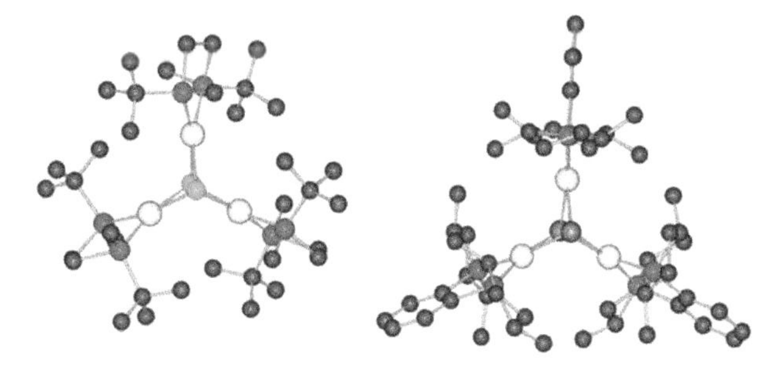

Figure 21. Molecular structures of the cations of [Rh$_3$((*R*,*R*)-*t*-Bu-BisP*)$_3$(μ_3-Cl)$_2$]$^+$, left, and [Rh$_3$((*S*,*S*)-Me-DuPhos)$_3$(μ_3-Br)$_2$]$^+$, right. Hydrogen atoms are omitted for clarity.

Such complexes are catalytically inactive; this was proven experimentally in the asymmetric hydrogenation of mac and dimethyl itaconate with the solvent complex [Rh(DIPAMP)(MeOH)$_2$]$^+$ in the presence of added sodium halides (Cl, Br, I) (Figure 22) [40].

Indeed, when [Rh(DIPAMP)(MeOH)$_2$]BF$_4$ and the prochiral olefin ((*Z*)-3-[1-(dimethylamino)-2-methylpent-2-en-3-yl]phenol) [217] are dissolved in MeOH-d_4 in a 1:10 ratio, the ^{31}P NMR spectrum of the resulting solution shows that almost 34% of the total signal intensity corresponds to the catalytically inactive trinuclear μ_3-chloro-bridged complex $[Rh_3(DIPAMP)_3(\mu_3\text{-}Cl)_2]BF_4$, resulting from traces of chloride left in the substrate after its synthesis (Figure 23) [40]. Similar to the competitive inhibition in enzymatic catalysis (Scheme 10) halides compete with the substrate for the solvent complex: the amount of trinuclear complex and the extent of deactivation it causes depend on the relative concentrations of halides and the substrate and on the stability constants of the complexes they form with the metal. To prevent deactivation, special attention should therefore be paid to substrate purification, especially on the industrial scale where, due to economical reasons, very high substrate to catalyst ratios are used.

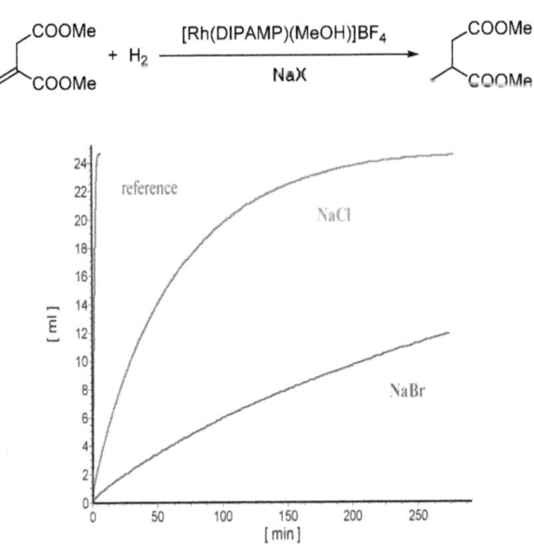

Figure 22. Hydrogen consumption (ml) for the hydrogenation of 1.0 mmol dimethyl itaconate with 0.01 mmol $[Rh(DIPAMP)(MeOH)_2]BF_4$ under addition of different sodium halides NaX: red: no additive (88% ee); green: 0.1 mmol NaCl, (82% ee); blue: 0.1 mmol NaBr, (80% ee); conditions: each 15 mL MeOH at 20 °C and 1 bar total pressure.

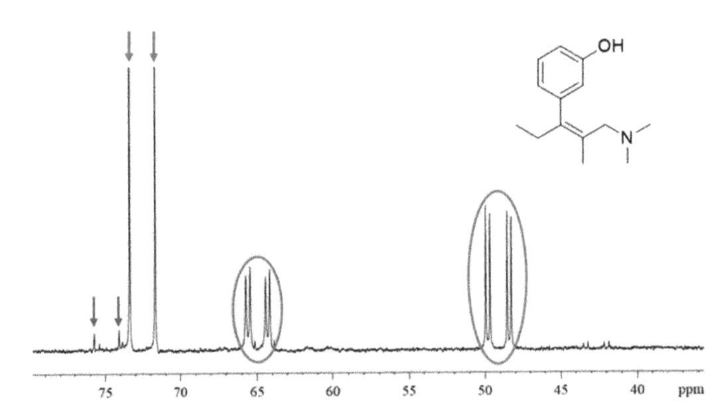

Figure 23. $^{31}P\{^1H\}$ NMR spectrum of a solution of $[Rh(DIPAMP)(MeOH\text{-}d_4)_2]BF_4$ and (3-[(2*R*,3*S*)-1-(dimethylamino)-2-methylpentan-3-yl]phenol (rhodium: substrate = 1:10). About 34% of the signal intensity (red arrows) is due to the doublet of the trinuclear complex $[Rh_3(DIPAMP)_3(\mu_3\text{-}Cl)_2]BF_4$. The signal set highlighted with blue arrows (2% of signal intensity) is due to the dinuclear complex $[Rh_2(DIPAMP)_2(\mu_2\text{-}Cl)_2])$. Highlighted in green is the substrate complex $[Rh(DIPAMP)(substrate)]BF_4$.

5. Catalyst Deactivation due to Irreversible Reactions of the Active Catalyst

Already in 1979 it was reported that the rate of itaconic acid hydrogenation with the rhodium–DIPAMP catalyst in methanol decreases with increasing substrate concentration [218]. Systematic investigations have shown that the substrate can react irreversibly with the catalyst to generate a Rh(III) alkyl complex. This represents a dead-end of the catalytic cycle because oxidative addition of molecular hydrogen is formally not possible on a Rh(III) complex. The inactive species has been thoroughly characterized and its molecular structure was unambiguously assigned by X-ray analysis [219,220]. Its rate of formation is influenced by substrate concentration, temperature, hydrogen pressure, and the presence of additives, either acidic or basic. The likely mechanism through which it is formed includes deprotonation of the β-carboxylic group of the coordinated substrate which leads to the formation of a rhodium–carboxylate bond, thus generating a neutral catalyst-substrate complex (Scheme 17). This step is obviously promoted by basic additives such as NEt_3. Next the OH of the α-carboxylic group oxidatively adds to rhodium generating a Rh(III) hydride species. Finally the double bond inserts into the Rh–H bond generating the Rh(III) alkyl complex.

BF_4 $-HBF_4$

orange diastereomeric **substrate complexes** → colorless **alkyl complex**

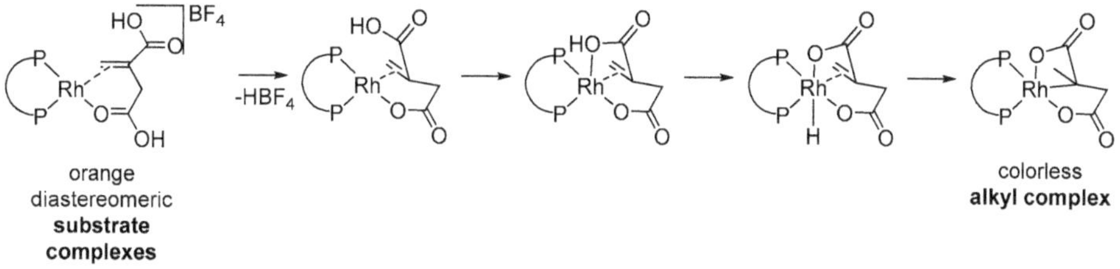

Scheme 17. Likely mechanism through which catalytically inactive Rh(III) alkyl complexes are formed during the hydrogenation of itaconic acid with cationic rhodium–diphosphine complexes.

This might be regarded as a general deactivation path which, regardless of the diphosphine ligand, could be open to other unsaturated substrates carrying carboxylic groups, as was shown for example for 2-acetamidoacrylic acid [221]. The catalytically inactive species resulting from the reaction of the latter with $[Rh(DPPE)(MeOH)_2]BF_4$ could be isolated and the presence of a Rh–C bond was confirmed by X-ray analysis: the Rh(III) alkyl complex, which is present in solution as a monomeric species as suggested by NMR investigations, crystallizes as a dimer with one carboxylate group acting as bridging ligand between the two rhodium centers. Interestingly this example demonstrates that one carboxylate group suffices to promote the formation of the inactive Rh(III) alkyl species (Figure 24).

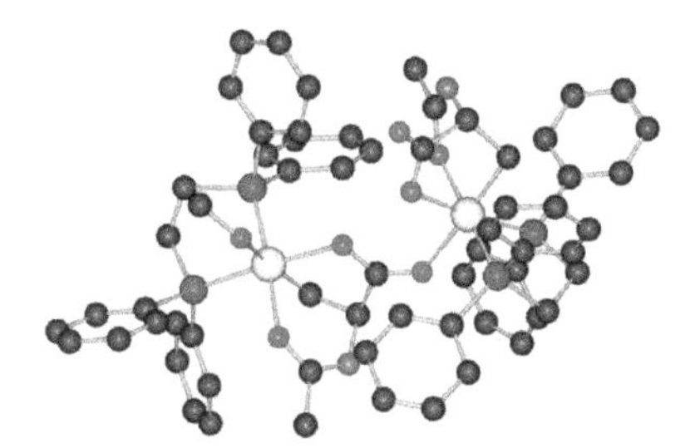

Figure 24. Solid state structure of the dimeric Rh(III)–alkyl complex arising from the reaction of 2-acetamidoacrylic acid with $[Rh(DPPE)(MeOH)_2]BF_4$, left. Hydrogen atoms are omitted for clarity.

Further investigations are needed to assess to what extent this deactivation path is open to other unsaturated substrates containing a carboxylic group. As to itaconic acid, it has been shown that the formation of inactive Rh(III) alkyl species represents a severe limitation [221] to the application of cloud point extraction as a methodology to extract the catalyst active form from micellar solutions in order to recycle it [222].

Not only the substrate but also other reaction components can irreversibly react with the catalytically active species. This is the case for halogenated solvents as Rh(I) complexes can activate C–X bonds (X = halogen) and hence undergo oxidative addition [223]. DCM is a solvent frequently applied in catalysis but a perusal of the literature shows how this might not always be a sensible choice. Examples of DCM oxidative addition to dimeric rhodium complexes of the type $[Rh_2(diphosphine)_2(\mu_2\text{-}Cl)_2]$ are scarce, and for a long time only one example was known for the ligand DPPE [224], but more have been reported for neutral monomeric rhodium complexes such as Rh(I)-β-diketonate/diphosphine derivatives with the ligands DPPM, DPPE, and DPPP [225] and for $[Rh(DMPE)_2Cl]$ [226]. Very electron-rich centers are required to promote breaking of the strong C–Cl bond. The activation of halobenzenes has been documented for cationic monomeric rhodium complexes with both di-[209,210] and monophosphines [197–200]. Several species can be formed following the oxidative addition of DCM to rhodium [227,228]: such species can be mononuclear and contain a terminal CH_2Cl group. They can be dinuclear and each rhodium center can have a terminal CH_2Cl group; they can be dinuclear and contain a bridging μ_2-CH_2 which results from the activation of both C–Cl bonds of CH_2Cl_2. The X-ray structures of two examples are reported in Figure 25 [229]

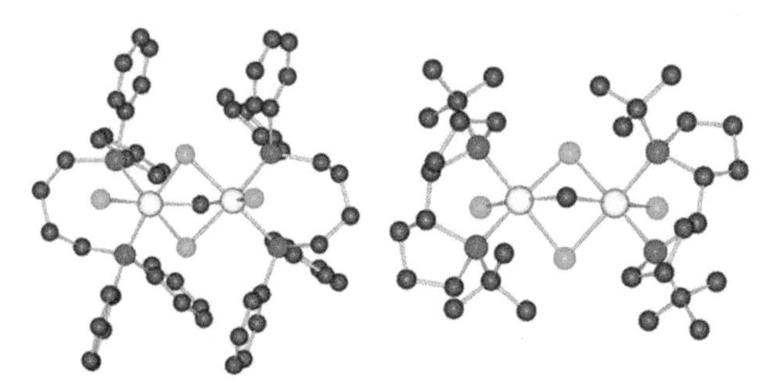

Figure 25. Molecular structure of $[Rh_2(DPPB)_2(\mu_2\text{-}Cl)_2(\mu_2\text{-}CH_2)Cl_2]$, left, and $[Rh_2((1S,1S',2R,2R')\text{-}TangPhos)_2(\mu_2\text{-}Cl)_2(\mu_2\text{-}CH_2)Cl_2]$, right. Hydrogen atoms are omitted for clarity.

The dinuclear species may also arise from further reaction of mononuclear rhodium complexes which have undergone DCM oxidative addition. The exact mechanism for the activation of DCM by neutral dimeric rhodium complexes stabilized by diphosphine ligands is not known but plausible pathways are shown in Scheme 18.

An issue, the consequences of which are often overlooked, is the establishment of the exact nature of the catalytically active species generated from the catalyst precursor. In the case of $[Rh_2(diphosphine)_2(\mu_2\text{-}Cl)_2]$, it is generally assumed that the dimeric precatalyst generates a highly reactive 14 electron monomeric species which initiates the catalytic cycle. This monomerization has been calculated for example for the ligand DPEPhos [64]. Simple metathesis experiments (Scheme 19) provided experimental evidence of the monomerization process [60,229].

On the other side in principle, a reversible equilibrium also exists between the classical cationic mononuclear Rh complexes $[Rh(diphosphine)(solvent)_2]^+$ and the dimeric neutral Rh complexes $[Rh_2(diphosphine)_2(\mu_2\text{-}Cl)_2]$, as described by the upper part of Scheme 16 [230]. With ammonia this reaction is irreversible (Scheme 20) [213]. This implies that, when discussing the mechanism of a catalytic cycle promoted by the dinuclear neutral species $[Rh_2(diphosphine)_2(\mu_2\text{-}Cl)_2]$, then the possibility of a mononuclear cationic species contributing to the observed activity should be taken into consideration.

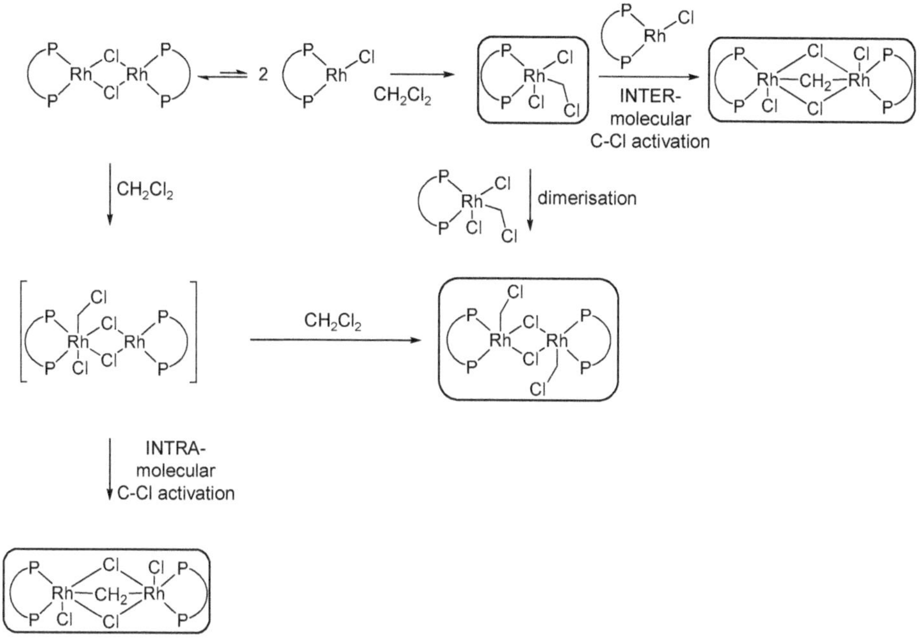

Scheme 18. Possible pathways for the oxidative addition of DCM to preformed neutral, dimeric Rh-complexes.

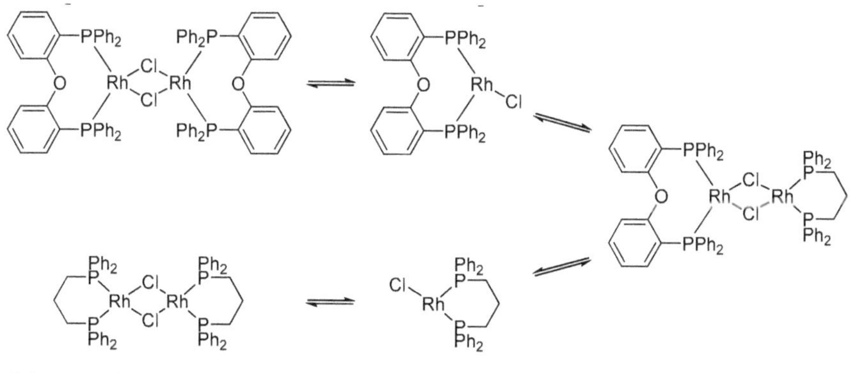

Scheme 19. Dimer–monomer equilibrium for complexes of the type $[Rh_2(diphosphine)_2(\mu_2\text{-}Cl)_2]$ for the ligands DPPP and DPEPhos. Experimental evidence based on ^{31}P NMR showing the formation of a dinuclear mixed species $[(DPPP)Rh((\mu_2\text{-}Cl)_2Rh(DPPE)]$ from the mononuclear fragments [(DPPP)RhCl] and [Rh(DPPE)Cl].

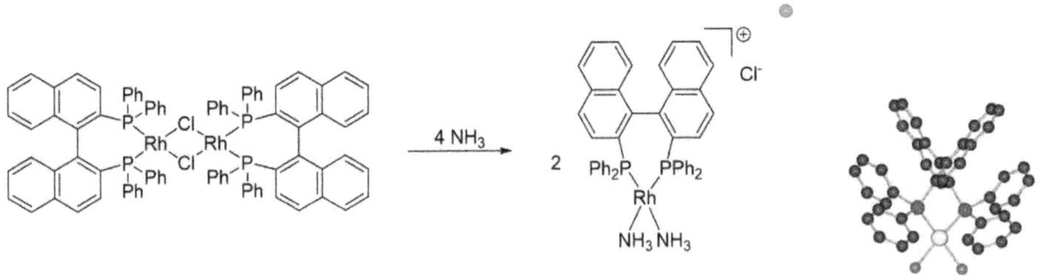

Scheme 20. Reaction scheme for the observed conversion of a dinuclear μ_2-chloro bridged complex into a monomeric cationic one, left, and the molecular structure of the resulting complex [Rh(BINAP)(NH_3)]Cl, right.

6. Conclusions

In the present work, the rich chemistry of rhodium/phosphine complexes, which are applied as homogeneous catalysts to promote a wide range of chemical transformations, has been used to showcase how the in situ generation of precatalysts, the conversion of precatalysts into the actually active species, as well as the reaction of the catalyst itself with other components in the reaction medium (substrates, solvents, additives) can lead to a number of deactivation phenomena. Such phenomena may go unnoticed or may be overlooked, thus preventing the full understanding of the catalytic process which is a prerequisite for its optimization.

As a summary of the discussion reported above, a few guidelines are presented here which may help the practitioner in the early stages of his/her investigation into a catalytic process promoted by rhodium–diphosphine catalysts.

The in situ preparation of a catalyst precursor from a suitable metal source and the desired ligand is very convenient in the screening of ligand libraries and reaction conditions. Yet fine-tuning of a catalytic process should include a detailed knowledge of the actual species generated during the in situ procedure. In the case of rhodium, the in situ generation of the catalyst precursor [Rh(diphosphine)(μ_2-X)]$_2$ (X = Cl−, OMe−, OH− ...) from [Rh(diolefin)(μ_2-X)]$_2$ and the corresponding diphosphine can in fact be far from selective and it is strongly dependent of the type of diphosphine, diolefin and reaction conditions. As a practical guide for the choice of proper reaction conditions, relevant information for the most common diphosphines can be found in reference [9a] and the relative supporting information.

The hydrogenation of prochiral olefins is promoted by catalyst precursors of the general formula [Rh(diphosphine)(diolefin)]$^+$ (dioelfin = COD, NBD). Generation of the active species [Rh(diphosphine)(solvent)$_2$]$^+$, in a proper solvent, requires hydrogenation of the diolefin. NBD-containing precursors are to be preferred as NBD is hydrogenated faster than COD. Prior to the addition of the substrate, the pre-hydrogenation time is then dictated by the used diphosphine. Data collected in Table A1 allow calculation of pre-hydrogenation time using the equation $t_{pre\text{-}hydrogenation} = 7 \times t_{1/2}$ with $t_{1/2} = \ln 2/k'$, seven half lives, in fact, correspond to 99.2% (practically complete) conversion of the diolefin in [Rh(diphosphine)(diolefin)]$^+$.

Knowledge of the correct pre-hydrogenation time is likewise important to prevent formation of polyhydride species, should hydrogenation of the catalyst precursor to generate the active species, in the absence of the substrate, last too long.

The choice of the solvent is also important in preventing deactivation phenomena: while a too strongly coordinating solvent may hamper substrate coordination, a moderately coordinating solvent such as MeOH can be easily displaced but at the same time prevent the aggregation of coordinatively unsaturated metal fragments to polynuclear inactive species. Aromatic solvents should be also avoided as they can sequester part of the active catalyst in the form of η^6-coordinated arene-Rh(I) complexes. Halogenated solvents like dichloromethane might not be a sensible choice either as rhodium can activate the rhodium–halide bond and generate rhodium (III)–halide species.

The use of purified substrates is also recommended as halides, which are common contaminants, may react with the rhodium catalyst generating polynuclear rhodium halide species.

Catalysis has been defined "a foundational pillar of green chemistry", and a sensible use of the catalyst to fully exploit its potential makes it even greener [231].

Funding: DFG (Deutsche Forschungsgemeinschaft) HE 2890/7-1 and HE 2890/7-2

Acknowledgments: We would like to sincerely thank PD Torsten Beweries and Detlef Selent for their helpful discussions and advice. In addition, we would like to thank PD Wolfgang Baumann for his support with analytical problems, especially those concerning NMR spectroscopy.

Conflicts of Interest: The authors declare no conflict of interest.

Appendix A

Table A1. Pseudo rate constants relative to diolefin hydrogenation for the hydrogenation of COD and NBD for different [Rh(diphosphine)(diolefin)]$^+$ complexes (25.0 °C; 1.0 bar total pressure). The values were obtained with MeOH as the solvent and with BF_4^- as the anion unless stated otherwise.

Diphosphine	Solvent	1st Order Hydrogenation Rate Constant (1/min)		Reference
		COD	NBD	
BINAP		$2.3 \cdot 10^{-1}$	26.8	[101]
	THF	$2.8 \cdot 10^{-1}$	20.5	[101]
	Propylene Carbonate	$1.4 \cdot 10^{-1}$	16.6	[101]
BPPM		$2.2 \cdot 10^{-1}$	1.2	[101]
CatASium®D(R)		ca. $1.7 \cdot 10^{-3}$	n.d.	[108]
CatASium®M(R)		$5.0 \cdot 10^{-2}$	25	[101]
	THF	$1.5 \cdot 10^{-1}$	12	[100]
	Propylene Carbonate	$8.5 \cdot 10^{-2}$	9.4	[100]
Chiraphos		$1.3 \cdot 10^{-3}$	3.0	[105]
Cyc-JaPhos		1.1	at least 700	[105]
DaniPhos		ca. 0.03	3.6	[100]
	THF	n.d.	4.8	[100]
DCPE		$7.5 \cdot 10^{-2}$	48.8	[105]
Difluorphos		$1.6 \cdot 10^{-1}$	28.6	[95]
DIOP		$2.3 \cdot 10^{-1}$	1.29	[104]
DIPAMP		ca. $2.9 \cdot 10^{-3}$	ca. 9	[173]
	EtOH	n.d.	13	[173]
	i-PrOH	n.d.	5.9	[173]
	THF	n.d.	5.0	[173]
	Trifluoroethanol	n.d.	3.9	[173]
	Propylene Carbonate	n.d.	4.8	[173]
DPOE		$2.5 \cdot 10^{-1}$	16.6	[105]
DPPB		$1.6 \cdot 10^{-1}$	1.25	[104]
DPPE		ca. $2.0 \cdot 10^{-3}$	17.3	[105]
DPPP		$2.4 \cdot 10^{-2}$	1.55	[105]
DTBM-SEGPhos		$6.3 \cdot 10^{-2}$	12.0	[95]
Duanphos		$6.8 \cdot 10^{-1}$	53.7	[95]
Et-Butiphane		$2.9 \cdot 10^{-2}$	n.d.	[91]
Et-DuPhos		$1.2 \cdot 10^{-1}$	52.2	[101,105]
Et-Ferrotane		$2.0 \cdot 10^{-1}$	19.3	[95]
H_8-BINAP		$8.5 \cdot 10^{-1}$	57.5	[95]
i-Pr-Butiphane		$1.0 \cdot 10^{-2}$	n.d.	[91]
JaPhos		7.15	230	[105]
Josiphos		$2.3 \cdot 10^{-2}$	ca. 34	[100]

Table A1. *Cont.*

Diphosphine	Solvent	1st Order Hydrogenation Rate Constant (1/min)		Reference
		COD	NBD	
Me-Butiphane		$1.0 \cdot 10^{-1}$	n.d.	[91]
Me-DuPhos		$1.1 \cdot 10^{-1}$	35.2	[101]
	THF	$1.6 \cdot 10^{-2}$	39	[101]
	Propylene Carbonate	$1.4 \cdot 10^{-2}$	18	[101]
Me-α-glup		$3.7 \cdot 10^{-1}$	13.4	[104]
Ph-β-glup-OH		$2.0 \cdot 10^{-1}$	9.5	[104]
Prophos		ca. $3.0 \cdot 10^{-3}$	9.2	[105]
Propraphosderivates				
R=2-pentyl		3.77	21.9	[104]
R=3-pentyl		4.09	21.4	[104]
R=cyclohexyl		5.44	20.2	[104]
R=cyclopentyl		2.94	18.4	[104]
R=methyl		$5.3 \cdot 10^{-1}$	8.2	[104]
SEGPhos		$4.7 \cdot 10^{-1}$	29.9	[95]
Synphos		$8.0 \cdot 10^{-1}$	67	[117]
Tangphos		$3.7 \cdot 10^{-1}$	194.4	[120]
t-Bu-BisP*		$2.1 \cdot 10^{-1}$	90	[120]
t-Bu-Ferrotane		$5.7 \cdot 10^{-1}$	ca. 50	[95]
3-Pen-SMS-Phos		n.d.	3.8	[23]

Table A2. Abbreviations including diphosphine ligands.

Abbreviations	
acac	acetylacetonateanion
COE	cyclooctene
COD	*cis,cis*-1,5-cyclooctadiene
DCM	dichloromethane
IUPAC	International Union of Pure and Applied Chemistry
mac	methyl-(*Z*)-α -acetamidocinnamate
NBD	bicyclo[2.2.1]hepta-2,5-diene
SHOP	Shell Higher Olefin Process
THF	tetrahydrofuran
Diphosphines	
BINAP	2,2′-bis(diphenylphosphino)-1,1′-binaphthyl
BPPM	2,3-bis(diphenylphosphino)-*N*-phenylmaleimide
CatASium® D(R)	*N*-Benzyl-(3*R*,4*R*)-bis(diphenylphosphino)pyrrolidine
CatASium® M(R)	3,4-Bis[(2*R*,5*R*)-2,5-dimethyl-1-phospholanyl]furan-2,5-dione
Chiraphos	2,3-bis(diphenylphosphino)butan
Cyc-JaPhos	1-(2-dicyclohexylphosphinophenyl)pyrol-2-dicyclohexylphosphine
DaniPhos	dicyclohexyl(1-(2-(diphenylphosphanyl)phenyl)ethyl)phosphane-(tricarbonyl)chrom
DCPB	1,4-bis(dicyclohexylphosphino)butane
DCPE	1,2-bis(dicyclohexylphosphino)ethane
Difluorphos	5,5′-bis(diphenylphosphino)-2,2,2′,2′-tetrafluoro-4,4′-bi-1,3-benzodioxole
DIOP	2,3-*O*-isopropylidene-2,3-dihydroxy-1,4-bis(diphenylphosphino)butane)
DIPAMP	1,2-bis[(2-methoxyphenyl)phenylphosphino]ethan
DM-SEGPhos	5,5′–bis[di(3,5-xylyl)phosphino]-4,4′-bi-1,3-benzodioxole
DPEPhos	bis[(2-diphenylphosphino)phenyl]ether
DPOE	1,2-bis(diphenylphosphinoxy)ethan
DPPB	1,4-bis(diphenylphosphino)butane
DPPE	1,2-bis(diphenylphosphino)ethan
DPPF	1,1′-bis(diphenylphosphino)ferrocene
DPPMP	2-[(diphenylphosphino)methyl]pyridine
DPPP	1,3-bis(diphenylphosphino)propane
DTBM-SEGPhos	5,5′-bis[di(3,5-di-*t*-butyl-4-methoxyphenyl)phosphino]-4,4′-bi-1,3-benzodioxole

Table A2. *Cont.*

Diphosphines	
Duanphos	(1,1′,2,2′)-2,2′-di-*t*-butyl-2,3,2′,3′-tetrahydro-1*H*,1′*H*(1,1′)biisophos-phindolyl
Et-ButiPhane	2,3-bis(2,5-diethylphospholanyl)benzo[b]thiophene
Et-DuPhos	1,2-bis-2,5-(diethylphospholano)benzene
Et-Ferrotane	1,1′-bis-(2,4-diethylphosphonato)ferrocene
H_8-BINAP	2,2′-bis(diphenylphospino)-5,5′,6,6′,7,7′,8,8′-octahydro-1,1′-binaphthyl
i-Pr-Butiphane	2,3-bis(2,5-diisopropylphospholanyl)benzo[b]thiophene
JaPhos	1-(2-dicyclohexylphosphino)butane
Josiphos	[2-(diphenylphosphino)ferrocenyl]-ethyldicyclohexylphosphine
Me-BPE	1,2-bis[(2,5)-2,5-dimethylphospholano]ethane
Me-Butiphane	2,3-bis(2,5-dimethylphospholanyl)benzo[b]thiophene
Me-DuPhos	1,2-bis-(2,5-dimethylphospholano)benzene
Me-α-glup	methyl-4,6-*O*-benzylidene-2,3-*0*-bis(diphenylphosphino)-α-*D*-glucopyranoside
MonoPhos	3,5-dioxa-4-phosphacyclohepta[2,1-*a*;3,4-*a*′]dinaphthalen-4-yl)-dimethylamine
Ph-β-glup-OH	Phenyl-2,3-*O*-bis(diphenylphosphino)-β-*D*-glucopyranoside
PPF-P(t-Bu)$_2$	1-[2-(diphenylphosphino)ferrocenyl]ethyl-di-*t*-butylphosphine
Prophos	1,2-bis(diphenylphosphino)propane
SEGPhos	5,5′-bis(diphenylphosphino)-4,4′-bi-1,3-benzodioxole
Synphos	[(5,6),(5′,6′)-bis(ethylenedioxy)biphenyl-2,2′-diyl]bis(diphenylphosphine)
Tangphos	1,1′-di-*t*-butyl-(2,2′)-diphospholane
t-Bu-BisP*	1,2-bis(*t*-butylmethylphosphanyl)ethane
t-Bu-Ferrotane	1,1′-Bis-(2,4-diethylphosphonato)ferrocene
3-Pen-SMS-Phos	1,2-bis[(*o*-3-pentyl-*O*-phenyl)(phenyl)phosphino]ethane

References and Notes

1. Berzelius, J.J. *Årsberättelse om Framstegen i Fysik och Kemi*; Royal Swedish Academy of Sciences, P.A. Norstedt & Söner: Stockholm, Sweden, 1835.
2. Berzelius, J.J. Quelques idees sur une nouvelle force agissant dans les combinaisons des corps organiques. *Ann. Chim.* **1836**, *61*, 146–151.
3. Ostwald, W. Über den Wärmewert der Bestandteile der Nahrungsmittel. *Z. Phys. Chem.* **1894**, *15*, 705–706.
4. Ostwald, W. Über Katalyse. *Ann. Nat.* **1910**, *9*, 1–25.
5. Steinborn, D. *Fundamentals of Organometallic Catalysis*; Wiley-VCH: Weinheim, Germany, 2012.
6. Wisniak, J. The History of Catalysis. From the Beginning to Nobel Prizes. *Educ. Chim.* **2010**, *21*, 60–69. [CrossRef]
7. Lindström, B.; Pettersson, L.J. A brief history of catalysis. *CATTECH* **2003**, *7*, 130–138. [CrossRef]
8. Laidler, K.J. The Development of Theories of Catalysis. *Arch. Hist. Exact Sci.* **1986**, *35*, 345–374. [CrossRef]
9. Laidler, K.J. A glossary of terms used in chemical kinetics, including reaction dynamics. *Pure Appl. Chem.* **1996**, *68*, 149–192. [CrossRef]
10. Kamer, P.C.J.; Vogt, D.; Thybaut, J. *Contemporary Catalysis: Science, Technology, and Applications*; Kamer, P.C.J., Vogt, D., Thybaut, J., Eds.; Royal Society of Chemistry: London, UK, 2017.
11. Hagen, J. *Industrial Catalysis: A Practical Approach*; Wiley-VCH Verlag GmbH & Co. KGaA: Weinheim, Germany, 2015.
12. Blaser, H.U. Looking Back on 35 Years of Industrial Catalysis. *CHIMIA Int. J. Chem.* **2015**, *69*, 393–406. [CrossRef]
13. Bartholomew, C.H.; Farrauto, R.J. *Fundamentals of Industrial Catalytic Processes*, 2nd ed.; John Wiley & Sons, Inc.: Hoboken, NJ, USA, 2005.
14. Cornils, B.; Herrmann, W.A.; Beller, M.; Paciello, R. (Eds.) *Applied Homogeneous Catalysis with Organometallic Compounds: A Comprehensive Handbook in Four Volumes*, 3rd ed.; Wiley-VCH Verlag GmbH & Co. KGaA: Weinheim, Germany, 2017.
15. Cornils, B.; Herrmann, W.A.; Zanthoff, H.; Wong, C.-H. (Eds.) *Catalysis from A to Z: A Concise Encyclopedia*, 4th ed.; John Wiley & Sons, Inc.: Hoboken, NJ, USA, 2013.
16. Slaugh, L.H.; Mullineaux, R.D. Novel Hydroformylation Catalysts. *J. Organomet. Chem.* **1968**, *13*, 469–477.

17. Osborn, J.A.; Jardine, F.H.; Young, J.F.; Wilkinson, G. The Preparation and Properties of Tris(triphenylphosphine)halogenorhodium(I) and Some Reactions thereof including Catalytic Homogeneous Hydrogenation of Olefins and Acetylenes and their Derivatives. *J. Chem. Soc. A* **1966**, 1711–1732. [CrossRef]
18. Scott, S.L. A Matter of Life(time) and Death. *ACS Catal.* **2018**, *8*, 8597–8599. [CrossRef]
19. Crabtree, R.H. Deactivation in Homogeneous Transition Metal Catalysis: Causes, Avoidance, and Cure. *Chem. Rev.* **2015**, *115*, 127–150. [CrossRef] [PubMed]
20. Argyle, M.D.; Bartholomew, C.H. Heterogeneous Catalyst Deactivation and Regeneration: A Review. *Catalysts* **2015**, *5*, 145–269. [CrossRef]
21. Van Leeuwen, P.W.N.M.; Chadwick, J.C. *Homogeneous Catalysts Activity-Stability-Deactivation*; Wiley-VCH: Weinheim, Germany, 2011
22. Heller, D.; de Vries, A.H.M.; de Vries, J.G. Catalyst Inhibition and Deactivation in Homogeneous Hydrogenation. In *Handbook of Homogeneous Hydrogenation*; de Vries, H.G., Elsevier, C., Eds.; Wiley-VCH Verlag GmbH & Co. KGaA: Weinheim, Germany, 2007; Chapter 44; pp. 1483–1516.
23. Meissner, A.; Alberico, E.; Drexler, H.-J.; Baumann, W.; Heller, D. Rhodium diphosphine complexes: a case study for catalyst activation and deactivation. *Catal. Sci. Technol.* **2014**, *4*, 3409–3425. [CrossRef]
24. Yuan, S.-W.; Han, H.; Li, Y.-L.; Wu, X.; Bao, X.; Gu, Z.-Y.; Xia, J.-B. Intermolecular C−H Amidation of (Hetero)arenes to Produce Amides through Rhodium-Catalyzed Carbonylation of Nitrene Intermediates. *Angew. Chem. Int. Ed.* **2019**, *131*, 8979–8984.
25. Oonishi, Y.; Masusaki, S.; Sakamoto, S.; Sato, Y. Rhodium(I)-Catalyzed Enantioselective Cyclization of Enynes by Intramolecular Cleavage of the Rh−C Bond by a Tethered Hydroxy Group. *Angew. Chem. Int. Ed.* **2019**, *131*, 8687.
26. Zheng, J.; Breit, B. Regiodivergent Hydroaminoalkylation of Alkynes and Allenes by a Combined Rhodium and Photoredox Catalytic System. *Angew. Chem. Int. Ed.* **2019**, *131*, 3430–3435. [CrossRef]
27. Ohmura, T.; Sasaki, I.; Suginome, M. Catalytic Generation of Rhodium Silylenoid for Alkene–Alkyne–Silylene [2 + 2 + 1] Cycloaddition. *Org. Lett.* **2019**, *21*, 1649–1653. [CrossRef]
28. Yang, X.-H.; Davison, R.T.; Dong, V.M. Catalytic Hydrothiolation: Regio- and Enantioselective Coupling of Thiols and Dienes. *J. Am. Chem. Soc.* **2018**, *140*, 10443–10446. [CrossRef]
29. Wu, S.T.; Luo, S.Y.; Guo, W.J.; Wang, T.; Xie, Q.X.; Wang, J.H.; Liu, G.Y. Direct Conversion of Ethyl Ketone to Alkyl Ketone via Chelation-Assisted Rhodium(I)-Catalyzed Carbon Carbon Bond Cleavage: Ligands Play an Important Role in the Inhibition of beta-Hydrogen Elimination. *Organometallics* **2018**, *37*, 2335–2341. [CrossRef]
30. Choi, K.; Park, H.; Lee, C. Rhodium-Catalyzed Tandem Addition–Cyclization–Rearrangement of Alkynylhydrazones with Organoboronic Acids. *J. Am. Chem. Soc.* **2018**, *140*, 10407–10411. [CrossRef] [PubMed]
31. Zheng, W.F.; Xu, Q.J.; Kang, Q. Rhodium/Lewis Acid Catalyzed Regioselective Addition of 1,3-Dicarbonyl Compounds to Internal Alkynes. *Organometallics* **2017**, *36*, 2323–2330. [CrossRef]
32. Yu, Y.; Xu, M. Chiral phosphorus-olefin ligands for asymmetric catalysis. *Huaxue Xuebao* **2017**, *75*, 655–670. [CrossRef]
33. Wang, H.W.; Lu, Y.; Zhang, B.; He, J.; Xu, H.J.; Kang, Y.S.; Sun, W.Y.; Yu, J.Q. Ligand-Promoted Rhodium(III)-Catalyzed ortho-C-H Amination with Free Amines. *Angew. Chem. Int. Ed.* **2017**, *56*, 7449–7453. [CrossRef] [PubMed]
34. Saito, H.; Nogi, K.; Yorimitsu, H. Rh/Cu-cocatalyzed Ring-opening Diborylation of Dibenzothiophenes for Aromatic Metamorphosis via Diborylbiaryls. *Chem. Lett.* **2017**, *46*, 1131–1134. [CrossRef]
35. Furusawa, T.; Tanimoto, H.; Nishiyama, Y.; Morimoto, T.; Kakiuchi, K. Rhodium-catalyzed Carbonylative Annulation of 2-Bromobenzylic Alcohols with Internal Alkynes Using Furfural via beta-Aryl Elimination. *Chem. Lett.* **2017**, *46*, 926–929. [CrossRef]
36. Loh, C.C.J.; Schmid, M.; Peters, B.; Fang, X.; Lautens, M. Benzylic Functionalization of Anthrones via the Asymmetric Ring Opening of Oxabicycles Utilizing a Fourth-Generation Rhodium Catalytic System. *Angew. Chem. Int. Ed.* **2016**, *55*, 4600–4604. [CrossRef] [PubMed]
37. Lee, T.; Hartwig, J.F. Rhodium-Catalyzed Enantioselective Silylation of Cyclopropyl C-H Bonds. *Angew. Chem. Int. Ed.* **2016**, *55*, 8723–8727. [CrossRef]

38. Meissner, A.; Preetz, A.; Drexler, H.-J.; Baumann, W.; Spannenberg, A.; Koenig, A.; Heller, D. In Situ Synthesis of Neutral Dinuclear Rhodium Diphosphine Complexes [{Rh(diphosphine)(μ_2-X)}$_2$]: Systematic Investigations. *Chem. Plus. Chem.* **2015**, *80*, 169–180.
39. Meissner, A.; Koenig, A.; Drexler, H.-J.; Thede, R.; Baumann, W.; Heller, D. New Pentacoordinated Rhodium Species as Unexpected Products during the In Situ Generation of Dimeric Diphosphine-Rhodium Neutral Catalysts. *Chem. Eur. J.* **2014**, *20*, 14721–14728. [CrossRef] [PubMed]
40. Preetz, A.; Kohrt, C.; Meissner, A.; Wei, S.; Buschmann, H.; Heller, D. Halide bridged trinuclear rhodium complexes and their inhibiting influence on catalysis. *Catal. Sci. Technol.* **2013**, *3*, 462–468. [CrossRef]
41. Lumbroso, A.; Vautravers, N.R.; Breit, B. Rhodium-Catalyzed Selective anti-Markovnikov Addition of Carboxylic Acids to Alkynes. *Org. Lett.* **2010**, *12*, 5498–5501. [CrossRef] [PubMed]
42. Koschker, P.; Lumbroso, A.; Breit, B. Enantioselective Synthesis of Branched Allylic Esters via Rhodium-Catalyzed Coupling of Allenes with Carboxylic Acids. *J. Am. Chem. Soc.* **2011**, *133*, 20746–20749. [CrossRef] [PubMed]
43. Lumbroso, A.; Koschker, P.; Vautravers, N.R.; Breit, B. Redox-Neutral Atom-Economic Rhodium-Catalyzed Coupling of Terminal Alkynes with Carboxylic Acids Toward Branched Allylic Esters. *J. Am. Chem. Soc.* **2011**, *133*, 2386–2389. [CrossRef] [PubMed]
44. Pritzius, A.B.; Breit, B. Asymmetric Rhodium-Catalyzed Addition of Thiols to Allenes: Synthesis of Branched Allylic Thioethers and Sulfones. *Angew. Chem. Int. Ed.* **2015**, *54*, 3121–3125. [CrossRef] [PubMed]
45. Beck, T.M.; Breit, B. Regio- and Enantioselective Rhodium-Catalyzed Addition of 1,3-Diketones to Allenes: Construction of Asymmetric Tertiary and Quaternary All Carbon Centers. *Angew. Chem. Int. Ed.* **2017**, *56*, 1903–1907. [CrossRef]
46. Parveen, S.; Li, C.K.; Hassan, A.; Breit, B. Chemo-, Regio-, and Enantioselective Rhodium-Catalyzed Allylation of Pyridazinones with Terminal Allenes. *Org. Lett.* **2017**, *19*, 2326–2329. [CrossRef] [PubMed]
47. Berthold, D.; Breit, B. Chemo-, Regio-, and Enantioselective Rhodium-Catalyzed Allylation of Triazoles with Internal Alkynes and Terminal Allenes. *Org. Lett.* **2018**, *20*, 598–601. [CrossRef]
48. Wei, S.; Pedroni, J.; Meissner, A.; Lumbroso, A.; Drexler, H.-J.; Heller, D.; Breit, B. Development of an Improved Rhodium Catalyst for Z-Selective Anti-Markovnikov Addition of Carboxylic Acids to Terminal Alkynes. *Chem. Eur. J.* **2013**, *19*, 12067–12076. [CrossRef]
49. Keller, E.; Pierrard, J.-S. (Eds.) *All Pictures Showing X-ray Structures Were Prepared Using the program SCHAKAL*; SCHAKAL99; University of Freiburg: Breisgau, Germany, 1999.
50. Möller, S.; Drexler, H.-J.; Heller, D. Two Precatalysts for Application in Propargylic CH Activation. *Acta Cryst. C* **2019**, submitted.
51. Moon, S.; Nishii, Y.; Miura, M. Thioether-Directed Peri-Selective C–H Arylation under Rhodium Catalysis: Synthesis of Arene-Fused Thioxanthenes. *Org. Lett.* **2019**, *21*, 233–236. [CrossRef] [PubMed]
52. Satake, S.; Kurihara, T.; Nishikawa, K.; Mochizuki, T.; Hatano, M.; Ishihara, K.; Yoshino, T.; Matsunaga, S. Pentamethylcyclopentadienyl Rhodium(III)–Chiral Disulfonate Hybrid Catalysis for Enantioselective C–H Bond Functionalization. *Nat. Catal.* **2018**, *1*, 585–591. [CrossRef]
53. Ghosh, K.; Mihara, G.; Nishii, Y.; Miura, M. Nondirected C-H Alkenylation of Arenes with Alkenes under Rhodium Catalysis. *Chem. Lett.* **2019**, *48*, 148–151. [CrossRef]
54. Fischer, C.; Koenig, A.; Meissner, A.; Thede, R.; Selle, C.; Pribbenow, C.; Heller, D. Quantitative UV/Vis Spectroscopic Investigations of the In Situ Synthesis of Neutral μ_2-Chloro-Bridged Dinuclear (Diphosphine)rhodium Complexes. *Eur. J. Inorg. Chem.* **2014**, *34*, 5849–5855. [CrossRef]
55. Duan, C.-L.; Tan, Y.-X.; Zhang, J.-L.; Yang, S.; Dong, H.-Q.; Tian, P.; Lin, G.-Q. Highly Enantioselective Rhodium-Catalyzed Cross-Addition of Silylacetylenes to Cyclohexadienone-Tethered Internal Alkynes. *Org. Lett.* **2019**, *21*, 1690–1693. [CrossRef]
56. Hilpert, L.J.; Sieger, S.V.; Haydl, A.M.; Breit, B. Palladium- and Rhodium-Catalyzed Dynamic Kinetic Resolution of Racemic Internal Allenes Towards Chiral Pyrazoles. *Angew. Chem. Int. Ed.* **2019**, *131*, 3416–3419. [CrossRef]
57. Tanaka, K. *Rhodium Catalysis in Organic Synthesis: Methods and Reactions*; Wiley-VCerlag GmbH & Co. KGaA: Weinheim, Germany, 2018.
58. Fairlie, D.P.; Bosnich, B. Homogeneous Catalysis - Conversion of 4-Pentenals to Cyclopentanones by Efficient Rhodium-Catalyzed Hydroacylation. *Organometallics* **1988**, *7*, 936–945. [CrossRef]

59. Fischer, C.; Beweries, T.; Preetz, A.; Drexler, H.-J.; Baumann, W.; Peitz, S.; Rosenthal, U.; Heller, D. Kinetic and Mechanistic Investigations in Homogeneous Catalysis Using Operando UV/vis Spectroscopy. *Catal. Today* **2010**, *155*, 282–288. [CrossRef]
60. Mannu, A.; Drexler, H.-J.; Thede, R.; Ferro, M.; Baumann, W.; Rüger, J.; Heller, D. Oxidative Addition of CH_2Cl_2 to Neutral Dimeric Rhodium Diphosphine Complexes. *J. Organomet. Chem.* **2018**, *871*, 178–184. [CrossRef]
61. Möller, S.; Drexler, H.-J.; Kubis, C.; Alberico, E.; Heller, D. Investigations into the Mechanism of the In Situ Formation of Neutral Dinuclear Rhodium Complexes. *J. Organomet. Chem.* **2019**. submitted.
62. Time for 98 % conversion under the following reaction conditions: Ca. $1.0 \cdot 10^{-2}$ mmol [Rh(diolefin)(μ_2-Cl)]$_2$ and $2.0 \cdot 10^{-2}$ mmol ligand in 5 ml solvent.
63. Lumbroso, A.; Cooke, M.L.; Breit, B. Catalytic Asymmetric Synthesis of Allylic Alcohols and Derivatives and their Applications in Organic Synthesis. *Angew. Chem. Int. Ed.* **2013**, *52*, 1890–1932. [CrossRef] [PubMed]
64. Gellrich, U.; Meissner, A.; Steffani, A.; Kahny, M.; Drexler, H.J.; Heller, D.; Plattner, D.A.; Breit, B. Mechanistic Investigations of the Rhodium Catalyzed Propargylic CH Activation. *J. Am. Chem. Soc.* **2014**, *136*, 1097–1104. [CrossRef] [PubMed]
65. Li, C.; Breit, B. Rhodium-Catalyzed Chemo- and Regioselective Decarboxylative Addition of β-Ketoacids to Allenes: Efficient Construction of Tertiary and Quaternary Carbon Centers. *J. Am. Chem. Soc.* **2014**, *136*, 862–865. [CrossRef] [PubMed]
66. Koschker, P.; Kaehny, M.; Breit, B. Enantioselective Redox-Neutral Rh-Catalyzed Coupling of Terminal Alkynes with Carboxylic Acids Toward Branched Allylic Esters. *J. Am. Chem. Soc.* **2015**, *137*, 3131–3137. [CrossRef] [PubMed]
67. James, B.R.; Mahajan, D. Bis(ditertiaryphosphine) Complexes of Rhodium(I). Synthesis, Spectroscopy, and Activity for Catalytic Hydrogenation. *Can. J. Chem.* **1979**, *57*, 180–187. [CrossRef]
68. Castellanos-Paez, A.; Thayaparan, J.; Castillon, S.; Claver, C. Reactivity of Tetracarbonyl Dithiolate-Bridged Rhodium(I) Complexes with Diphosphines. *J. Organomet. Chem.* **1998**, *551*, 375–381. [CrossRef]
69. Slack, D.A.; Baird, M.C. Investigations of Olefin Hydrogenation Catalysts. The Major Species Present in Solutions Containing Rhodium(I) Complexes of Chelating Diphosphines. *J. Organomet. Chem.* **1977**, *142*, C69–C72. [CrossRef]
70. Crosman, A.; Hoelderich, W.F. Enantioselective Hydrogenation over Immobilized Rhodium Diphosphine Complexes on Aluminated SBA-15. *J. Catal.* **2005**, *232*, 43–50. [CrossRef]
71. Van Haaren, R.J.; Zuidema, E.; Fraanje, J.; Goubitz, K.; Kamer, P.C.J.; van Leeuwen, P.W.N.M.; van Strijdonck, G.P.F. Probing the Mechanism of Rhodium (I) Catalyzed Dehydrocoupling of di-n-hexylsilane. *C. R. Chim.* **2002**, *5*, 431–440. [CrossRef]
72. Wagner, H.H.; Hausmann, H.; Hoelderich, W.F. Immobilization of Rhodium Diphosphine Complexes on Mesoporous Al-MCM-41 Materials: Catalysts for Enantioselective Hydrogenation. *J. Catal.* **2001**, *203*, 150–156. [CrossRef]
73. Sinou, D.; Bakos, J. (S,S)-2,3-Bis[Di(m-Sodiumsulfonatophenyl)-Phosphino]Butane (Chiraphosts) and (S,S)-2,4-Bis[Di(m-Sodiumsulfonatophenyl)- Phosphino]Pentane (BDPPts). *Inorg. Synth.* **1998**, *32*, 36–40.
74. Baxley, G.T.; Weakley, T.J.R.; Miller, W.K.; Lyon, D.K.; Tyler, D.R. Synthesis and Catalytic Chemistry of Two New Water-soluble Chelating Phosphines. Comparison of Ionic and Nonionic Functionalities. *J. Mol. Catal. A Chem.* **1997**, *116*, 191–198. [CrossRef]
75. Bartik, T.; Bunn, B.B.; Bartik, B.; Hanson, B.E. Synthesis, Reactions, and Catalytic Chemistry of the Water-Soluble Chelating Phosphine l,2-Bis[bis(m-sodiosulfonatophenyl)phosphino]ethane (DPPETS). Complexes with Nickel, Palladium, Platinum, and Rhodium. *Inorg. Chem.* **1994**, *33*, 164–169. [CrossRef]
76. Bakos, J.; Tóth, I.; Heil, B.; Szalontai, G.; Párkányi, L.; Fülöp, V. Catalytic and Structural Studies of Rh^I Complexes of (−)-(2S,4S)-2,4-bis(diphenylphosphino)pentane. Asymmetric Hydrogenation of Acetophenonebenzylimine and Acetophenone. *J. Organomet. Chem.* **1989**, *370*, 263–276. [CrossRef]
77. Meng, G.; Szostak, M. Rhodium-Catalyzed C–H Bond Functionalization with Amides by Double C–H/C–N Bond Activation. *Org. Lett.* **2016**, *18*, 796–799. [CrossRef] [PubMed]
78. Mizukami, A.; Ise, Y.; Kimachi, T.; Inamoto, K. Rhodium-Catalyzed Cyclization of 2-Ethynylanilines in the Presence of Isocyanates: Approach toward Indole-3-carboxamides. *Org. Lett.* **2016**, *18*, 748–751. [CrossRef] [PubMed]

79. Zhao, C.; Liu, L.-C.; Wang, J.; Jiang, C.; Zhang, Q.-W.; He, W. Rh(I)-Catalyzed Insertion of Allenes into C–C Bonds of Benzocyclobutenols. *Org. Lett.* **2016**, *18*, 328–331. [CrossRef]
80. Chen, D.; Zhang, X.; Qi, W.-Y.; Xu, B.; Xu, M.-H. Rhodium(I)-Catalyzed Asymmetric Carbene Insertion into B–H Bonds: Highly Enantioselective Access to Functionalized Organoboranes. *J. Am. Chem. Soc.* **2015**, *137*, 5268–5271. [CrossRef]
81. Lim, D.S.W.; Lew, T.T.S.; Zhang, Y. Direct Amidation of N-Boc- and N-Cbz-Protected Amines via Rhodium-Catalyzed Coupling of Arylboroxines and Carbamates. *Org. Lett.* **2015**, *17*, 6054–6057. [CrossRef]
82. Gopula, B.; Yang, S.-H.; Kuo, T.-S.; Hsieh, J.-C.; Wu, P.-Y.; Henschke, J.P.; Wu, H.-L. Direct Synthesis of Chiral 3-Arylsuccinimides by Rhodium-Catalyzed Enantioselective Conjugate Addition of Arylboronic Acids to Maleimides. *Chem. Eur. J.* **2015**, *21*, 11050–11055. [CrossRef]
83. Lee, S.; Lee, W.L.; Yun, J. Rhodium-Catalyzed Addition of Alkyltrifluoroborate Salts to Imines. *Adv. Synth. Catal.* **2015**, *357*, 2219–2222. [CrossRef]
84. Rao, H.; Yang, L.; Shuai, Q.; Li, C.-J. Rhodium-Catalyzed Aerobic Coupling between Aldehydes and Arenesulfinic Acid Salts: A Novel Synthesis of Aryl Ketones. *Adv. Synth. Catal.* **2011**, *353*, 1701–1706. [CrossRef]
85. Morimoto, T.; Yamasaki, K.; Hirano, A.; Tsutsumi, K.; Kagawa, N.; Kakiuchi, K.; Harada, Y.; Fukumoto, Y.; Chatani, N.; Nishioka, T. Rh(I)-Catalyzed CO Gas-Free Carbonylative Cyclization Reactions of Alkynes with 2-Bromophenylboronic Acids Using Formaldehyde. *Org. Lett.* **2009**, *11*, 1777–1780. [CrossRef] [PubMed]
86. Cornils, B.; Herrmann, W.A.; Schlögl, R.; Wong, C.-H. (Eds.) *Catalysis from A to Z*, 2nd ed.; Wiley-VCH: Weinheim, Germany, 2003.
87. Temkin, O.N. Mechanisms of Formation of Catalytically Active Metal Complexes. In *Homogeneous Catalysis with Metal Complexes: Kinetic Aspects and Mechanisms*, 1st ed.; Temkin, O.N., Ed.; John Wiley & Sons Ltd.: Chichester, UK, 2012; Chapter 5, pp. 453–544.
88. Evans, D.; Osborn, J.A.; Wilkinson, G. Hydroformylation of Alkenes by Use of Rhodium Complex Catalysts. *J. Chem. Soc. A* **1968**, 3133–3142. [CrossRef]
89. Brown, C.K.; Wilkinson, G. Homogeneous Hydroformylation of Alkenes with Hydridocarbonyltris-(triphenylphosphine)rhodium(i) as Catalyst. *J. Chem. Soc. A* **1970**, 2753–2764. [CrossRef]
90. Adams, G.M.; Ryan, D.E.; Beattie, N.A.; McKay, A.I.; Lloyd-Jones, G.C.; Weller, A.S. Dehydropolymerization of H3B·NMeH2 Using a [Rh(DPEphos)]+ Catalyst: The Promoting Effect of NMeH2. *ACS Catal.* **2019**, *9*, 3657–3666. [CrossRef] [PubMed]
91. Fischer, C.; Schulz, S.; Drexler, H.-J.; Selle, C.; Lotz, M.; Sawall, M.; Neymeyr, K.; Heller, D. The Influence of Substituents in Diphosphine Ligands on the Hydrogenation Activity and Selectivity of the Corresponding Rhodium Complexes as Exemplified by ButiPhane. *ChemCatChem* **2012**, *4*, 81–88. [CrossRef]
92. 0.01 mmol rhodium complex and 1.0 mmol prochiral olefin in 15.0 ml MeOH at 25.0 °C and 1 bar total pressure
93. Nagel, U.; Kinzel, E.; Andrade, J.; Prescher, G. Enantioselektive Katalyse, 4. Synthese N-substituierter (R,R)-3,4-Bis(diphenylphosphino)-pyrrolidine und Anwendung ihrer Rhodiumkomplexe zur Asymmetrischen Hydrierung von α-(Acylamino)acrylsäure-Derivaten. *Chem. Ber.* **1986**, *119*, 3326–3343. [CrossRef]
94. Nagel, U.; Krink, T. Catalytic Hydrogenation with Rhodium Complexes Containing dipamp-pyrphos Hybrid Ligands. *Angew. Chem. Int. Ed. Engl.* **1993**, *32*, 1052–1054. [CrossRef]
95. Thiel, I.; Horstmann, M.; Jungk, P.; Keller, S.; Fischer, F.; Drexler, H.-J.; Heller, D.; Hapke, M. Insight into the Activation of *In Situ*-Generated Chiral Rh(I)-Catalysts and their Application in Cyclotrimerizations. *Chem. Eur. J.* **2017**, *23*, 17048–17057. [CrossRef]
96. Alberico, E.; Baumann, W.; de Vries, J.G.; Drexler, H.-J.; Gladiali, S.; Heller, D.; Henderickx, H.J.W.; Lefort, L. Unravelling the Reaction Path of Rhodium–MonoPhos-Catalysed Olefin Hydrogenation. *Chem. Eur. J.* **2011**, *17*, 12683–12695. [CrossRef] [PubMed]
97. Van den Berg, M. Rhodium-Catalyzed Asymmetric Hydrogenation Using Phosphoramidite Ligands. Ph.D. Thesis, University of Groningen, Groningen, The Netherlands, 2006. Available online: http://hdl.handle.net/11370/c6d065b0-5bc1-4a2a-a452-fb237d5a4e6e (accessed on 5 June 2019).
98. Esteruelas, M.A.; Herrero, J.; Martin, M.; Oro, M.L.A.; Real, V.M. Mechanism of the Hydrogenation of 2,5-Norbornadiene Catalyzed by $[Rh(NBD)(PPh_3)_2]BF_4$ in Dichloromethane: a Kinetic and Spectroscopic Investigation. *J. Organomet. Chem.* **2000**, *599*, 178–184. [CrossRef]

99. Heller, D.; Kortus, K.; Selke, R. Kinetische Untersuchungen zur Ligandenhydrierung in Katalysatorvorstufen für die Asymmetrische Reduktion prochiraler Olefine. *Liebigs Ann.* **1995**, 575–581. [CrossRef]
100. Braun, W.; Salzer, A.; Drexler, H.-J.; Spannenberg, A.; Heller, D. Investigations into the Hydrogenation of Diolefins and Prochiral Olefins Employing the "Daniphos"-type Ligands. *Dalton Trans.* **2003**, 1606–1613. [CrossRef]
101. Preetz, A.; Drexler, H.-J.; Fischer, C.; Dai, Z.; Börner, A.; Baumann, W.; Spannenberg, A.; Thede, R.; Heller, D. Rhodium Complex Catalyzed Asymmetric Hydrogenation - Transfer of Pre-Catalysts Into Active Species. *Chem. Eur. J.* **2008**, *14*, 1445–1451. [CrossRef]
102. Selent, D.; Heller, D. *In-Situ Techniques for Homogeneous Catalysis, in Catalysis From Principle to Application;* Beller, M., Renken, A., van Santen, R., Eds.; Wiley-VCH: Weinheim, Germany, 2012; Chapter 23; pp. 465–492.
103. Drexler, H.-J.; Preetz, A.; Schmidt, T.; Heller, D. Kinetics of Homogeneous Hydrogennations: Measurement and Interpretation. In *Handbook of Homogeneous Hydrogenation;* de Vries, H.G., Elsevier, C., Eds., Wiley-VCH: Weinheim, Germany, 2007; Chapter 10; pp. 257–293.
104. Heller, D.; Borns, S.; Baumann, W.; Selke, R. Kinetic Investigations of the Hydrogenation of Diolefin Ligands in Catalyst Precursors for the Asymmetric Reduction of Prochiral Olefins, II. *Chem. Ber.* **1996**, *129*, 85–89. [CrossRef]
105. Drexler, H.-J.; Baumann, W.; Spannenberg, A.; Fischer, C.; Heller, D. COD- versus NBD-Precatalysts. Dramatic Difference in the Asymmetric Hydrogenation of Prochiral Olefins with Five Membered Diphosphine Rh-Hydrogenation Catalysts. *J. Organomet. Chem.* **2001**, *621*, 89–102. [CrossRef]
106. Baseda Krüger, M.; Selle, C.; Heller, D.; Baumann, W. Determination of Gas Concentrations in Liquids by Nuclear Magnetic Resonance: Hydrogen in Organic Solvents. *J. Chem. Eng. Data* **2012**, *57*, 1737–1744. [CrossRef]
107. Krüger, M.B. Bestimmung von Gaskonzentrationen in Flüssigen Medien Mittels NMR-Spektroskopie: Eine Methode für Kinetische und in-situ Studien. Ph.D. Thesis, University of Rostock, Rostock, Germany, 2013.
108. Greiner, L.; Ternbach, M.B. Kinetic Study of Homogeneous Alkene Hydrogenation by Model Discrimination. *Adv. Synth. Catal.* **2004**, *346*, 1392–1396. [CrossRef]
109. Drexler, H.-J.; Zhang, S.; Sun, A.; Spannenberg, A.; Arrieta, A.; Preetz, A.; Heller, D. Cationic Rh-Bisphosphane-Diolefin Complexes as Precatalysts for Enantioselective Catalysis - What Informations Do Single Crystal Structures Contain Regarding Product Chirality? *Tetrahedron Asymmetry* **2004**, *15*, 2139–2150. [CrossRef]
110. Brown, J.M.; Chaloner, P.A. The Mechanism of Asymmetric Hydrogenation Catalysed by Rhodium (I) Dipamp Complexes. *Tetrahedron Lett.* **1978**, *19*, 1877–1880. [CrossRef]
111. Brown, J.M.; Chaloner, P.A. The Mechanism of Asymmetric Homogeneous Hydrogenation. Rhodium (I) Complexes of Dehydroamino Acids Containing Asymmetric Ligands Related to Bis(1,2-diphenylphosphino)ethane. *J. Am. Chem. Soc.* **1980**, *102*, 3040–3048. [CrossRef]
112. De Vries, J.G.; Lefort, L. High-Throughput Experimentation and Ligand Libraries. In *Handbook of Homogeneous Hydrogenation;* de Vries, H.G., Elsevier, C., Eds.; Wiley-VCH: Weinheim, Germany, 2007; Chapter 36, pp. 1245–1278.
113. Preetz, A.; Drexler, H.-J.; Schulz, S.; Heller, D. BINAP: Rhodium-Diolefin Complexes in Asymmetric Hydrogenation. *Tetrahedron Asymm.* **2010**, *21*, 1226–1231. [CrossRef]
114. Cobley, C.J.; Lennon, I.C.; McCague, R.; Ramsden, J.A.; Zanotti-Gerosa, A. On the Economic Application of DuPHOS Rhodium(I) Catalysts: A Comparison of COD versus NBD Precatalysts. *Tetrahedron Lett.* **2001**, *42*, 7481–7483. [CrossRef]
115. Moxham, G.L.; Randell-Sly, H.E.; Brayshaw, S.K.; Woodward, R.L.; Weller, A.S.; Willis, M.C. A Second-Generation Catalyst for Intermolecular Hydroacylation of Alkenes and Alkynes Using β-S-Substituted Aldehydes: The Role of a Hemilabile P-O-P Ligand. *Angew. Chem. Int. Ed.* **2006**, *45*, 7618–7622. [CrossRef] [PubMed]
116. Preetz, A.; Fischer, C.; Kohrt, C.; Drexler, H.-J.; Baumann, W.; Heller, D. Cationic Rhodium-BINAP Complexes: Full Characterization of Solvate- and Arene Bridged Dimeric Species. *Organometallics* **2011**, *30*, 5155–5159. [CrossRef]
117. Meißner, A.; Drexler, H.-J.; Keller, S.; Selle, C.; Ratovelomanana-Vidal, V.; Heller, D. Synthesis and Characterisation of Cationic Synphos-Rhodium Complexes. *Eur. J. Inorg. Chem.* **2014**, 4836–4842. [CrossRef]

118. Webster, R.; Bçing, C.; Lautens, M. Reagent-Controlled Regiodivergent Resolution of Unsymmetrical Oxabicyclic Alkenes Using a Cationic Rhodium Catalyst. *J. Am. Chem. Soc.* **2009**, *131*, 444–445. [CrossRef]
119. Preetz, A.; Kohrt, C.; Drexler, H.-J.; Torrens, A.; Buschmann, H.; Lopez, M.G.; Heller, D. Asymmetric Ring Opening of Oxabicyclic Alkenes With *Cationic* Rhodium Complexes. *Adv. Synth. Catal.* **2010**, *352*, 2073–2080. [CrossRef]
120. Fischer, C.; Kohrt, C.; Drexler, H.-J.; Baumann, W.; Heller, D. Trinuclear Hydride Complexes of Rhodium. *Dalton Trans.* **2011**, *40*, 4162–4166. [CrossRef]
121. Kohrt, C.; Hansen, S.; Drexler, H.-J.; Rosenthal, U.; Schulz, A.; Heller, D. Molecular Vibration Spectroscopy Studies on Novel Trinuclear Rhodium-7-Hydride Complexes of the General Type $\{[Rh(PP^*)X]_3(\mu_2\text{-}X)_3(\mu_3\text{-}X)\}(BF_4)_2$ (X = H, D). *Inorg. Chem.* **2012**, *51*, 7377–7383. [CrossRef] [PubMed]
122. Kohrt, C.; Baumann, W.; Spannenberg, A.; Drexler, H.-J.; Gridnev, I.; Heller, D. Formation of Trinuclear Rhodium-Hydride Complexes $\{[Rh(PP^*)H]_3(\mu_2\text{-}H)_3(\mu_3\text{-}H)\}(Anion)_2$ - During Asymmetric Hydrogenation? *Chem. Eur. J.* **2013**, *19*, 7443–7451. [CrossRef] [PubMed]
123. Kohrt, C. Mehrkernige Hydridkomplexe des Rhodiums mit Bisphosphanliganden: Charakterisierung, Bildung und Anwendung. Ph.D. Thesis, University of Rostock, Rostock, Germany, 2013.
124. Gridnev, I.D.; Imamoto, T. Mechanism of Enantioselection in Rh-Catalyzed Asymmetric Hydrogenation. The origin of Utmost Catalytic performance. *Chem. Comm.* **2009**, 7447–7464. [CrossRef] [PubMed]
125. Indeed a mechanism for the formation of the trinuclear rhodium (III) polyhydride species $\{[Rh(diphosphine)H]_3(\mu_2\text{-}H)_3(\mu_3\text{-}H)\}^{2+}$ from the rhodium(I) solvate dihydride $[Rh(III)(diphosphine)(H)_2(solvent)_2]^+$ has been put forward, based on experimental evidence, which includes the intermediacy of a dinuclear rhodium (III) polyhydride species of general formula $\{[Rh(diphosphine)H]_2(\mu_2\text{-}H)_3\}[X]$. The equilibria, the species present in solution and their relative concentrations, depends on the properties of the diphosphine ligand, the temperature and the solvent.
126. The hydrogenation of diolefins in complexes with diphosphines which do not contain aryl groups in nonpolar solvents such as DCM is even more complicated: With the ligand *t*-BuBisP* $^1H\text{-}^{103}Rh$-HMQC-NMR spectra show the formation of three polyhydrido species. One of them has been assigned the structure $\{[Rh(t\text{-}Bu\text{-}BisP^*)H]_3(\mu_2\text{-}H)_3(\mu_3\text{-}Cl)\}_2{}^+$, while that of the other two remain unclear.
127. Cornish-Bowden, A. *Fundamentals of Enzyme Kinetics*, 4th ed.; Wiley-VCH Verlag & Co. KgaA: Weinheim, Germany, 2012.
128. Kollár, L.; Törös, S.; Heil, B.; Markó, L. Phosphinerhodium Complexes as Homogeneous Catalysts: XI. Decarbonylation of Primary Alcohols Used as Solvents Under Conditions of Olefin Hydrogenation; a Side Reaction Leading to Catalyst Deactivation. *J. Organomet. Chem.* **1980**, *192*, 253–256.
129. Benn, R.; Rufińska, A. High-Resolution Metal-NMR Spectroscopy of Organometallic Compounds. *Angew. Chem. Int. Ed. Engl.* **1986**, *25*, 861–881. [CrossRef]
130. Von Philipsborn, W. Probing Organometallic Structure and Reactivity by Transition Metal NMR Spectroscopy. *Chem. Soc. Rev.* **1999**, *28*, 95–105. [CrossRef]
131. Ernsting, J.M.; Elsevier, C.J.; de Lange, W.G.J.; Timmer, K. Inverse Two-Dimensional ^{31}P, $^{103}Rh\{^1H\}$ NMR of Cationic Rhodium (1) Complexes Containing Chelating Diphosphines. *Magn. Reson. Chem.* **1991**, *29*, S118–S124. [CrossRef]
132. Leitner, W.; Bühl, M.; Fornika, R.; Six, C.; Baumann, W.; Dinjus, E.; Kessler, M.; Krüger, C.; Rufińska, A. ^{103}Rh Chemical Shifts in Complexes Bearing Chelating Bidentate Phosphine Ligands. *Organometallics* **1999**, *18*, 1196–1206. [CrossRef]
133. Halpern, J.; Riley, D.P.; Chan, A.S.C.; Pluth, J.J. Novel Coordination Chemistry and Catalytic Properties of Cationic 1,2-Bisjdiphenylphosphino)ethanerhodium(I) Complexes. *J. Am. Chem. Soc.* **1977**, *99*, 8055–8057. [CrossRef]
134. Halpern, J.; Chan, A.S.C.; Riley, D.P.; Pluth, J.J. Some Aspects of the Coordination Chemistry and Catalytic Properties of Cationic Rhodium-Phosphine Complexes. *Adv. Chem. Ser.* **1979**, *173*, 16–25.
135. Landis, C.R.; Halpern, J. Homogeneous Catalysis of Arene Hydrogenation by Cationic Rhodium Arene Complexes. *Organometallics* **1983**, *2*, 840–842. [CrossRef]
136. Fischer, C.; Thede, R.; Drexler, H.-J.; König, A.; Baumann, W.; Heller, D. Investigations into the Formation and Stability of Cationic Rhodium Diphosphane η6-Arene Complexes. *Chem. Eur. J.* **2012**, *18*, 11920–11928. [CrossRef] [PubMed]

137. Polster, J.; Lachmann, H. *Spectrometric Titrations: Analysis of Chemical Equilibria*; Wiley-VCH: Weinheim, Germany, 1989.
138. Benesi, H.A.; Hildebrand, J.H. A Spectrophotometric Investigation of the Interaction of Iodine with Aromatic Hydrocarbons. *J. Am. Chem. Soc.* **1949**, *71*, 2703–2707. [CrossRef]
139. Scott, R.L. Some Comments on the Benesi-Hildebrand Equation. *Rec. Trav. Chim.* **1956**, *75*, 787–789. [CrossRef]
140. Scatchard, G. The Attractions of Proteins for Small Molecules and Ions. *Ann. N. Y. Acad. Sci.* **1949**, *51*, 660–672. [CrossRef]
141. Gampp, H.; Maeder, M.; Meyer, C.J.; Zuberbühler, A.D. Calculation of Equilibrium Constants from Multiwavelength Spectroscopic Data—I: Mathematical Considerations. *Talanta* **1985**, *32*, 95–101. [CrossRef]
142. Gampp, H.; Maeder, M.; Meyer, C.J.; Zuberbühler, A.D. Calculation of Equilibrium Constants from Multiwavelength Spectroscopic data—II132, 95.: Specfit: Two User-Friendly Programs in Basic and Standard Fortran 77. *Talanta* **1985**, *32*, 257–264. [CrossRef]
143. Gampp, H.; Maeder, M.; Meyer, C.J.; Zuberbühler, A.D. Calculation of Equilibrium Constants From Multiwavelength Spectroscopic Data—III: Model-Free Analysis of Spectrophotometric and ESR Titrations. *Talanta* **1985**, *32*, 1133–1139. [CrossRef]
144. Gampp, H.; Maeder, M.; Meyer, C.J.; Zuberbühler, A.D. Calculation of Equilibrium Constants from Multiwavelength Spectroscopic Data—IV: Model-Free Least-Squares Refinement by Use of Evolving Factor Analysis. *Talanta* **1986**, *33*, 943–951. [CrossRef]
145. ReactLAB^{TM} Equilibria. Available online: http://jplusconsulting.com/products/reactlab-equilibria/ (accessed on 5 June 2019).
146. Fischer, C. UV-vis-Spektroskopie in der homogenen Katalyse – Komplexchemische Untersuchungen an Rhodium-Katalysatoren. Ph.D. Thesis, University of Rostock, Rostock, Germany, 2010.
147. Gridnev, I.D.; Higashi, N.; Asakura, K.; Imamoto, T. Mechanism of Asymmetric Hydrogenation Catalyzed by a Rhodium Complex of (S,S)-1,2-Bis(tert-butylmethylphosphino)ethane. Dihydride Mechanism of Asymmetric Hydrogenation. *J. Am. Chem. Soc.* **2000**, *122*, 7183–7194. [CrossRef]
148. Gridnev, I.D.; Yasutake, M.; Higashi, N.; Imamoto, T. Asymmetric Hydrogenation of Enamides with Rh-BisP* and Rh-MiniPHOS Catalysts. Scope, Limitations, and Mechanism. *J. Am. Chem. Soc.* **2001**, *123*, 5268–5276. [CrossRef] [PubMed]
149. Heller, D.; Drexler, H.-J.; Spannenberg, A.; Heller, B.; You, J.; Baumann, W. The Inhibiting Influence of Aromatic Solvents on the Activity of Asymmetric Hydrogenations. *Angew. Chem. Int. Ed.* **2002**, *41*, 777–780. [CrossRef]
150. Sablong, R.; van der Vlugt, J.I.; Thormann, R.; Mecking, S.; Vogt, D. Disperse Amphiphilic Submicron Particles as Non-Covalent Supports for Cationic Homogeneous Catalysts. *Adv. Synth. Catal.* **2005**, *347*, 633–636. [CrossRef]
151. Balué, J.; Bayón, J.C. Hydroformylation of Styrene Catalyzed by a Rhodium Thiolate Binuclear Catalyst Supported on a Cationic Exchange Resin. *J. Mol. Catal. A* **1999**, *137*, 193–203. [CrossRef]
152. Šebesta, R. (Ed.) *Enantioselective Homogeneous Supported Catalysis*; RSC: Cambridge, UK, 2012.
153. Pugin, B.; Blaser, H.-U. The Immobilization of Rhodium-4-(diphenylphosphino)-2-(diphenylphosphinomethyl)-pyrrolidine (Rh-PPM) Complexes: A Systematic Study. *Adv. Synth. Catal.* **2006**, *348*, 1743–1751. [CrossRef]
154. Leadbeater, N.E.; Marco, M. Preparation of Polymer-Supported Ligands and Metal Complexes for Use in Catalysis. *Chem. Rev.* **2002**, *102*, 3217–3274. [CrossRef] [PubMed]
155. Chapuis, C.; Barthe, M.; de Saint Laumer, J.-Y. Synthesis of Citronellal by RhI-Catalysed Asymmetric Isomerization of N,N-Diethyl-Substituted Geranyl- and Nerylamines or Geraniol and Nerol in the Presence of Chiral Diphosphino Ligands, under Homogeneous and Supported Conditions. *Helv. Chim. Acta* **2001**, *84*, 230–242. [CrossRef]
156. Bayston, D.J.; Fraser, J.L.; Ashton, M.R.; Baxter, A.D.; Polywka, M.E.C.; Moses, E. Preparation and Use of a Polymer Supported BINAP Hydrogenation Catalyst. *J. Org. Chem.* **1998**, *63*, 3137–3140. [CrossRef]
157. Renaud, E.; Baird, M.C. Effects of Catalyst Site Accessibility on Catalysis by Rhodium(I) Complexes of Amphiphilic Ligands $[Ph_2P(CH_2)nPMe_3]^+$ (n = 2, 3, 6 or 10) tethered to a cation-exchange resin. *J. Chem. Soc. Dalton Trans.* **1992**, 2905–2906. [CrossRef]

158. Kalck, P.; de Oliveira, E.L.; Queau, R.; Peyrille, B.; Molinier, J. Dinuclear Rhodium Complexes Immobilized on Functionalized Diphenylphosphino-(Styrene-Divinylbenzene) Resins Giving High Selectivities for Linear Aldehydes in Hydroformylation reactions. *J. Organomet. Chem.* **1992**, *433*, C4–C8. [CrossRef]
159. Fischer, C.; Thede, R.; Baumann, W.; Drexler, H.-J.; König, A.; Heller, D. Investigations into Metal Leaching from Polystyrene-Supported Rhodium Catalysts. *ChemCatChem* **2016**, *8*, 352–356. [CrossRef]
160. Schwarze, M.; Milano-Brusco, J.-S.; Strempel, V.; Hamerla, T.; Wille, S.; Fischer, C.; Baumann, W.; Arlt, W.; Schomäcker, R. Rhodium Catalyzed Hydrogenation Reactions in Aqueous Micellar Systems as Green Solvents. *RSC Adv.* **2011**, *1*, 474–483.
161. Mikami, K.; Kataoka, S.; Yusa, Y.; Aikawa, K. Racemic but Tropos (Chirally Flexible) BIPHEP Ligands for Rh(I)-Complexes: Highly Enantioselective Ene-Type Cyclization of 1,6-Enynes. *Org. Lett.* **2004**, *6*, 3699–3701. [CrossRef] [PubMed]
162. Mikami, K.; Yusa, Y.; Hatano, M.; Wakabayashi, K.; Aikawa, K. Highly Enantioselective Spiro Cyclization of 1,6-Enynes Catalyzed by Cationic Skewphos Rhodium(I) Complex. *Chem. Commun.* **2004**, 98–99. [CrossRef] [PubMed]
163. Barnhart, R.W.; Wang, X.; Noheda, P.; Bergens, S.H.; Whelan, J.; Bosnich, B. Asymmetric Catalysis. Asymmetric Catalytic Intramolocular Hydrosilation and Hydroacylation. *Tetrahedron* **1994**, *50*, 4335–4346. [CrossRef]
164. Allen, D.G.; Wild, S.B.; Wood, D.L. Catalytic Asymmetric Hydrogenation of Prochiral Enamides by Rhodium(I) Complexes Containing the Enantiomers of (R*,R*)-(.+-.)-1,2-Phenylenebis(methylphenyl-phosphine) and its Arsenic Isosteres. *Organometallics* **1986**, *5*, 1009–1015. [CrossRef]
165. Miyashita, A.; Takaya, H.; Souchi, T.; Noyori, R. 2, 2′-Bis(diphenylphosphino)-1, 1′-binaphthyl(binap): A New Atropisomeric Bis(triaryl)phosphine. Synthesis and its Use in the Rh(l)-Catalyzed Asymmetric Hydrogenation of α-(Acylamino)acrylic Acids. *Tetrahedron* **1984**, *40*, 1245–1253. [CrossRef]
166. Riley, D.P.J. Solution Studies of the Asymmetric Hydrogenation Catalyst System Derived from The [Rhodium (1,2-Bis(diphenylphosphino)-1-cyclohexylethane)] Moiety. *Organomet. Chem.* **1982**, *234*, 85–97. [CrossRef]
167. Miyashita, A.; Yasuda, A.; Takaya, H.; Toriumi, K.; Ito, T.; Souchi, T.; Noyori, R. Synthesis of 2,2′-Bis(diphenylphosphino)-1,1′-binaphthyl (BINAP), an Atropisomeric Chiral Bis(triaryl)phosphine, and its Use in the Rhodium(I)-Catalyzed Asymmetric Hydrogenation of Alpha-(acylamino)acrylic Acids. *J. Am. Chem. Soc.* **1980**, *102*, 7932–7934. [CrossRef]
168. Gridnev, I.D.; Alberico, E.; Gladiali, S. Captured at Last: a Catalyst–Substrate Adduct and a Rh-Dihydride Solvate in the Asymmetric Hydrogenation by a Rh-Monophosphine Catalyst. *Chem. Commun.* **2012**, *48*, 2186–2188. [CrossRef]
169. Gridnev, I.D.; Fan, C.; Pringle, P.G. New Insights Into the Mechanism of Asymmetric Hydrogenation Catalysed by Monophosphonite–Rhodium Complexes. *Chem. Commun.* **2007**, 1319–1321. [CrossRef] [PubMed]
170. Rifat, A.; Patmore, N.J.; Mahon, M.F.; Weller, A.S. Rhodium Phosphines Partnered with the Carborane Monoanions $[CB_{11}H_6Y_6]^-$ (Y = H, Br). Synthesis and Evaluation as Alkene Hydrogenation Catalysts. *Organometallics* **2002**, *21*, 2856–2865. [CrossRef]
171. Marcazzan, P.; Ezhova, M.B.; Patrick, B.O.; James, B.R. Synthesis and Structure of Dimeric Rh-Bis(tertiary phosphine) Complexes, Exceptionally Useful Synthetic Precursors. *C. R. Chim.* **2002**, *5*, 373–378. [CrossRef]
172. Singewald, E.T.; Mirkin, C.A.; Stern, C.L. A Redox-Switchable Hemilabile Ligand: Electrochemical Control of the Coordination Environment of a RhI Complex. *Angew. Chem. Int. Ed.* **1995**, *35*, 1624–1627. [CrossRef]
173. Preetz, A.; Baumann, W.; Fischer, C.; Drexler, H.-J.; Schmidt, T.; Thede, R.; Heller, D. Asymmetric Hydrogenation. Dimerization of Solvate Complexes: Synthesis and Characterization of Dimeric $[Rh(DIPAMP)]_2^{2+}$, a Valuable Catalyst Precursor. *Organometallics* **2009**, *28*, 3673–3677. [CrossRef]
174. Fairlie, D.P.; Bosnich, B. Homogeneous Catalysis. Mechanism of Catalytic Hydroacylation: the Conversion of 4-Pentenals to Cyclopentanones. *Organometallics* **1988**, *7*, 946–954. [CrossRef]
175. Barnhart, R.W.; Xianqi, W.; Noheda, P.; Bergens, S.H.; Whelan, J.; Bosnich, B. Asymmetric Catalysis. Asymmetric Catalytic Intramolecular Hydroacylation of 4-Pentenals Using Chiral RhodiumDiphosphine Catalysts. *J. Am. Chem. Soc.* **1994**, *116*, 1821–1830. [CrossRef]
176. Masutomi, K.; Sugiyama, H.; Uekusa, H.; Shibata, Y.; Tanaka, K. Asymmetric Synthesis of Protected Cyclohexenylamines and Cyclohexenols by Rhodium-Catalyzed [2+2+2] Cycloaddition. *Angew. Chem. Int. Ed.* **2016**, *55*, 15373–15376. [CrossRef]

177. Aida, Y.; Sugiyama, H.; Uekusa, H.; Shibata, Y.; Tanaka, K. Rhodium-Catalyzed Asymmetric [2 + 2 + 2] Cycloaddition of α,ω-Diynes with Unsymmetrical 1,2-Disubstituted Alkenes. *Org. Lett.* **2016**, *18*, 2672–2675. [CrossRef]
178. Yoshizaki, S.; Nakamura, Y.; Masutomi, K.; Yoshida, T.; Noguchi, K.; Shibata, Y.; Tanaka, K. Rhodium-Catalyzed Asymmetric [2 + 2 + 2] Cycloaddition of 1,6-Enynes with Cyclopropylideneacetamides. *Org. Lett.* **2016**, *18*, 388–391. [CrossRef]
179. Shintani, R.; Takano, R.; Nozaki, K. Rhodium-Catalyzed Asymmetric Synthesis of Silicon-Stereogenic Silicon-Bridged Arylpyridinones. *Chem. Sci.* **2016**, *7*, 1205–1211. [CrossRef] [PubMed]
180. Torres, A.; Roglans, A.; Pla-Quintana, A. An Enantioselective Cascade Cyclopropanation Reaction Catalyzed by Rhodium(I): Asymmetric Synthesis of Vinylcyclopropanes. *Adv. Synth. Catal.* **2016**, *358*, 3512–3516. [CrossRef]
181. Fernandez, M.; Parera, M.; Parella, T.; Lledj, A.; le Bras, J.; Muzart, J.; Pla-Quintana, A.; Roglans, A. Rhodium-Catalyzed [2+2+2] Cycloadditions of Diynes with Morita–Baylis–Hillman Adducts: A Stereoselective Entry to Densely Functionalized Cyclohexadiene Scaffolds. *Adv. Synth. Catal.* **2016**, *358*, 1848–1853. [CrossRef]
182. Amatore, M.; Leboeuf, D.; Malacria, M.; Gandon, V.; Aubert, C. Highly Enantioselective Rhodium-Catalyzed [2+2+2] Cycloaddition of Diynes to Sulfonimines. *J. Am. Chem. Soc.* **2013**, *135*, 4576–4579. [CrossRef] [PubMed]
183. Araki, T.; Noguchi, K.; Tanaka, K. Enantioselective Synthesis of Planar-Chiral Carba-Paracyclophanes: Rhodium-Catalyzed [2+2+2] Cycloaddition of Cyclic Diynes with Terminal Monoynes. *Angew. Chem. Int. Ed.* **2013**, *52*, 5617–5621. [CrossRef] [PubMed]
184. Augé, M.; Barbazanges, M.; Tran, A.T.; Simonneau, A.; Elley, P.; Amouri, H.; Aubert, C.; Fensterbank, L.; Gandon, V.; Malacria, M.; et al. Atroposelective [2+2+2] Cycloadditions Catalyzed by a Rhodium(I)–Chiral Phosphate System. *Chem. Commun.* **2013**, *49*, 7833–7835. [CrossRef]
185. Lu, Y.; Woo, S.K.; Krische, M.J. Total Synthesis of Bryostatin 7 via C–C Bond-Forming Hydrogenation. *J. Am. Chem. Soc.* **2011**, *133*, 13876–13879. [CrossRef]
186. Kong, J.R.; Krische, M.J. Catalytic Carbonyl Z-Dienylation via Multicomponent Reductive Coupling of Acetylene to Aldehydes and α-Ketoesters Mediated by Hydrogen: Carbonyl Insertion into Cationic Rhodacyclopentadienes. *J. Am. Chem. Soc.* **2006**, *128*, 16040–16041. [CrossRef]
187. Kong, J.-R.; Cho, C.-W.; Krische, M.J. Hydrogen-Mediated Reductive Coupling of Conjugated Alkynes with Ethyl (N-Sulfinyl)iminoacetates: Synthesis of Unnatural α-Amino Acids via Rhodium-Catalyzed C–C Bond Forming Hydrogenation. *J. Am. Chem. Soc.* **2005**, *127*, 11269–11276. [CrossRef]
188. Rhee, J.U.; Krische, M.J. Highly Enantioselective Reductive Cyclization of Acetylenic Aldehydes via Rhodium Catalyzed Asymmetric Hydrogenation. *J. Am. Chem. Soc.* **2006**, *128*, 10674–10675. [CrossRef]
189. Jang, H.-Y.; Hughes, F.W.; Gong, H.; Zhang, J.; Brodbelt, J.S.; Krische, M.J. Enantioselective Reductive Cyclization of 1,6-Enynes via Rhodium-Catalyzed Asymmetric Hydrogenation: C–C Bond Formation Precedes Hydrogen Activation. *J. Am. Chem. Soc.* **2005**, *127*, 6174–6175. [CrossRef] [PubMed]
190. Masutomi, K.; Sakiyama, N.; Noguchi, K.; Tanaka, K. Rhodium-Catalyzed Regio-, Diastereo-, and Enantioselective [2+2+2] Cycloaddition of 1,6-Enynes with Acrylamides. *Angew. Chem. Int. Ed.* **2012**, *51*, 13031–13035. [CrossRef] [PubMed]
191. Sakiyama, N.; Noguchi, K.; Tanaka, K. Rhodium-Catalyzed Intramolecular Cyclization of Naphthol- or Phenol-Linked 1,6-Enynes Through the Cleavage and Formation of sp^2 C-O Bonds. *Angew. Chem. Int. Ed.* **2012**, *51*, 5976–5980. [CrossRef] [PubMed]
192. Ishida, M.; Shibata, Y.; Noguchi, K.; Tanaka, K. Rhodium-Catalyzed Asymmetric [2+2+2] Cyclization of 1,6-Enynes and Aldehydes. *Chem. Eur. J.* **2011**, *17*, 12578–12581. [CrossRef] [PubMed]
193. Miyauchi, Y.; Kobayashi, M.; Tanaka, K. Rhodium-Catalyzed Intermolecular [2+2+2] Cross-Trimerization of Aryl Ethynyl Ethers and Carbonyl Compounds To Produce Dienyl Esters. *Angew. Chem. Int. Ed.* **2011**, *50*, 10922–10926. [CrossRef] [PubMed]
194. Jang, H.-Y.; Krische, M.J. Rhodium-Catalyzed Reductive Cyclization of 1,6-Diynes and 1,6-Enynes Mediated by Hydrogen: Catalytic C–C Bond Formation via Capture of Hydrogenation Intermediates. *J. Am. Chem. Soc.* **2004**, *126*, 7875–7880. [CrossRef] [PubMed]
195. Preetz, A. Rhodium-Präkatalysatoren in der asymmetrischen Katalyse. Ph.D. Thesis, University of Rostock, Rostock, Germany, 2008.

196. König, A.; Fischer, C.; Alberico, E.; Selle, C.; Drexler, H.-J.; Baumann, W.; Heller, D. Oxidative Addition of Aryl Halides to Cationic Bis(phosphane)rhodium Complexes: Application in C–C Bond Formation. *Eur. J. Inorg. Chem.* **2017**, 2040–2047. [CrossRef]
197. Jiao, Y.; Brennessel, W.W.; Jones, W.D. Oxidative Addition of Chlorohydrocarbons to a Rhodium Tris(pyrazolyl)borate Complex. *Organometallics* **2015**, *34*, 1552–1566. [CrossRef]
198. Townsend, N.S.; Chaplin, A.B.; Naser, M.A.; Thompson, A.L.; Rees, N.H.; Macgregor, S.A.; Weller, A.S. Reactivity of the Latent 12-Electron Fragment $[Rh(P\textit{i}Bu_3)_2]^+$ with Aryl Bromides: Aryl-Br and Phosphine Ligand C-H Activation. *Chem. Eur. J.* **2010**, *16*, 8376–8389. [CrossRef]
199. Chen, S.; Li, Y.; Zhao, J.; Li, X. Chelation-Assisted Carbon-Halogen Bond Activation by a Rhodium(I) Complex. *Inorg. Chem.* **2009**, *48*, 1198–1206. [CrossRef]
200. Douglas, T.M.; Chaplin, A.B.; Weller, A.S. Dihydrogen Loss from a 14-Electron Rhodium(III) Bis-Phosphine Dihydride To Give a Rhodium(I) Complex That Undergoes Oxidative Addition with Aryl Chlorides. *Organometallics* **2008**, *27*, 2918–2921. [CrossRef]
201. Pike, S.D.; Weller, A.S. C–Cl activation of the weakly coordinating anion $[B(3,5\text{-}Cl_2C_6H_3)_4]^-$ at a Rh(I) centre in solution and the solid-state. *Dalton Trans.* **2013**, *42*, 12832–12835. [CrossRef] [PubMed]
202. Curto, S.G.; Esteruelas, M.A.; Olivan, M.; Onate, E.; Velez, A. Selective C–Cl Bond Oxidative Addition of Chloroarenes to a POP–Rhodium Complex. *Organometallics* **2017**, *36*, 114–128. [CrossRef]
203. Puri, M.; Gatard, S.; Smith, D.A.; Ozerov, O.V. Competition Studies of Oxidative Addition of Aryl Halides to the (PNP)Rh Fragment. *Organometallics* **2011**, *30*, 2472–2482. [CrossRef]
204. Gatard, S.; Guo, C.; Foxman, B.M.; Ozerov, O.V. Thioether, Dinitrogen, and Olefin Complexes of (PNP)Rh: Kinetics and Thermodynamics of Exchange and Oxidative Addition Reactions. *Organometallics* **2007**, *26*, 6066–6075. [CrossRef]
205. Gatard, S.; Celenligil-Cetin, R.; Guo, C.; Foxman, B.M.; Ozerov, O.V. Carbon−Halide Oxidative Addition and Carbon−Carbon Reductive Elimination at a (PNP)Rh Center. *J. Am. Chem. Soc.* **2006**, *128*, 2808–2809. [CrossRef]
206. Timpa, S.D.; Pell, C.J.; Zhou, J.; Ozerov, O.V. Fate of Aryl/Amido Complexes of Rhodium(III) Supported by a POCOP Pincer Ligand: C–N Reductive Elimination, β-Hydrogen Elimination, and Relevance to Catalysis. *Organometallics* **2014**, *33*, 5254–5262. [CrossRef]
207. Timpa, S.D.; Fafard, C.M.; Herberta, D.E.; Ozerov, O.V. Catalysis of Kumada–Tamao–Corriu Coupling by a (POCOP)Rh Pincer Complex. *Dalton Trans.* **2011**, *40*, 5426–5429. [CrossRef]
208. Ito, J.-I.; Miyakawa, T.; Nishiyama, H. Amine-Assisted C−Cl Bond Activation of Aryl Chlorides by a (Phebox)Rh-Chloro Complex. *Organometallics* **2008**, *27*, 3312–3315. [CrossRef]
209. Qian, Y.Y.; Lee, M.H.; Yang, W.; Chan, K.S. Aryl Carbon–Chlorine (Ar–Cl) and Aryl Carbon–Fluorine (Ar–F) Bond Cleavages by Rhodium Porphyrins. *J. Organomet. Chem.* **2015**, *791*, 82–89. [CrossRef]
210. Willems, S.T.H.; Budzelaar, P.H.M.; Moonen, N.N.P.; de Gelder, R.; Smits, J.M.M.; Gal, A.W. Coordination and Oxidative Addition at a Low-Coordinate Rhodium(I) β-Diiminate Centre. *Chem. Eur. J.* **2002**, *8*, 1310–1320. [CrossRef]
211. Yamagata, T.; Tani, K.; Tatsuno, Y.; Saito, T. A New Rhodium Trinuclear Complex Containing Highly Protected Hydroxo Groups, $[\{Rh(binap)\}_3(\mu_3\text{-}OH)2]ClO_4$, Responsible for Deactivation of the 1,3-Hydrogen Migration Catalyst of Allylamine [Binap = 2,2′-Bis(diphenylphosphino)-1,1′-binaphthyl]. *J. Chem. Soc. Chem. Commun.* **1988**, 466–468. [CrossRef]
212. Preetz, A.; Baumann, W.; Drexler, H.-J.; Fischer, C.; Sun, J.; Spannenberg, A.; Zimmer, O.; Hell, W.; Heller, D. Trinuclear Rhodium Complexes and Their Relevance for Asymmetric Hydrogenation. *Chem. Asian. J.* **2008**, *3*, 1979–1982. [CrossRef] [PubMed]
213. Horstmann, M.; Drexler, H.-J.; Baumann, W.; Heller, D. Ammine and Amido Complexes of Rhodium: Synthesis, Application and Contributions to Analytics. Unpublished work. 2019.
214. [2-(3-Methoxy-phenyl)-cyclohex-1-enylmethyl]-dimethylamine, which can be transformed by a stereoselective hydrogenation to Tramadol, an important analgesic drug used world-wide, is a member of a class of 1,2-disubstituded cyclohexene derivatives which are important intermediates in the synthesis of pharmaceutically active compounds acting on the central nervous system.
215. The hydrogenation product is a key intermediate in the synthesis of 4-amino-2-(*R*)-methylbutan-1-ol en route towards the receptor antagonist TAK-637.

216. Cobley, C.J.; Lennon, I.C.; Praquin, C.; Zanotti-Gerosa, A. Highly Efficient Asymmetric Hydrogenation of 2-Methylenesuccinamic Acid Using a Rh-DuPHOS Catalyst. *Org. Process Res. Dev.* **2003**, *7*, 407–411. [CrossRef]
217. The prochiral olefin (Z)-3-[1-(dimethylamino)-2-methylpent-2-en-3-yl]phenol is an important precursor of the new analgesic drug Tapentadol which is as an open-chain analogue of Tramadol. The (*R,R*) stereoisomer of Tapentadol is a novel, centrally acting analgesic with a dual mode of action: M-opioid receptor (MOR) agonism and norepinephrine reuptake inhibition.
218. Christopfel, W.C.; Vineyard, B.D. Catalytic Asymmetric Hydrogenation With a Rhodium(I) Chiral Bisphosphine System. A Study of Itaconic Acid and Some of its Derivatives and Homologs. *J. Am. Chem. Soc.* **1979**, *101*, 4406–4408. [CrossRef]
219. Schmidt, T.; Drexler, H.-J.; Sun, J.; Dai, Z.; Baumann, W.; Preetz, A.; Heller, D. Unusual Deactivation in the Asymmetric Hydrogenation of Itaconic Acid. *Adv. Synth. Catal.* **2009**, *351*, 750–754. [CrossRef]
220. Schmidt, T.; Baumann, W.; Drexler, H.-J.; Heller, D. Unusual Deactivation in the Asymmetric Hydrogenation of Itaconic Acid. *J. Organomet. Chem.* **2011**, *696*, 1760–1767. [CrossRef]
221. Schmidt, M.; Schreiber, S.; Franz, L.; Langhoff, H.; Farhang, A.; Horstmann, M.; Drexler, H.-J.; Heller, D.; Schwarze, M. Hydrogenation of Itaconic Acid in Micellar Solutions: Catalyst Recycling with Cloud Point Extraction? *Ind. Eng. Chem. Res.* **2019**, *58*, 2445–2453. [CrossRef]
222. Other possibilities are extraction with organic solvents, micellar-enhanced ultrafiltration (MEUF) and phase separation.
223. Friedrich, H.B.; Moss, J.R. Halogenoalkyl Complexes of Transition Metals. *Adv. Organomet. Chem.* **1991**, *33*, 235–290.
224. Ball, G.E.; Cullen, W.R.; Fryzuk, M.D.; James, B.R.; Rettig, S.J. Oxidative Addition of Dichloromethane to $[(dppe)Rh]_2(\mu\text{-}Cl)_2$ (dppe = $Ph_2PCH_2CH_2PPh_2$). X-ray structure of $[(dppe)RhCl]_2(\mu\text{-}Cl)_2(\mu\text{-}CH_2)$. *Organometallics* **1991**, *10*, 3767–3769. [CrossRef]
225. Fennis, P.J.; Budzelaar, P.H.M.; Frijns, J.H.G. Dichloromethane Addition to Rhodium-β-Diketonate Complexes of Diphosphines and Pyridyl-Substituted Diphosphines. *J. Organomet. Chem.* **1990**, *393*, 287–298. [CrossRef]
226. Marder, T.B.; Fultz, W.C.; Calabrese, J.C.; Harlow, R.L.; Milstein, D. Activation of Dichloromethane by Basic Rhodium(I) and Iridium(I) Phosphine Complexes. Synthesis and Structures of *fac*-$[Rh(PMe_3)_3Cl_2(CH_2PMe_3)]Cl\cdot CH_2Cl_2$ and *trans*-$[Rh(Me_2PCH_2CH_2PMe_2)_2Cl(CH_2Cl)]Cl$. *J. Chem. Soc. Chem. Commun.* **1987**, 1543–1545. [CrossRef]
227. Blank, B.; Glatz, G.; Kempe, R. Single and Double C-Cl-Activation of Methylene Chloride by P,N-ligand Coordinated Rhodium Complexes. *Chem. Asian J.* **2009**, *4*, 321–327. [CrossRef] [PubMed]
228. Ciriano, M.A.; Tena, M.A.; Oro, A. Reactions of Chloroform and *gem*-Dichlorocarbons with Binuclear Rhodium Complexes Leading to Functionalized Methylene-Bridged Compounds. *J. Chem. Soc. Dalton Trans.* **1992**, 2123–2124. [CrossRef]
229. Mannu, A.; Ferro, M.; Möller, S.; Heller, D. Monomerisation of $[Rh_2(1,3\text{-Bis-(Diphenylphosphino)-Propane})_2(\mu_2\text{-}Cl)_2]$ Detected by Pulsed Gradient Spin Echo Spectroscopy and ^{31}P Nmr Monitoring of Metathesis Experiments. *J. Chem. Res.* **2018**, *42*, 402–404. [CrossRef]
230. This reaction has been known for a long time and is used in the synthesis of mononuclear cationic complexes from neutral μ2-chloro brigded precursors by chloride abstraction with a silver salt.
231. Anastas, P.T.; Kirchhoff, M.M.; Williamson, T.C. Catalysis as a Foundational Pillar of Green Chemistry. *Appl. Catal. A Gen.* **2001**, *221*, 3–13. [CrossRef]